2nd International Conference on Vocational Education and Electrical Engineering (ICVEE 2017)

IOP Conference Series: Materials Science and Engineering
Volume 336

Surabaya, Indonesia
9 November 2017

Editors:

Lusia Rakhmawati
Hapsari Peni
Pradini Puspitaninagayu

ISBN: 978-1-5108-6157-2
ISSN: 1757-8981

Printed from e-media with permission by:

Curran Associates, Inc.
57 Morehouse Lane
Red Hook, NY 12571

Some format issues inherent in the e-media version may also appear in this print version.

Printed by Curran Associates, Inc. (2018)

For permission requests, please contact the Institute of Physics
at the address below.

Institute of Physics
Dirac House, Temple Back
Bristol BS1 6BE UK

Phone: 44 1 17 929 7481
Fax: 44 1 17 920 0979

techtracking@iop.org

Additional copies of this publication are available from:

Curran Associates, Inc.
57 Morehouse Lane
Red Hook, NY 12571 USA
Phone: 845-758-0400
Fax: 845-758-2633
Email: curran@proceedings.com
Web: www.proceedings.com

Table of contents

Volume 336

The 2nd International Conference on Vocational Education and Electrical Engineering (ICVEE)

9 November 2017, Surabaya, Indonesia

Accepted papers received: 12 March 2018
Published online: 16 April 2018

Preface

Papers

Electricals

Instrumentation and control

Engineering Education

IOP Conf. Series: Materials Science and Engineering **336** (2018) 011001 doi:10.1088/1757-899X/336/1/011001

The 2nd International Conference on Vocational Education and Electrical Engineering (ICVEE) 2017

Preface

The 2nd International Conference on Vocational Education and Electrical Engineering (ICVEE) 2017 was a scientific forum where all of us could meet colleagues and friends of broad areas, discuss and disseminate research findings and discoveries as well as to develop knowledge, technology, arts, and sustainable research networks, particularly in vocational education, electrical engineering, and telematics. This year conference raises a very important theme on "Enhancing nation competitiveness thought Innovation in Electrical Engineering and Education". Therefore, it was a great pleasure and beneficial for all of us to take this advantage to build strong sustainable networks among researchers in order to develop knowledge, technology through recent research and innovation. By hosting this conference, Department of Electrical Engineering, Faculty of Engineering Universitas Negeri Surabaya was not only gaining the advancement of science and technology from all the findings and discoveries delivered during the conference, but also fruitful to encourage and enhance the arts and cultural values that would further dignify our nation and country among other worldwide.

This seminar can be accomplished, for the help of various parties. On this occasion, let us express our appreciation and gratitude to the key note speakers, the speakers, both from within and outside the country, so this seminar was warm and lively. To all those who have helped so that this international seminar to run smoothly in accordance with the intent and purpose, we extend our appreciation and gratitude. Not to forget we convey an apology, if in the organization of this seminar there are many shortcomings, and it all happened, solely because of the limitations that exist in us. Thus, thank you for all the attention.

The Editors
Lusia Rakhmawati, S.T., M.T.
Hapsari Peni A. T., S.Si, M.T.
Pradini Puspitaningayu, S.T., M.T.

IOP Conf. Series: Materials Science and Engineering **336** (2018) 011001 doi:10.1088/1757-899X/336/1/011001

LIST OF COMMITTEE

Conference Chair:
Prof. Dr. Munoto, M.Pd.

Co- Conference Chair:
Lusia Rakhmawati, S.T., M.T.

International Advisory Boards:
Oenardi Lawanto, Ph.D – Utah State University, Amerika Serikat
Prof. Madya Ir. Dr.Herlina Abdul Rahim – Universiti Teknologi Malaysia, Malaysia
Prof. Dr. Abdullah Bin Gani – University of Malaya, Malaysia
Prof. Chi-Jo Wang – Southern Taiwan University of Science and Technology, Taiwan
Prof. Dr. Mochamad Ashari, M. Eng. – Telkom University, Institut Teknologi Sepuluh Nopember, Indonesia
Prof. Dr. Ir. Gamantyo Hendrantoro, Ph.D. – Institut Teknologi Sepuluh Nopember, Indonesia

Scientific Committee:
Prof. Dr. Supari Muslim – Unesa, Indonesia
Prof. Dr. Bambang Suprianto, M.T. – Unesa, Indonesia
Dr. Wirawan, DEA – ITS, Indonesia
Dr. Ade Gafar Abdullah – UPI, Indonesia
Ir. Wijono, Ph.D – Universitas Brawijaya, Indonesia
Dr. Eng. Adi Wibowo, S.Si., M.Kom – Undip, Indonesia
Dr. IGP Asto Buditjahjanto – Unesa, Indonesia
Dr. Son Kuswadi – PENS, Indonesia
Iwan Kustiawan, Ph. D – UPI, Indonesia
Eppy Yundra, Ph. D – Unesa, Indonesia
Didin Wahyudin, Ph.D, – UPI, Indonesia
Dr. Lilik Anifah – Unesa, Indonesia
Krismadinata, Ph. D – Universitas Negeri Padang, Indonesia
Heri Suryoatmojo, ST., MT., Ph.D – ITS, Indonesia
Dr. Eng. Asep Bayu Dani Nandiyanto – UPI, Indonesia
Dr. Elly Matul Imah, S.Si., M.Kom. – Unesa, Indonesia

Organizing Committee:

Hapsari Peni A. T., S.Si, M.T. – Universitas Negeri Surabaya
Yulia Fransisca, S.Pd., M.Pd. – Universitas Negeri Surabaya
Pradini Puspitaningayu, S.T., M.T. – Universitas Negeri Surabaya
Reza Rahmadian, S.ST., MengSc – Universitas Negeri Surabaya
Mahendra Widyartono, S.T., M.T.- Universitas Negeri Surabaya
Widi Ariwibowo, S.T., M.T.- Universitas Negeri Surabaya
Dr. Euis Ismayati, M.Pd. – Universitas Negeri Surabaya
Ibrohim S.T., M.T. – Universitas Negeri Surabaya
Subuh Isnur Haryudo, S.T., M.T. – Universitas Negeri Surabaya

IOP Conf. Series: Materials Science and Engineering **336** (2018) 011001 doi:10.1088/1757-899X/336/1/011001

Farid Baskoro, ST., MT. – Universitas Negeri Surabaya
Rifqi Firmansyah, S.T., M.T. – Universitas Negeri Surabaya
Arif Widodo, S.T., M.Sc. – Universitas Negeri Surabaya
Ir. Achmad Imam Agung, M. Pd – Universitas Negeri Surabaya
Nur Kholis, S.T., M.T. – Universitas Negeri Surabaya
Endryansyah, S.T., M.T. – Universitas Negeri Surabaya
Dr. Wiryanto, M.Si – Universitas Negeri Surabaya
Aditya Chandra H., S.T., M.T – Universitas Negeri Surabaya

IOP Conf. Series: Materials Science and Engineering **336** (2018) 011001 doi:10.1088/1757-899X/336/1/011001

CONFERENCE PHOTOGRAPH

IOP Conf. Series: Materials Science and Engineering **336** (2018) 011001 doi:10.1088/1757-899X/336/1/011001

IOP Conf. Series: Materials Science and Engineering **336** (2018) 011001 doi:10.1088/1757-899X/336/1/011001

IOP Conf. Series: Materials Science and Engineering **336** (2018) 011002 doi:10.1088/1757-899X/336/1/011002

Peer review statement

All papers published in this volume of *IOP Conference Series: Materials Science and Engineering* have been peer reviewed through processes administered by the proceedings Editors. Reviews were conducted by expert referees to the professional and scientific standards expected of a proceedings journal published by IOP Publishing.

Implementation of Particle Swarm Optimization Method for Voltage Stability Analysis in 150 kV Sub System Grati – Paiton East Java

A B Kusumaningtyas*, M N Hidayat, and F Ronilaya

Electrical Engineering Department, Malang State Polytechnic . JL Soekarno-Hatta No. 9 Malang 65141 East Java Indonesia

*ajeng_bening@yahoo.com

Abstract. Based on the data from State Electric Company on 15 January 2013, the undistributed power in the 150 kV sub system Grati-Paiton Region IV, that consist of 26 bus 150 kV and 2 bus generation 500 kV system, was recorded 3.286,00 MW. At the same time, the frequency of the system was down to 49 Hz. This lead to a deficit generation and unstable voltage condition in the system. Fast Voltage Stability Index (FVSI) method is used in this research to analyze the voltage stability of the buses. For buses with unstable voltage condition, reactive power will be injected through capacitor installation. The site where the capacitor will be installed is determined using the Fast Voltage Stability Index (FVSI) method while the size of the capacitor is determined using the Particle Swarm Optimization (PSO) method. The PSO method has been applied in some researches, such as to determine optimal placement and sizing in radial distribution network as well as in transmission network.. In this research, the PSO method is used to find the Q_{loss} of an interconnection transmission system, which in turn, the value of the Q_{loss} is used to determine the capacitance of the capacitor needed by the system.

1. Introduction

Paiton is a one of the largest generation units to supply the electrical energy needed for Java, Bali, and Madura island. Based on the data taken from State Electric Company on 15 January 2013, the undelivered power in sub system Grati Paiton Region IV 150 kV was recorded at 3.286,00 MW and the frequency system was down to 49 Hz. This leads to unstable voltage condition and cause the system to blackout. Voltage stability is one of important parameters in a system. Voltage stability is an ability of a bus or transmission line to maintain the nominal voltage in a bus after disturbance occurs [1]. There are many methods that used to analyze the voltage stability such as PV curve, QV curve, Line Stability Index, Voltage Stability Index (VSI), L Index, and Fast Voltage Stability Index (FVSI) [3,4,5]. In this research, the voltage stability is analyzed using Fast Voltage Stability Index (FVSI) method. The purpose of this method is to indicate the condition of a bus. If the bus has the stability index close to unity its reach the limit level of instability condition. If the bus has the stability index close to zero this bus is in stable condition [3].

The bus that indicates in unstable condition will be injected with reactive power through the capacitor in order to improve voltage stability in a bus.

2. Reactive Power Loss

There is some method that usually used to reactive power injection that is through STATCOM, SVC, Shunt capacitor series, and capacitor installation [7,8,9]. In this research using a capacitor as a reactive power injection that installs in switchgear. The method that used to determine the capacitance of the capacitor is Particle Swarm Optimization method. The purpose of this method is to calculate reactive power loss (Q loss) in the transmission line, so from the reactive power loss value can determine the capacitance of the capacitor. The equation of load flow in transmission line is

$$S_R = P_R + jQ_R = E_R I^* = E_R\left[\frac{Es\ COS\delta + jEs \sin\delta - ER}{jX}\right] \quad (1)$$

Where
S_R = Apparent power of receiving bus
P_R = Active power in receiving bus
Q_R = Reactive power in receiving bus
E_R = voltage in receiving bus
E_S = voltage in sending bus

$$Q_S - Q_R = (E_S - E_R)^2 X^{-1} = XI^2 \quad (2)$$

Equation 2 is illustrated by the relation between reactive power, voltage, and impedance in transmission line system. The deviation between Qs and Qr is a reactive loss in line. To control the value of reactive loss in the transmission line can control the value of impedance.

3. Fast Voltage Stability Index

Fast Voltage Stability Index (FVSI) is a method that used to calculate the index of voltage stability. if the bus has the stability index close to unity (1) its reach the limit level of instability condition but if the bus has the stability index close to zero this bus is in stable condition [5].

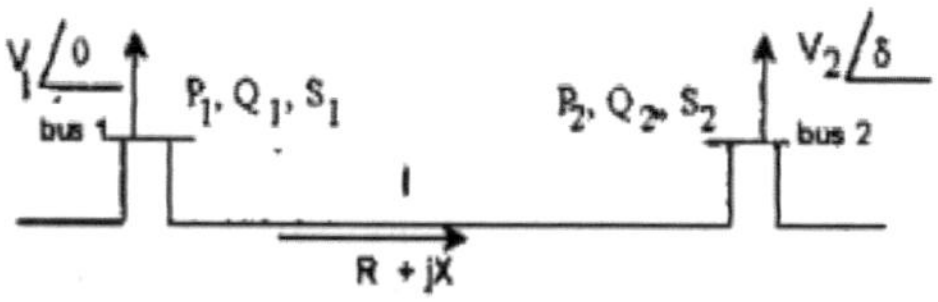

Figure 1. Radial system of 2 bus

From this figure the equation in receiving bus is

$$V_2^2 - \left[\frac{R}{X}\sin\delta + \cos\delta\right] V_1 V_2 + \left(X + \frac{R^2}{X}\right) Q_2 = 0 \quad (3)$$

$$\frac{4Z^2QX}{V1^2(R\sin\delta + X\cos\delta)^2} < 1 \quad (4)$$

"i" symbol is for sending bus and "j" symbol is for receiving bus so the equation for Fast Voltage stability method is

$$FVSIij = \frac{4Z^2Qj}{Vi^2X} \quad (5)$$

Where
Z = impedance of line (ohm)
X = reactance of line (ohm)
Qj = Reactive power in receiving bus (Var)
Vi = voltage in sending bus (Volt)

4. Particle Swarm Optimization

Particle Swarm Optimization is a one of a method that used to optimation calculated. This method firstly introduces by Kennedy and Eberhart in 1995. This method adapts from the behavior of bird in a swarm to collect the food [8,9,10]. The first step that must do in this method is the initialization of population which are created and each particle will be updated in every iteration and used to find the value of objective function. In each iteration, there are Pbest value and Gbest in a swarm. To update the velocity and new position can calculate with this equation

$$V_i\,(k+1) = W * V_i\,(k) + C_1{*}rand_1{*}(Pbesti - Sik + \; C_2{*}\,rand_2{*}\,Gbestik - Sik\,) \quad (6)$$

$$Si\,(\,k+1) = Si\,(k) + Vi\,(k+1) \quad (7)$$

Where
Vi(k) = velocity of particle I in kth iteration
Vi(k+1) = velocity of particle I in (k+1)th iteration
W = inertia weight
C1, C2 = Weighfting factor
Si(k) = particle position in kth iteration
Si(k+1) = particle position in (k+1)th iteration
Pbest = personal best of particle i
Gbest = global best of particle i

The value of weight factor can determine from this equation

$$W = Wmax - \frac{Wmax-Wmin}{iter\ max}\ x\ iter \quad (8)$$

Range value of weight factor is 0.9 to maximum and 0.4 to a minimum. Iter max is maximum iteration and iter is the number of iteration from this program.

5. Problem Formulation

The purpose of this research, in order to know the reactive power losses in transmission line wich, in turn, can determine the capacitance of a capacitor that will be used to reactive power injection to improve voltage stability and voltage profile in the system.If in transmission line illustrated with two bus that is bus I as a sending bus and bus j as a receiving bus and the impedance of transmission line is Z = R+jX per phase in line so the equation to calculate the reactive power loss is

$$Q_{ij} = \frac{1}{X^2+R^2}\left[Vi^2 + Vj^2 - 2ViVj\cos(\delta i - \delta j)\right]X \quad (9)$$

Total reactive power losses from system determine with add all of the losses in every branch. The range of voltage limit on a bus is

$$Vmin < Vi < Vmax \quad (10)$$

IOP Conf. Series: Materials Science and Engineering **336** (2018) 012001 doi:10.1088/1757-899X/336/1/012001

After knowing the value of Q loss in transmission line so that can determine the sizing of capacitance capacitor with this equation

$$Qc < Qloss \quad (11)$$

Qc is a capacitance of capacitor and Qloss is a reactive power loss in the transmission line. The value of the capacitor is smallest than reactive power loss in order to in normal condition (not in peak load) reactive power in the system not over, it can cause overvoltage.To calculated load flow analyze between two buses represented by this equation

$$Pij = \frac{1}{R^2+X^2}(R\,[Vi]^2 - R[Vi][Vj]\cos\gamma - X[Vi][Vj]\sin\gamma) \quad (12)$$

$$Qij = \frac{1}{R^2+X^2}(X\,[Vi]^2 - X[Vi][Vj]\cos\gamma - R\,[Vi][Vj]\sin\gamma) \quad (13)$$

6. Implementation of PSO

Particle Swarm Optimization is a one of a method that used to optimization calculation. PSO of this research in order to know the reactive power losses in transmission line which, in turn, can determine the capacitance of a capacitor that will be used to reactive power injection to improve voltage stability and voltage profile in the system.

Step 1: initialized data system that is transmission line data, voltage data in each bus, range value of voltage limit, entry the data of PSO parameters such as the number of particles, number of iteration, weight factor, acceleration factor (C1, C2) etc.

Step 2: location of capacitor installation predefine with FVSI method

Step 3: after that calculate the objective function to determine Qloss value with equation (9)

Step 4: Update the value of Pbest and Gbest

$$V_i\,(k+1) = W * V_i\,(k) + C_1{*}rand_1{*}(Pbesti - Sik + C_2{*}\,rand_2{*}\,Gbestik - Sik\,)$$

Step 5: Update the value of particle velocity with equation

$$Vmin < V < Vmax \quad (14)$$

Step 6: Update the particle position

$$Xij^{\,t} = Xij^{t-1} + Vij^{t}$$

$$Xmin < X < X\,max \quad (15)$$

Step 7: after getting the optimal result, replace the impedance line with line impedance after compensation through capacitor in objective function and repeat the iteration

7. Result and discussion

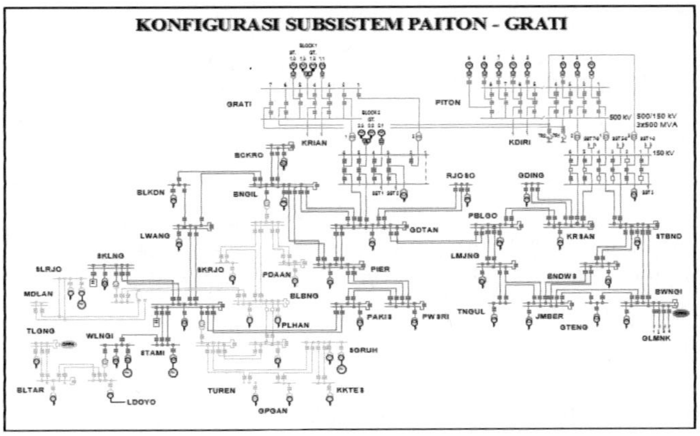

Figure 2 . Single line diagram Paiton – Grati

Electrical energy distribution system Paiton – Gratis consists of 26 bus 150 kV and 2 bus generation 500 kV. Based on the calculation of FVSI in sub system Paiton – Grati, the bus that has the highest FVSI index is in Situbondo – Banyuwangi line 0.313 Paiton – Situbondo 0.1602, Genteng – Banyuwangi line 0.15167, Probolinggo – Gondangwetan line 0.134869 (see appendix 1). The transmission line that has the highest index indicate that this bus in unstable voltage condition. In this bus will be installing the capacitor as a reactive power injection to improve voltage stability and voltage profile. After analyzing the voltage stability the value of Q loss can calculate with PSO method.This table represents the value of reactive power loss before installing the capacitor.

Table 1. Voltage of the bus and reactive power before compensation

Line transmission	Before installing capacitor		
	V(i) (kV)	V(j) (kV)	Qloss (kvar)
Probolinggo - Gondangwetan	143.8	143.8	38.3
Situbondo- Banyuwangi	144,4	143,5	323
Paiton – Situbondo	145,2	144,4	1281,3
Genteng - Banyuwangi	143	143,5	602,5

V(i) is a sending voltage bus and V(j) is a receiving voltage bus. Based on table 1 reactive power loss in probolinggo – the gondangwetan network is 38.3 kvar so that to decide the capacity of the capacitor as a reactive power injection used PSO method with the value from this parameter XL, Vi, and Vj.

Table 2. Parameter of PSO for Probolinggo – Gondangwetan network

Vi (kV)	Vj(kV)	$\delta 1$	$\delta 2$	XL(up)
143.8	148.3	-5.29	-5.58	239

This parameter will be calculated in the objective function (equation 9) so that can obtain reactive power loss with PSO method. Reactive power loss in Probolinggo-Gondangwetan network is 193

MVAR. The value of the Q loss is used to determine of capacitor needed, so with Q loss 193 MVAR capacitor that used is 150 MVAR. After installing capacitor do the simulation again so that obtained Q loss after compensation.

Table 3. Voltage of the bus and reactive power after compensation

Line transmission	Before installing capacitor		
	V(i) (kV)	V(j) (kV)	Qloss (kvar)
Probolinggo - Gondangwetan	153	153	29,2
Situbondo- Banyuwangi	150,6	149,8	274,5
Paiton – Situbondo	148,9	148,3	1113,5
Genteng - Banyuwangi	147	147,6	548,2

The capacitor that used as a reactive power injection installed in three bus wich is Probolinggo bus, Gondangwetan bus, and Banyuwangi bus. This bus is a bus that has the highest FVSI value. Based on reactive power loss calculated the capacity of capacitor needed is 150 MVAR in Gondangwetan bus, 100 MVAR in Probolinggo bus and 50 MVAR in Banyuwangi bus. From table 1 and 3 show that the value of reactive power loss decrease and the voltage value at the bus increase after compensation.

Table 4. FVSI of the bus after compensation

Line transmission	Before installing capacitor		After installing capacitor	
	Q (j)	FVSI	Q (i)	FVSI
Probolinggo- Gondangwetan	50.96	0.135	14.73	0.034
Situbondo- Banyuwangi	56.232	0.194	27.531	0.0875
Paiton- Situbondo	57.7	0.1412	57.7	0.1298
Genteng - Banyuwangi	56.232	0.1516	27.531	0.068

From table 4 show that after installing capacitor reactive power loss in bus decrease so that the value of FVSI decrease.

8. Conclusion

The main purpose of this research is to determine the capacity of the capacitor by calculated the reactive power loss in a transmission line using the Particle Swarm Optimization (PSO) method. This research applied in interconnection network Paiton – Grati 150 kV east java. The result shows that the capacity capacitor installed at gondangwetan bus is 150 Mvar, at Probolinggo bus is 100 Mvar and at Banyuwangi bus is 50 Mvar. After installing the capacitor, the reactive power loss at Probolinggo – Gondangwetan line decreases from 38.3 kvar to 29.2 kvar, so that the voltage of bus increases and the FVSI index decrease close to zero. This condition indicates that the bus is in stable condition after installing the capacitor. This research proves that the PSO can be used to determine reactive power loss and the capacity of capacitor that needed so that the voltage value in bus increase and voltage system be stable.

9. Acknowledgment

The author desire to gratefull to Mr. Nur as a staff in State electricity company (PLN) for the data and also author feel like gratitude to Mr. M Noor Hidayat, Ph.D. and Mr Ferdian Ronilaya, Ph.D. for their support, comment and suggestion with respect to this research.

10. Appendices

Table A1. Voltage stability Index based on FVSI

No	Sending bus	Receiving bus	V(j)	FVSI
1	PAITON	SITUBONDO	144.4	0,160221
2	PAITON	SITUBONDO	144.4	0,141227
3	SITUBONDO	BANYUWANGI	143.5	0,194443
4	SITUBONDO	BANYUWANGI	143.5	0,313891
5	SITUBONDO	BONDOWOSO	143.5	0,021024
6	SITUBONDO	BONDOWOSO	143.5	0,017146
7	BONDOWOSO	JEMBER	142.8	0,045072
8	BONDOWOSO	JEMBER	142.8	0,015463
9	JEMBER	GENTENG	143	0,063648
10	GENTENG	BANYUWANGI	143.5	0,15167
11	PAITON	KRAKSAN	145	0,014371
12	PAITON	KRAKSAN	145	0,014371
13	KRAKSAN	GENDING	145	0,010253
14	KRAKSAN	GENDING	145	0,010253
15	KRAKSAN	PROBOLINGGO	143.8	0,023407
16	KRAKSAN	PROBOLINGGO	143.8	0,016237
17	PROBOLINGGO	GONDANGWETAN	143.8	0,134869
18	PROBOLINGGO	GONDANGWETAN	143.8	0,114758
19	PROBOLINGGO	LUMAJANG	142.7	0,032083
20	PROBOLINGGO	LUMAJANG	142.7	0,027119
21	LUMAJANG	TANGUL	142.7	0,009698
22	TANGUL	JEMBER	142.8	0,034247
23	GRATI	GONDANGWETAN	143.8	0,042718
24	GRATI	GONDANGWETAN	143.8	0,036348
25	GONDANGWETAN	REJOSO	143.4	0,019137
26	GONDANGWETAN	REJOSO	143.4	0,019137
27	GONDANGWETAN	PIER	143	0,017886
28	GONDANGWETAN	PIER	143	0,016122
29	PIER	BANGIL	143	0,00369
30	PIER	BANGIL	143	0,003208
31	GONDANGWETAN	BANGIL	143	0,017003
32	GONDANGWETAN	BANGIL	143	0,014782
33	BANGIL	BUMICOKRO	142.9	0,000301
34	BANGIL	BUMICOKRO	142.9	0,000155
35	BANGIL	LAWANG	141.8	0,024532
36	LAWANG	KEBONAGUNG	141.2	0,055076

References

[1] Kundur, P., Balu, N. J., & Lauby, M. G. (1994). *Power system stability and control* (Vol. 7). New York: McGraw-hill.

[2] Taylor, C. W. (1994). *Power system voltage stability*. McGraw-Hill.

[3] Yazdanpanah-Goharrizi, A., & Asghari, R. (2007, September). A novel line stability index (NLSI) for voltage stability assessment of power systems. In *7th WSEAS International Conference on Power Systems, Beijing, China* (pp. 15-1).

[4] Kowsalya, M., Ray, K. K., & Kothari, D. P. (2008, November). Voltage stability enhancement through the optimal location of var compensator. In Sustainable Energy Technologies, 2008. ICSET 2008. IEEE International Conference on (pp. 360-365). IEEE

[5] Vadivelu, K. R., & Marutheswar, G. V. (2014). Fast voltage stability index based optimal reactive power planning using differential evolution. Electr. Electron. Eng. Int. J, 3(1), 51-

IOP Conf. Series: Materials Science and Engineering **336** (2018) 012001 doi:10.1088/1757-899X/336/1/012001

60.
[6] Wang, K., Ye, M., Xiong, W., Wang, F., & Hou, J. (2016, September). Coordinated control of STATCOM and mechanically switched capacitors to improve short-term voltage stability. In Power System Technology (POWERCON), 2016 IEEE International Conference on (pp. 1-5). IEEE.
[7] Dixit, M., Kundu, P., & Jariwala, H. R. (2016, March). Optimal allocation and sizing of shunt capacitor in the distribution system for power loss minimization. In Electrical, Electronics and Computer Science (SCEECS), 2016 IEEE Students' Conference on (pp. 1-6). IEEE.
[8] Shivashankar, S. (2013, February). Optimal location of TCSC in transmission lines using contingency severity index and performance index methods for single contingency using PSO. In Power, Energy and Control (ICPEC), 2013 International Conference on (pp. 135-138). IEEE
[9] Sonwane, P. M., & Kushare, B. E. (2014, April). Optimal capacitor placement and sizing for enhancement of distribution system reliability and power quality using PSO. In Convergence of Technology (I2CT), 2014 International Conference for (pp. 1-7). IEEE.
[10] Prakash, K., & Sydulu, M. (2007, June). Particle swarm optimization based capacitor placement on radial distribution systems. In Power Engineering Society General Meeting, 2007. IEEE (pp. 1-5). IEEE.
[11] Zhang, P., Teng, S., Zhang, K., Yu, X., Chen, Y., & Chen, Y. (2016, October). Optimal allocation method of dynamic VAR compensation based on a novel transient voltage stability index. In Power and Energy Engineering Conference (APPEEC), 2016 IEEE PES Asia-Pacific (pp. 785-789). IEEE
[12] Mosbah, M., Mohammedi, R. D., Arif, S., & Hellal, A. (2016, November). Optimal of shunt capacitor placement and size in Algerian distribution network using particle swarm optimization. In Modelling, Identification, and Control (ICMIC), 2016 8th International Conference on (pp. 192-197). IEEE
[13] Lotfi, H., Samadi, M., & Dadpour, A. (2016, April). Optimal capacitor placement and sizing in radial distribution system using an improved Particle Swarm Optimization algorithm. In Electrical Power Distribution Networks Conference (EPDC), 2016 21st Conference on (pp. 147-152). IEEE.

IOP Conf. Series: Materials Science and Engineering **336** (2018) 012002 doi:10.1088/1757-899X/336/1/012002

Fixed Nadir Focus Concentrated Solar Power Applying Reflective Array Tracking Method

B Setiawan*, A M DAMayanti, A Murdani, I I A Habibi, and R N Wakidah

Electrical Engineering Departement, Malang State Polytechnic. Jl. Soekarno-Hatta No. 9, Malang 65141, Indonesia

*stbuddy@yahoo.com

Abstract. The Sun is one of the most potential renewable energy develoPMent to be utilized, one of its utilization is for solar thermal concentrators, CSP (Concentrated Solar Power). In CSP energy conversion, the concentrator is as moving the object by tracking the sunlight to reach the focus point. This method need quite energy consumption, because the unit of the concentrators has considerable weight, and use large CSP, means the existence of the usage unit will appear to be wider and heavier. The addition of weight and width of the unit will increase the torque to drive the concentrator and hold the wind gusts. One method to reduce energy consumption is direct the sunlight by the reflective array to nadir through CSP with Reflective Fresnel Lens concentrator. The focus will be below the nadir direction, and the position of concentrator will be fixed position even the angle of the sun's elevation changes from morning to afternoon. So, the energy concentrated maximally, because it has been protected from wind gusts. And then, the possibility of dAMage and changes in focus construction will not occur. The research study and simulation of the reflective array (mechanical method) will show the reflective angle movement. The distance between reflectors and their angle are controlled by mechatronics. From the simulation using fresnel 1m^2, and efficiency of solar energy is 60.88%. In restriction, the intensity of sunlight at the tropical circles 1KW/peak, from 6 AM until 6 PM.

1. Introduction

Solar energy is one of the energies exploited by many people because of its abundant availability, but its utilization is also not optimal. This is because the nature of solar energy is affected by the rotation and revolution of the earth. The observation that can be done is to see the position of the sun as if moving from the equator to 23.5° LU back to the equator, continue to 23.5° LS, and back again to the equator [1].

Total energy produced by the sun has an average power of 1 KW/m^2 on the surface of the earth with energy up to 7 kWh/m^2/day on average per year [2]. And it has sunshine hours annually about 2975 hours or 124 days while the average length of irradiation is about 8.2 hours per day [3]. Although the solar energy reaches the Earth is only 30%, the energy radiated within 30 minutes is capable of energy for a year [4]. The intensity of solar radiation averaged about 4.8 kWh/m^2 per day in all parts of Indonesia. One tool that can utilize solar energy is CSP (Concentrated Solar Power) using a Fresnel lens.

CSP's main constraint is its relatively low efficiency. So that the intensity of the CSP is always optimal, it is necessary to adjust the CSP position to the sun position automatically. John A. Duffie [5] suggests that efforts to minimize angles mean the effort to position the solar collector perpendicular to

IOP Conf. Series: Materials Science and Engineering **336** (2018) 012002 doi:10.1088/1757-899X/336/1/012002

the reflection of solar radiation, so as to maximize the energy of solar radiation. The tracking system is classified by its rotational motion.

CSP panel solar light direction is less effective if it is done manually by humans. Due to the ever-changing position of the sun, it is necessary to create a control system that can adjust the direction of the CSP panel by using a solar or sensor algorithm.

2. Modelling of the System

2.1. The Solar Coordinates of the Solar Path

The solar time clock, solar hour, the one-day period at the coordinates of the earth's surface, moves from east to west, parallel to latitude, from an average assume of 6 AM to 18 PM on the equator track. While the solar declination trajectory per year moves from the north, -23.45° North Latitude, December 21, southbound 23.45° south latitude, June 21, and turns north; a period of one year [6].

In the calculation, the sun hour angle to the coordinate point is expressed in an equation, with the 'degree: minute: second' value. The angle of the sun (elevation), obtained from the equation :

$$h_s = 15^o h_{sn} \quad (1)$$

Where,
h_s = angle of the sun
h_{sn} = time angle

While the declination angle, as Figure 1a, is obtained by the equation [7] :

$$\delta_s = 23.45^o \sin([360\frac{(284+n)}{365}]^o) \quad (2)$$

Where,
δ_s = n-day declination angle
n = n-day in one year
n = 1 is January 1st

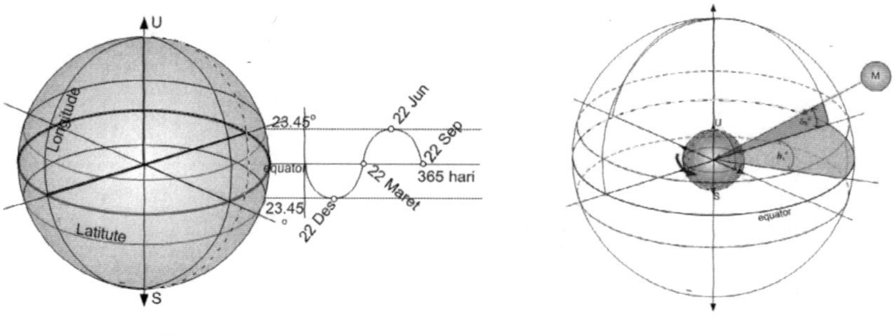

Figure 1. a) Trajectory sun angle declination δ_s a year
b) Sun time angle h_s and declination δ_s sun

Changes the time-angle trajectory and the sun's declination of a tracker coordinate can be illustrated as shown in Figure 1.

Whereas, the azimuth angle on day n in a year is expressed by equation (2), by the AMount of 'degree: minutes: seconds'; with the description, δ_s = n-day declination angle; n = day to n in one year, n = 1 is January 1st. Latitude (L).

2.2. Concentrated Solar Power

The solar concentration is a way to increase the heat of the sun by using an additional equiPMent. The principle of CSP is illustrated in Figure 2.

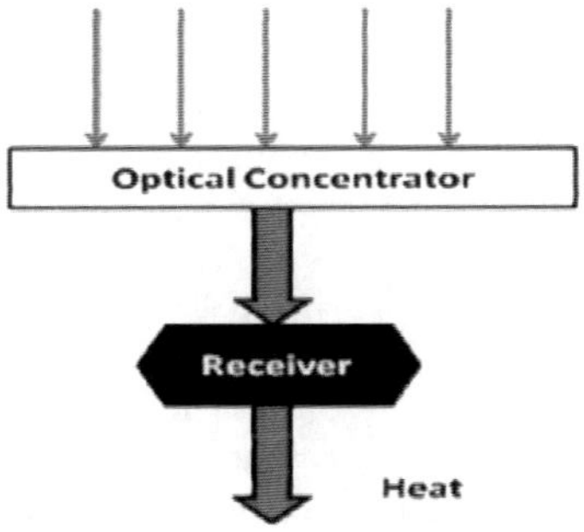

Figure 2. Principle of concentrating solar energy [8]

In solar thermal concentrating, solar energy is optically focused before being transferred into heat. This concentration is obtained by reflecting solar radiation using a mirror or lens. The reflected light will be concentrated in the focus area [4]. There are two types of Fresnel: the refractive lens and the reflective mirrors, as shown in Figure 3. Fresnel bias lenses are mostly used in photovoltaic applications whereas many reflective mirrors are applied in solar thermal. Optical design Fresnel lens is more flexible and can produce a uniform flux density on the absorber [9].

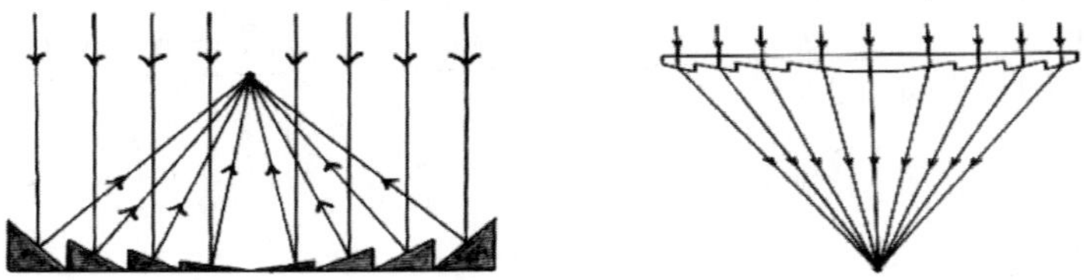

Figure 3. a) Reflective fresnel lens [8]
b) Refractive Fresnel lens [8]

The efficiency of concentrator can be calculated its efficiency with equation (3) and (4), and the focus of the heat in equation (5).

$$\eta_{receive} = \frac{Q_{absorbed} - Q_{lost}}{Q_{solar}} \quad (3)$$

Where,

$\eta_{receive}$ = Efficiency
$Q_{absorbed}$ = *Absorbed energy*
Q_{lost} = Lost energy
Q_{solar} = Energy of the sun

$$Q_{solar} = h_{optic} ICA$$

(4)

h_{optic} = efficiency
I = Solar power (Watt/m^2)
C = kali konsentrasi
A = Area(m^2)

Focus temperature, T_{max}

$$T_{max} = \left(\frac{IC}{\sigma}\right)^{0.25} \tag{5}$$

σ = absorbtivity

2.3. Electronic Solar Tracker

The Axis tracker is aimed at tracking the elevation of the east-west trajectory; by using analogous control hardware in a concentrated solar power system, parabolic trough concentrator [10]. And a declination axis tracker to track the coming sun every year from north to south and south to north.

Where the reflective equations of the elevation axis tracker are as follows:

$$\boldsymbol{\rho(t)} = \frac{\boldsymbol{(h_s-90)}}{\boldsymbol{2}}\boldsymbol{(t)} \tag{6}$$

ρ = reflector angle {-45° < ρ <45°}
h_s = angle of the sun{0° < h_s <180°}
t = time {6AM < t < 18PM}

Figure 4 shows the range of angular motion of the tracker to the change in the angle of the solar elevation as shown in equation (6).

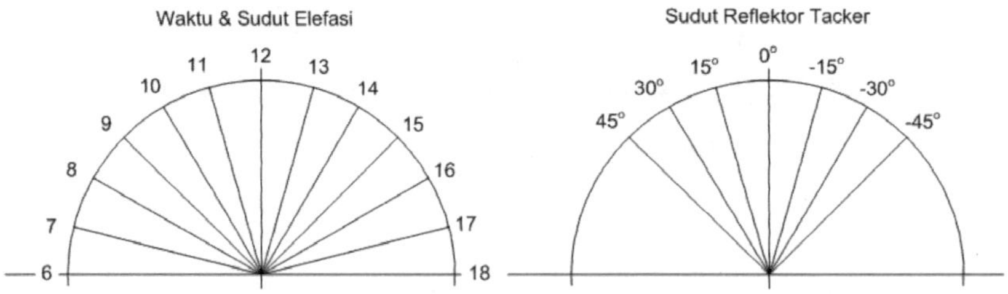

Figure 4. Reflector angle to elevation time

And the reflective equations of the declination axis tracker are as follows:

$$\boldsymbol{\tau(t)} = \frac{\boldsymbol{\delta_s-90}}{\boldsymbol{2}}\boldsymbol{(t)} \tag{7}$$

τ = reflector angle {-11,72° < τ <11,72°}
δ_s = angle of the sun {0° < δ_s <180°}
t = day {1 < t < 364}

Declination Tracker to track sunlight north and south, +/- 23.45° in north and south latitude in one year a period.

2.4. Proposed Array Solar Tracking

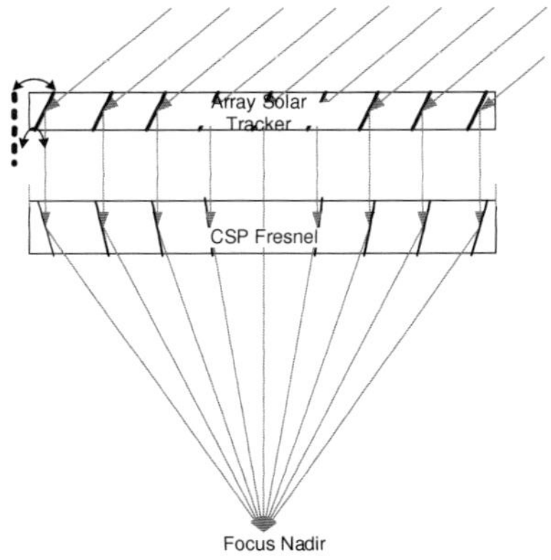

Figure 5. CSP Focus Nadir

The solar light direction system is driven by a solar tracker, then directed toward the Fresnel lens. And the Fresnel lens will focus the sunlight on one point called focus nadir, so the high temperature occur. The direction of the sun's rays is accomplished by utilizing a linear mirror array per axis (reflector). When the sun's rays come angled α^o to the reflector, the angular reflected angle of α^o nadir rays is also against the reflector on the sAMe coming and leaving lines.

Figure 6. Reflection of sunlight by the reflector. (a) Elevation (b) Declination

3. Simulation and Result

The solar cycle at 6 AM until 6 PM is illustrated with a sine signal as above with a maximum intensity is 120000 lux which is equivalent to 1KW. However such power is not entirely acceptable to the flat collector because the flat plate collectors cannot fully capture the sun's heat at a given sun. To see the heat-receiving power of the equation below:

$$P_f = P_s \cos \Theta \quad (8)$$

P_f = Power received flat collector (Watt)
P_s = Solar Power (Watt)
Θ = The resulting angle between the position of the current sun with the top position
The equation above can be made a graph of solar power to flat collector with cycle start at 6 AM until 18 PM expressed in figure 7:

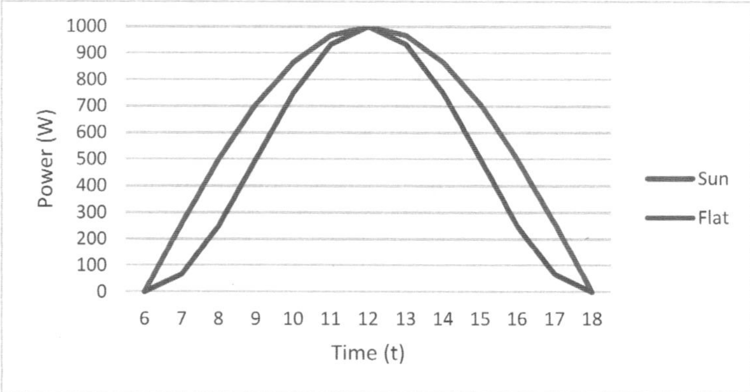

Figure 7. Graph of flat power efficiency to solar power

So the graph it can be seen that the energy efficiency of flat collector is

$$\boldsymbol{\eta flat}(\%) = \sum \frac{\boldsymbol{Pflat}}{\boldsymbol{Ps}} \times \mathbf{100} \quad (9)$$

ηflat = Flat efficiency (%)
The above equation it can be seen that the solar collector with flat plate has an efficiency energy 78.99%. Before the solar light reflect to Fresnel, solar light through the reflector. So, the light will be the direction. The reflector has an efficiency 82%, efficiency depends on the materials used (stainless still).
After sunlight is passed through the reflector, then Fresnel which serves to focus sunlight into heat at one point. From the efficiency can know how much power Fresnel with the following equation:

$$\boldsymbol{Pfr} = \boldsymbol{Pf} \times \boldsymbol{\eta reflector} \quad (10)$$

Pf_r = Power fresnel (Watt)
Pf = Power flat (Watt)
ηreflector= Reflector efficiency (%)

To see how much power on the nadir, the efficiency value of Fresnel is required. Fresnel used is made of PMMA (Polymethyl-methacrylates) which has the efficiency of up to 94% sun absorption. So the nadir power is known from the following equation:

$$\boldsymbol{Pn} = \boldsymbol{Pfr} \times \boldsymbol{\eta fresnel} \quad (11)$$

P_n = Power nadir (Watt)
ηfresnel = Fresnel efficiency (%)

The above equation it can be seen the value of flat power, fresnel power, and power at nadir. From these data it can be graphed as below:

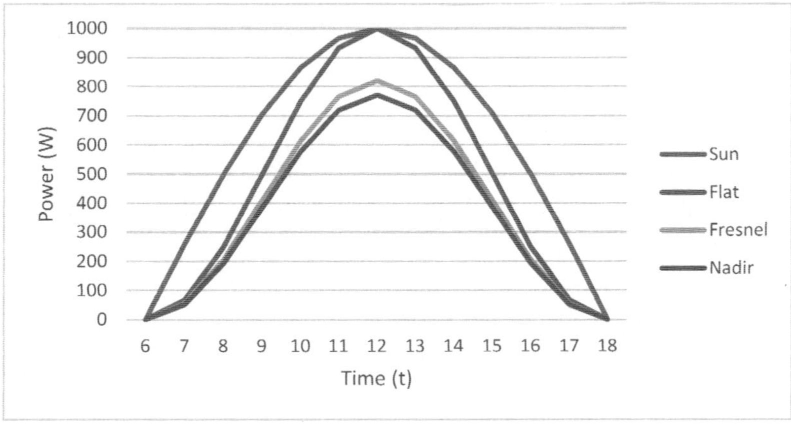

Figure 8. Comparison of overall efficiency

$$\eta Nadir(\%) = \sum \frac{Pn}{Ps} \times 100 \quad (12)$$

ηnadir = Nadir efficiency (%)

The result of the equation shows that the utilization of efficiency solar energy is 60.88%.

The temperature of nadir obtained in the research with Fresnel utilization up to 789°C at 12 AM. High or low temperatures that are focused on the nadir point are obtained from the movement of the sun angle radiated to the reflector that is forwarded to the Fresnel. The reflector follows the direction of the sun assumed that at 12 AM with the flat reflector angle is 90°.

4. Conclusion

From the simulation results with the utilization of concentrated solar power by using fresnel lens is an effective way that can absorb solar energy well. The Fresnel lens of PMMA (polymethyl-methacrylate) is the best Fresnel lens because it has the highest efficiency of 94%. With the utilization of Fresnel, the efficiency of sunlight utilization is 60.88%. With the temperature concentrated at 12 AM is 789°C with a reflector angle of 90°.

References

[1] Hardianto, H. E., & Rinaldi, R. S. (2012). Perancangan Prototype Penjejak Cahaya Matahari Pada Aplikasi Pembangkit Listrik Tenaga Surya. In *FORISTEK: Forum Teknik Elektro dan Teknologi Informasi* (Vol. 2, No. 2).

[2] Mousazadeh, H., Keyhani, A., Javadi, A., Mobli, H., Abrinia, K., & Sharifi, A. (2009). A review of principle and sun-tracking methods for maximizing solar systems output. *Renewable and sustainable energy reviews*, *13*(8), 1800-1818.

[3] Kartiasiha, F., Syaukat, Y., & Anggraenib, L. Determinan Intensitas Energi di Indonesia The Determinants of Energy Intensity in Indonesia.

[4] Kalogirou, S. A. (2004). Solar thermal collectors and applications. *Progress in energy and combustion science*, *30*(3), 231-295.

[5] Beckman, W. A., & Duffie, J. A. (1991). *Solar engineering of thermal processes.* ISBN 0-471-51056-4.

IOP Conf. Series: Materials Science and Engineering **336** (2018) 012002 doi:10.1088/1757-899X/336/1/012002

[6] Setiawan, B., Mauridhi, H., & MochAMad, A. (2013). A High Precision Angle Compensation Controller for Dish Solar Tracker Installed on a Moving Large Ship. *International Review of Automatic Control (IREACO)*, *6*(6), 695-700.Koyuncu, B., & BalasubrAManian, K. (1991). A microprocessor controlled automatic sun tracker. *IEEE Transactions on Consumer Electronics*, *37*(4), 913-917.

[7] Akinjiola, O. P., & Balachandran, U. B. (2012). Concentrated Solar Thermal (CST) System for Fuelwood Replacement and Household Water Sanitation in Developing Countries. *Journal of Sustainable develoPMent*, *5*(6), 25.

[8] Menghani, P. D., Udawant, R. R., Funde, A. M., & Dingare, S. V. (2013). Low-pressure steAM generation by solar energy with fresnel lens: a review. *IOSR Journal of Mechanical and Civil Engineering*, *5*, 60-63.

[9] Lee, C. Y., Chou, P. C., Chiang, C. M., & Lin, C. F. (2009). Sun tracking systems: a review. *Sensors*, *9*(5), 3875-3890.

IOP Conf. Series: Materials Science and Engineering **336** (2018) 012003 doi:10.1088/1757-899X/336/1/012003

Design of Stand-Alone Hybrid Power Generation System at Brumbun Beach Tulungagung East Java

A N Rahmat*, M N Hidayat, F Ronilaya, A Setiawan

Electrical Engineering Department, State Polytechnic Of Malang, Malang, Indonesia

*Angg4.nr@gmail.com

Abstract. Indonesian government insists to optimize the use of renewable energy resources in electricity generation. One of the efforts is launching Independent Energy Village plan. This program aims to fulfill the need of electricity for isolated or remote villages in Indonesia. In order to support the penetration of renewable energy resources in electricity generation, a hybrid power generation system is developed. The simulation in this research is based on the availability of renewable energy resources in Brumbun beach, Tulungagung, East Java. Initially, the electricity was supplied through stand-alone electricity generations which are installed at each house. Hence, the use of electricity between 5 p.m. – 9 p.m. requires high operational costs. Based on the problem above, this research is conducted to design a stand-alone hybrid electricity generation system, which may consist of diesel, wind, and photovoltaic. The design is done by using HOMER software to optimize the use of electricity from renewable resources and to reduce the operation of diesel generation. The combination of renewable energy resources in electricity generation resulted in NPC of $ 44.680, COE of $ 0,268, and CO_2 emissions of 0,038 % much lower than the use of diesel generator only.

1. Introduction

Increased consumption of electrical energy needs is followed by an increase in population and electrical energy needs. Currently, electricity energy has become a primary requirement. since Indonesia, electricity development projects are still focused on urban and industrial areas because the distribution of power grids in remote areas get less attention so that the electricity supply is not reliable. Independent Energy of Village is one of the government's programs to meet the needs of electric energy by optimizing available renewable energy to be able to reduce people's dependence on the use of non-renewable energy sources [1].

Many countries have developed hybrid power generations and their numbers are increasing now [2]. The use of hybrid power generations can reduce the use of expensive fossil fuels and can replace them with clean and environmentally friendly electrical energy. Which in turn, it will improve the quality of the lives of people in remote areas [3].

2. System Description

A hybrid power generation system consists of two or more renewable energy sources that are operated together to obtain an efficient and optimal system [3]. The combination of different energy sources provides the balance and stability of a power plant [3]. The advantages of hybrid power generation system [2][4-6], among others: (1) improving the reliability of the system in supplying the load, (2) reducing the emission gas from the use of fossil fuels, (3) providing a 24-hour power supply, (4) extend the life of the system, and (5) improve the effectiveness and efficiency of the use of electrical energy optimally. Hybrid power generation system configuration [2][5][6], among others: (1) series hybrid system, (2) hybrid switch system, and (3) parallel hybrid system. The advantages of parallel hybrid configuration [6], among others: (1) more optimal load fulfillment, (2) operational efficiency and maintenance of diesel generator, and (3) the other component size is more minimalist thus reducing investment cost.

2.1. Renewable Energy Sources

Various types of renewable energy sources that can be converted into electrical energy, such as wind, solar, water and geothermal. Renewable energy is energy that is in nature free to be utilized. Indonesia is an archipelago country and passed by the equator so that the source of renewable energy is very abundant. Wind is an easily accessible source of renewable energy in coastal areas. Brumbun Beach ($8^0$18' LS,$111^0$094' BT) is located in Ngrejo Village, Tangunggunung District, Tulungagung District, East Java Province. It can be accessed by passing through damaged and impassable roads, the last distance between power lines and residential areas is around 5 km. The data of solar intensity and wind velocity is obtained through hybrid power plant optimization software, by filling in the latitude and longitude coordinates of the location to be studied.

The characteristics and potential of solar energy in selected locations are analyzed based on global solar radiation with monthly averages as well as monthly brightness [7]. Global Horizontal Irradiance (GHI) is the total solar radiation incident on a horizontal surface [8]. At the Brumbun site, it known that the minimum value of solar radiation in January at 4.850 kWh/m^2/day with a brightness index of 0.448 and in September is a maximum value of 6.220 kWh/m^2/day with a brightness index of 0.613, while the average annual solar radiation is 5.37 kWh/m^2/day.

Wind speed parameters are needed in designing wind turbines. The data of wind resources is equipped with a high anemometer in which wind speed is measured [8]. At Brumbun beach area, the recorded maximum wind velocity is in August, i.e.: 6.1 m/s and the minimum wind speed is in December, i.e.: 2.95 m/s, while the average wind speed per year is 4.43 m/s.

Electric energy consumptive behavior for 24 hours on the consumer of electrical energy. Some facilities that require electrical energy, among others: homes, mosques, schools, and shops. The peak load profile occurring at 7 pm for 3.62 kW.

2.2. System Configuration

The configuration of stand-alone or off-grid hybrid power plants is particularly suitable for areas far from the grid's reach, such as remote areas or small populated islands, apart from costly network expansion as well as to minimize electricity generation from non-renewable sources. Here are the components used in the design of power generation systems in this study.

Genset is a power plant to meet the electrical load with a small capacity and to increase electrical power when the peak load. The use of generators is suitable for isolated areas and not yet covered by the power grid. A diesel generator is a model based on its Fuel Consumption (FG) pattern, which is proportional to its output power [12]:

$$F_G = B_G \, x \, P_{G\text{-}rated} + A_G \, x \, P_{G\text{-}out} \tag{1}$$

Photovoltaic is a power generation technology that converts photon from solar radiation into electrical energy [13][14]. In this research, the PV panels used were monocrystalline type because it has the highest efficiency with 24.1% [16]. HOMER modeled the composition of PV as a device that produces dc electricity proportional to the incidence of global solar radiation present, regardless of the temperature and the affected voltage [7]. The equation for determining the output power is [8]:

$$P_{PV} = f_{PV} \cdot Y_{PV} \cdot \frac{I_T}{I_S} \tag{2}$$

Where, f_{PV} is the derating factor of PV, Y_{PV} is the capacity value of the PV module (kW), I_T is GHI (kW/m^2), whereas I_S is 1 kW/m^2.

Electrical energy is generated from wind blows that rotate turbine connected to the generator. By using this software can be determined the use of the optimum amount of wind turbines. Four steps determining the output power of wind turbine [7], are: a) It determines the average wind speed for the hour at the anemometer height by referring to the wind resource data, b) It calculates the corresponding wind speed at the turbine's hub using the logarithmic law or the power law, c) It refers to the turbine's power curve to calculate its power output at that wind speed assuming standard air density, d) It multiplies that power output value by the air density ratio, which is the ratio of actual water density to the standard water density. The equation determines the output power of the wind turbine [8]:

$$P_{WT} = ½ \rho A v^3 C_{pmax} \tag{3}$$

Where PWT is the wind turbine output power (kW), ρ is the real air density (kg/m^3), v is the wind velocity (m/s) while C_{pmax} is the wind velocity coefficient typically is 0.59. The battery is an equipment used to store direct DC electric charge. To improve the efficiency of the system by reducing shortage factor, energy storage is required [10]. The battery storage capacity is given as [13]:

$$CWh = (EL \text{ x } AD)/(\eta_{inv} \text{ x } \eta_{batt} \text{ x } DOD) \tag{4}$$

Where EL is the average energy needed every day (kWh/day), the AD is the battery Autonomy Day, DOD is the depth of discharge, while η inv and η Batt represent each inverter and battery efficiently. The bidirectional converter is a device used to convert DC power source generated from a power source into an AC power source (inverter) and otherwise (rectifier). When the output load is closer to the converter's loaded workload the efficiency is better and otherwise. The efficiency of the converter is better when the output load is close to the workload shown in the converter. The inverter efficiency equation is:

$$\eta_{inv} = Cbatt + [(100\% - \eta_{inv})]C_{batt} \tag{5}$$

2.3. *Cost Optimization*

Economic evaluation of the entire hybrid system is achieved by optimizing the total life-cycle cost of the system configurations [17]. The Net Present Cost (NPC) of a component is the present value of all the costs of installing and operating that component over the project lifetime, minus the present value of all the revenues that it earns over the project lifetime [8].

$$C_{NPC} = \frac{Cann\ tot}{CRF\ (i\ N)} \tag{6}$$

HOMER defines the level of Cost of Energy (COE) as the average cost per kWh of the useful electrical energy produced by the system [8]. To determine the value of COE, HOMER compares the value of the total annual cost (NPC) to the actual electrical load by the hybrid system (kWh/year).

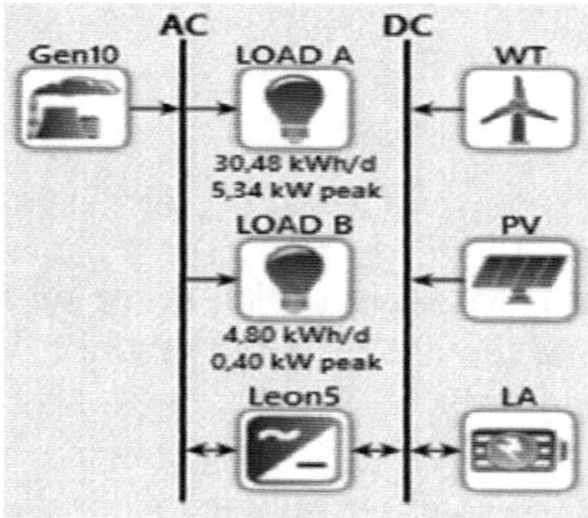

Figure 1. The proposed hybrid configuration

The equation of COE is:

$$COE = \frac{Cann\ tot}{Eload\ serve} \tag{7}$$

3. Method

In this research, simulation and design of hybrid power plant system are done using HOMER software. It is a simulation tool developed by NREL that is used to design a hybrid power plant system [7]. It simulates and optimizes off-grid and grid-connected power generation systems [4][9][11], which can consist of a combination of wind turbines, photovoltaic, micro-hydro, biomass, generators (diesel/gasoline), microturbine, fuel-cell, battery, and hydrogen storage, also serving both electrical and thermal loads. Its algorithm can optimization and sensitivity analysis facilitates the evaluation of various possible system configurations [8]. To make the design of optimum hybrid power plant systems many references suggest using this software [3-5][7][12-14].

The stand-alone hybrid power plant design in this study uses a parallel configuration as shown in Figure 1. Selection of wind turbine with output DC voltage due to the voltage generated by fluctuating conditions. A diesel generator will be used as back-up voltage source if PV panels and wind turbines cant supply batteries and at that time the battery is empty. A diesel generator is also switched on maximum for 1 hr per day during peak load to keep diesel generator always in good condition also keep stable voltage.

Table 1. The input window of hybrid system components

Quantity	PV	WT	Converter	Genset	Battery
Size (kW)	8	6	5	10	20
Capital ($)	7800	3000	2000	5750	4000
Replacement ($)	7800	3000	2000	5750	4000
O&M ($/yr)		100			10
Lifetime (year)	25	25	10	15000 hour	10

The sizing of a hybrid system as shown in Figure 1 is as follows:

- Determine power consumption demand. Total Appliance use = 35,28 kWh/day
- Output Power of PV = P_{PV} = 0,8.4.5,37/1 = 17,2 kWh
- Output Power of WT = P_{WT} = ½ .1,225.1,07.2,29 3.0,59 = 4,64 kWh
- Inverter sizing. Total load of equipment = 3620 W. To maintain the security and stability of the inverter should be considered to be 25-30% larger than the total load. The inverter that should be installed with a minimum specification is 5 kW.
- Battery storage capacity CWh = (21,620x3)/(0,85x0,6x12) = 3532 Ah. So the capacity of batteries to be installed voltage of 24 V with 4000 Ah with 32 hr autonomy.

4. Result and discussion

The parameters used in determining the optimal power system configuration using HOMER software are the smallest values of Net Present Cost (NPC), Cost Of Energy (COE), and CO_2 emissions gas from some of the power plant configurations offered. The optimization analysis is divided into four kinds of system configuration: Diesel Generator, Diesel Generator, and PV also battery, Diesel Generator and wind turbine (WT) also the battery, and Diesel Generator PV and WT are also batteries.

Architecture						Cost				System
						COE ($)	NPC ($)	Operating cost ($)	O&M ($)	CO_2 (kg/yr)
	✓	✓	✓	✓	✓	$0,268	$44.680	$1.402	$358,70	1.077
	✓		✓	✓	✓	$0,370	$61.652	$2.947	$528,90	3.749
		✓	✓	✓	✓	$0,556	$92.518	$6.016	$1.237	9.306
			✓	✓	✓	$0,811	$134.952	$9.530	$1.844	15.166
⚠	✓	✓	✓		✓	$1,07	$177.550	$12.299	$3.800	17.152
⚠	✓		✓		✓	$1,14	$190.418	$13.527	$4.054	19.176
⚠		✓	✓		✓	$1,51	$252.127	$18.672	$5.748	25.849
			✓			$1,60	$267.023	$20.211	$6.132	28.462

Figure 2. Simulation results of the hybrid configuration system

Case 1: Diesel System Generation. From Figure 2 it is shown that the use of power generators only from diesel generators has a COE value of $ 1.60 of NPC value of $ 67,023 and CO_2 emissions of 28,462 kg/yr. The absence of Battery causes utilization-focused only on diesel. This condition is a current problem in the residents of Brumbun Beach, where to meet the need for electricity must generate diesel generators with operational tariffs and very expensive maintenance also energy utilization that has not been 24 hours

Case 2: Diesel – PV System Generation. From Figure 2 it is shown that the use of diesel generator, PV and battery storage power generators has a more optimum value than an only diesel generator. With a COE value of $ 0.370 the NPC value of $ 61,562 and CO_2 emissions levels of 3,749 kg/ yr. With the PV is able to reduce the use of diesel generators to supply the utility voltage during the day. Battery life is required to maintain the stability of this generating system with 32 hr autonomous clock.

Case 3: Diesel – WT System Generation. Figure 2 shows that the use of diesel generator, WT and battery storage media, has a COE value of $ 0.566.00, an NPC value of $ 92,518.00, and a CO_2 emissions level of 9,306 kg/yr. With the WT is able to reduce the use of diesel generator but for the stability of battery filling is very little, because the topography of the coast Brumbun makes little wind gusts. To maximize the electricity generated from wind turbines, a low-speed turbine type is required. Batteries are required to maintain the stability of this generating system with 16 hours less autonomous clock than case 2. The length of autonomous batteries is due to the fluctuating levels of renewable energy sources.

Case 4: Diesel – PV – WT System Generation. From Figure 2 it is shown that the use of diesel generator, WT, PV and battery storage power generators have the smallest value of COE of $ 0.268 the smallest NPC value of $ 44,460 and CO_2 emissions of 1,077 kg/yr. With the combination of two sources of renewable energy PV and wind turbine capable of providing optimal.

5. Conclusion

The aim of this research is to design and simulate the design of hybrid power generation in community settlement at Brumbun Beach. From the design and simulation of the proposed power generation system, optimal results are obtained. Power generation with centralized system configurations has many advantages over-dispersed systems. The results obtained from the design and simulation of hybrid power plants using HOMER software have the value of Cost Of Energy (COE) for 0,268 $, Net Present Cost (NPC) value of 44,460 $, and with CO_2 emissions of 1.077 kg/yr. The hybrid power generation designed is capable of supplying power for 24 hours at a cost per kWh.

6. Acknowledgment
This research project was supported by The State Polytechnic of Malang, Indonesia

References
[1] Juwito, A. F., Pramonohadi, S., & Haryono, T. (2015). Optimalisasi energi terbarukan pada pembangkit tenaga listrik dalam menghadapi desa mandiri energi di Margajaya. Jurnal Semesta Teknika, 15(1).
[2] Kunaifi, K. (2011). PROGRAM HOMER UNTUK STUDI KELAYAKAN PEMBANGKIT LISTRIK HIBRIDA DI PROPINSI RIAU. Telematika, (3).
[3] Magarappanavar, U. S., & Koti, S. (2016). Optimization of Wind-Solar-Diesel Generator Hybrid Power System using HOMER. Optimization, 3(06).
[4] Sopian, K., Othman, M. Y., & Rahman, M. A. A. (2005). Performance of a photovoltaic diesel hybrid system in Malaysia. ISESCO Sci Technol Vision, 1, 37-39.
[5] Sari, D. P., & Nazir, R. (2015). Optimalisasi desain sistem pembangkit listrik tenaga hybrid diesel generator-photovoltaic array menggunakan HOMER (studi kasus: Desa Sirilogui, Kabupaten Kepulauan Mentawai). Jurnal Nasional Teknik Elektro, 4(1).
[6] Nayar, C. V., Phillips, S. J., James, W. L., Pryor, T. L., & Remmer, D. (1993). Novel wind/diesel/battery hybrid energy system. Solar energy, 51(1), 65-78.
[7] Olatomiwa, L., Mekhilef, S., Huda, A. S. N., & Sanusi, K. (2015). Techno economic analysis of hybrid PV–diesel–battery and PV–wind–diesel–battery power systems for mobile BTS: the way forward for rural development. Energy Science & Engineering, 3(4), 271-285.
[8] HOMER Manual Book
[9] Energy, H. Global Horizontal Irradiance (GHI). (2017, September 2017). [Online]. Available: http://www.homerenergy.com
[10] Lambert, T., Gilman, P., & Lilienthal, P. (2006). Micropower system modeling with HOMER. Integration of alternative sources of energy, 1(15), 379-418
[11] Energy, H. (2012). Energy modeling software for hybrid renewable energy systems. Overview [online], available: http://homerenergy. com/[accessed 2 nd April 2017].
[12] Ismail, M. S., Moghavvemi, M., & Mahlia, T. M. I. (2013). Techno-economic analysis of an optimized photovoltaic and diesel generator hybrid power system for remote houses in a tropical climate. Energy Conversion and Management, 69, 163-173.
[13] Quaschning, V. (2016). Understanding renewable energy systems. Routledge.
[14] Sitompul, R. (2011). Manual Pelatihan Teknologi Energi Terbarukan yang Tepat Untuk Aplikasi di Masyarakat Pedesaan. Jakarta: PNPM Support Facility (PSF), 45-75.
[15] Handbook for Solar Photovoltaic (PV) Systems, Grenzone Pte Ltd, Phoenix Solar Pte Ltd, Singapore Polytechnic, Solar Energy, Research Institute of Singapore (SERIS), SP PowerGrid, Urban Redevelopment Authority, Singapore, ISBN: 978-981-08-4462-2, hal. 4-15.
[16] GSES. 2016. A Practical Guide to PV Efficiency. GSES Technical Paper.

IOP Conf. Series: Materials Science and Engineering **336** (2018) 012004 doi:10.1088/1757-899X/336/1/012004

Breakdown Voltage of CF_3CHCl_2 gas an Alternative to SF_6 Gas using HV Test and Bonding Energy Methods

Tedy Juliandhy[1*], T. Haryono[1], Suharyanto[1], Indra Perdana[2]

[1]Departement of Electrical Engineering and Information Technology,
[2]Departement of Chemical Engineering,
Engineering Faculty, GadjahMada University

* tedy.s3te14@mail.ugm.ac.id

Abstract. For more than two decades of Sulphur Hexafluoride (SF6) gases is used as a gas insulation in high voltage equipment especially in substations. In addition to getting an advantage as an insulating gas. SF6 gas is recognized as one of the greenhouse effect gases that cause global warming. Under the Kyoto Protocol, SF6 gas is one of those gases whose use is restricted and gradually reduced to the presence of a replacement gas for SF6 gas. One of the alternative gas alternatives which have the potential of replacing SF6 gas as an insulating gas in Gas Insulated Switchgear (GIS) equipment in the substation is Dichlorotrifluoroethane (CF3CHCl2) gas. The purpose of this paper is to enable a comparison of breakdown voltage with high voltage test and method of calculating Bonding energy to Dichlorotrifluoroethane gas as substitute gas for SF6 gas. At 0.1 bar gas pressure obtained an average breakdown voltage of 18.68 kV / mm at 25oC chamber temperature and has the highest breakdown voltage at 50oC with a breakdown voltage of 19.56 kV / mm. The CF3CHCl2 gas has great potential as an insulating gas because it has more insulation ability high of SF6 gas, and is part of the gas recommended under the Kyoto Protocol. Gas CF3CHCl2 has the capacity to double the value of electronegativity greater than SF6 gas as a major requirement of gas isolation and has a value of Global Warming Potential (GWP) and Ozone Depleting lower than from SF6 gas.

1. Introduction

Gas Insulated Switchgear (GIS) on high voltage equipment is grounded by the power system and the transmission system is expected to function reliably and efficiently[1]–[3]. The ability of gas insulation in GIS equipment plays an important role in performing this function. SF6 gas insulation assists the reliability of the system with its ability to isolate the gas and as an electric arc extinguisher during switching.

GIS is an electrical substation that uses SF_6 gas as its insulating medium. SF_6 gas in addition to its advantages as a gas insulation used in high voltage equipment also has a deficiency as a gas that is limited or reduced its use in Kyoto Protocol conventions because it causes environmental damage such as greenhouse effect, global warming and the occurrence of acid rain in certain areas. One of the contents of the amendment of the Kyoto Protocol is the utilization of alternative gas alternatives as a more environmentally friendly gas insulation[4]–[9]. Under the Kyoto Protocol, the use of SF_6 gas as an insulating medium for high voltage electrical appliances and other applications is recommended for restriction and ultimately terminated until the discovery of a replacement gas that does not pollute the environment and damage the environment. A large amount of research and research investigating new gases as an alternative SF_6 gas as gas insulation in substation equipment shows a positive response to

IOP Conf. Series: Materials Science and Engineering **336** (2018) 012004 doi:10.1088/1757-899X/336/1/012004

the Kyoto Protocol Convention[10]. The response form of the researchers can be seen in the number of researchers who explored the substitute gas of SF_6 gas at most and by mixing the SF_6 gas with other gases.

Monitoring and maintenance of the quality of gas insulation are essential to ensure reliable operation of electrical appliances. In addition to having a reliable capability as a gas insulation, SF_6 is also the largest contributor to environmental damage with global warming, greenhouse effect, and acid rain. Global warming and acid rain are one of the causes of environmental damage caused by SF_6 gas usage[11]–[14]. A more serious characteristic of damage is the formation of acid rain, caused by chemical reactions in the air during leakage of high voltage equipment installation using SF_6 gas as an insulating medium.

The issue of global warming and the depletion of the ozone layer at Kyoto's climate convention resulted in the Kyoto Protocol as a basis for restrictions on the use of gases that contribute to global warming and environmental degradation. It cannot be ignored that the utilization of SF_6 gas in High Voltage equipment especially in a substation of Gas Insulated Switchgear (GIS) equipment is not separated from SF_6 capability as proven gas isolation. The ability of SF_6 as gas insulation because it has the ability to extinguish the electric arc when switching on high equipment also has the value of isolation (dielectric strength) of 90 kV / mm bar. This capability is used on SF_6 gas at the base of gas utilization SF_6 Gas Insulated Switchgear (GIS). The specification of SF_6 gas so that it is used as gas insulation in high voltage equipment can be seen in the table below / Table I:

Table 1. Specifications Sulphur Hexafluoride Gas

Specifications	**Sulphur hexafluoride (SF_6)**
Breakdown Voltage	90 kV/mm
Electronegativity	3.98
Relative Dielectric Strength	2.5
GWP	23900
ODP	0.08
Atmosphere live-time	3200 Year

In table 1, the outline of SF6 gas has good isolation capability in high voltage equipment usage, its ability to withstand breakdown voltage, its electronegativity, and its insulating strength value compared to O2 gas. As for the eco-friendly gas requirements required by the Kyoto Protocol, SF6 gases including environmentally damaging gas are seen at a value of Global Warming Potential (GWP) of 23900 times compared to CO gas, depleting the ozone layer or Ozone Depleting Potential (ODP) of 0.08, and the residence time in the atmosphere reaches 3200 years (Atmosphere Live Time value)[15]–[18]. The existence of the Kyoto Protocol was positively responded by the researchers to develop a gas insulation study in accordance with the requirements of the Kyoto Protocol. The researcher's response to finding alternative gas as an SF6 gas replacement has been largely undertaken by the following research using either new gas or SF6 gas mixed with other gases in order to reduce the released concentrate in the air as shown in the research in Table II [19]–[22].

Table 2. SF_6 gas replacement research

Researcher/Year	**Research Title**	**Research methods**
GenyoUeta, et al (2011)	Evaluation of Breakdown Characteristics of CO2 Gas for Non-standard Lightning Impulse Waveforms Breakdown Characteristics under Single frequency Oscillation Waveforms of 1.3 MHz to 4.0 MHz	Testing CO2 gas as an alternative to SF6 gas with Non-standard Lightning Impulse Waveforms
Junichi Wada, et al (2013)	Evaluation of Breakdown Characteristics of N2 Gas for Non-standard Lightning Impulse	Testing N2 gas as an alternative to SF6 gas with

	Waveforms - Breakdown Characteristics in the Presence of Bias Voltages under Non-uniform Electric Field	Non-standard Lightning Impulse Waveforms
Zoran Rajovi´c, et al (2013)	Influence of SF6–N2 Gas Mixture Parameters on the Effective Breakdown Temperature of the Free Electron Gas	Test the voltage across the SF6-N2 gas mixture and calculate the temperature of the free electrons
Li Kang, et al (2013)	Study of the Decomposition of c-C4F8 and its Mixture with N2 under 50Hz ac Corona Discharge	Test gas c-C4F8 and mixed with N2 gas

One of the potential gases used as an alternative to SF_6 gas is CF_3CHCl_2 (Dichlorotrifluoroethane) gas. Use of CF_3CHCl_2 gas as an alternative gas and has great potential to replace SF_6 gas because CF_3CHCl_2 gas has properties and characteristics similar to SF_6 but in accordance with the insulation gas as required by the Kyoto Protocol. Characteristics of dichlorotrifloroethane gas that has almost the same characteristics as in the forming elements bonding compounds. If theSF_6 gas has the element F (Flour) as an electronegative element and belongs to the halogen group in the chemical element CF_3CHCl_2 gas has elements F and Cl as elements that are electronegative. Because of its specific electronegative properties, the CF_3CHCl_2 gas also has the ability to extinguish the electric arc. The ability of gas insulation owned by SF_6 gas is related to chemical bond with its compounds. Theoretically, the stronger the chemical compound bond, the greater the energy required to release the bonds of the compound. The energy required to release the bonding chemical compounds in the electrical world is referred to as the breakdown voltage. Translucent voltage is a voltage that can penetrate chemical bonds on compounds that are given the minimum voltage or tension required to decompose or release bonds to a chemical compound. In bond chemistry in an interconnected molecule is called binding energy. The binding energy itself is defined as the energy required to release the bond between the forming atoms in the molecule of a compound. Considering the characteristics and capabilities of CF_3CHCl_2 gas as a substitute for SF_6 alternative gas the authors will try to compare the capability of CF_3CHCl_2 gas by using High Voltage (HV) test and binding energy calculation method to get the estimation or prediction of the penetration value of the material used as the insulator.

In this paper, we discussed the comparison of CF_3CHCl_2 gas breakdown voltage by using high voltage test and penetrating voltage value by calculation or prediction of binding energy calculation. If the penetrating voltage value of the binding energy can be used as a prediction to determine the penetrating voltage value of a substance, this method can be recommended as a way to predict the penetrating voltage value of a compound to obtain the insulating material.

2. Experimental setup

2.1. High Voltage testand model

Testing using high voltage using equipment: Vacuum Pump brand Value, Testing transformer type JEC-120 with primary voltage 100-200 AC and secondary voltage 100 kV AC. A test tube gas chamber with a diameter of 30cm and a length of the 40cm tube as showed in figure1and figure 2.

Figure 1. Chamber Gas

While the series of test equipment shown in Figure 2

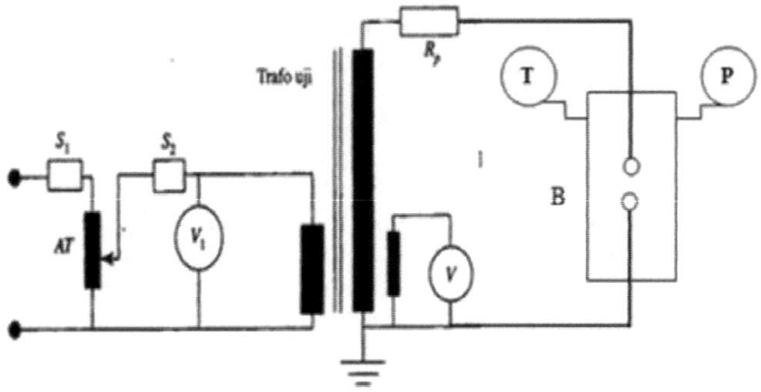

Figure 2. Circuit of HV Test

Before the test, any gas inside the chamber tube is emptied using a vacuum pump. After the chamber tube is empty or vacuum condition then continued by filling the chamber with test gas / CF_3CHCl_2. Gas pressure is included in the chamber at 0.1 bar pressure, electrode gap set at 1mm and 2mm gap distances, the temperature in the chamber set at 25°C, 30°C, 40°C, and 50°C. Increased temperature in the chamber using 100-watt heaters and controller with a thermocouple. After the initial treatment is carried out according to the setting, the CF_3CHCl_2 gas test is conducted to determine the penetrating voltage value of the CF_3CHCl_2 gas material when applied. For each test, sampling test of the breakdown voltage is 5 times. From the results of the 5-time test taken the average value of the test gas voltage in kV / mm Bar. The result of this mean breakdown value will be compared with the value of the penetrating voltage that can be obtained from the prediction method of the energy value of the bond and the value of the voltage required to achieve the penetrating voltage value in the test gas material. Table III shows the breakdown voltage of high voltage test results.

Table 3. Breakdown Voltage CF_3CHCl_2 gas

P (Bar)	d (mm)	T (Celsius)	Breakdown Voltage (kV)					Average (kV)
			1	2	3	4	5	
0.1	1	25	19.4	18.2	19.5	17.9	18.4	18.68
0.1	1	30	19.4	17.9	17	16.6	16.3	17.44
0.1	1	40	16.7	16.7	17.7	16.3	16.5	16.76
0.1	1	50	19.6	18.2	20.9	20.5	18.6	19.56
0.1	2	25	37.3	37.3	38.4	34.3	34.9	36.44

IOP Conf. Series: Materials Science and Engineering **336** (2018) 012004 doi:10.1088/1757-899X/336/1/012004

0.1	2	30	32.6	32.6	31.4	33.3	32.2	32.42
0.1	2	40	32.6	27.9	30.2 9	29	34.9	30.9
0.1	2	50	39.6	39.6	38.4	40.8	36.8	39.04

2.2. *Bond Energy*

Bond energy is the energy needed to break the bonds between atoms in a molecule. The SI unit of bond energy is kilojoules per mole. The greater the binding energy in a compound the greater the energy required to release the binding of the compound. The following table shows the binding energy of the various bonds that have been recorded in the binding energy of a compound/Table IV:

Table 4. Bond Energy

Bond	**Energy (kJoule/mol)**	**Bond**	**Energy (kJoule/mol)**
C--C	348	H--I	299
C--H	413	H--C	413
C--N	293	H--N	391
C--O	358	H--O	366
C--F	485	H--H	436
C—Cl	328	O—O	145
C—Br	276	O—H	463
C—I	240	O—F	190
C—S	259	O—Cl	203
Si—H	323	O—I	234
Si—Si	226	S—H	339
Si—C	301	S—F	327
Si—O	368	S—Cl	253
N—H	391	S—Br	218
N—N	170	S—S	266
N—O	201	F—F	158
N—F	272	Cl-Cl	243
N—Cl	200	Cl—F	253
N—Br	243	Br-Br	193
I----Cl	208	Br---F	237
I---Br	175	Br---Cl	218
I----I	151	Br---P	65

Comparison bonding energy calculation on SF_6 gas and CF_3CHCl_2 gas below is obtained the value of binding energy equal to:
Using the binding energy table in accordance with table 3 was obtained:

SF6 ----------> S + 6F (1)

S-F = (6x327) = 1962 kJ/mol (2)

The binding energy of the chemical reaction of SF_6 is 1962 kJoule/mol. Dichlorotrifluoroethane (CF_3CHCl_2) gas binding energy. Using the binding energy table in accordance usingTable IV was obtained:

F Cl

IOP Conf. Series: Materials Science and Engineering **336** (2018) 012004 doi:10.1088/1757-899X/336/1/012004

F--C--C--H

F Cl

The binding energy of the chemical reaction of CF_3CHCl_2 is:

$$3\ (\text{C---F}) = 3\ (485\ \text{kJ/mol}) = 1455\ \text{kJ/mol} \quad (3)$$

$$2\ (\text{C---Cl}) = 2\ (328\ \text{kJ/mol}) = 656\ \text{kJ/mol} \quad (4)$$

$$\text{C---H} = 1\ (413\ \text{kJ/mol}) = 413\ \text{kJ/mol} \quad (5)$$

Bond Energy CF_3CHCl_2 = (1455 + 656 +413) kJ/mol = 2524 kJ/mol.

3. Result and Discussion

High voltage tests the value of gas penetration voltage CF3$CHCl_2$ after testing 5 times obtained the value of translucent voltage average 18.68 kV / mm or 186.6 kV / mm bar in temperature condition 25°C. While the average value of test breakdown voltage of the test gas at a temperature of 50°C obtained value 19.56 kV / mm or 195.6 kV / mm bar. As the temperature is increased then the gas in the chamber space will undergo development so as to increase the density of the test gas in the chamber. Due to the gas that expands in the chamber, there steps up the pressure, and the greater the pressure in the chamber the test gas bond will be compressed in other words the stronger the bonding of the molecule in the chemical compound. The stronger the chemical bond, the greater the energy needed to release the chemical bonds on the compound forming or the greater the bonding energy, the greater the ability of the test gas to withstand the penetration voltage/insulation properties the greater. The percentage increase in the capability of the gas holds a breakdown voltage of 4.5% so that at 50°C for the capability of the gas in holding the penetrating voltage inside the chamber is higher than the CF_3CHCl_2 gas capability at 25°C. When the chamber was given a temperature condition at 30°C - 40°C the CF_3CHCl_2 gas capability suppressed a translucent voltage ranging from 16-17 kV. The value of the translucent voltage used as a reference then took the lowest penetration value of 16.76 kV or 167.6 kV / mm bar at a temperature of 40°C. GIS equipment at the substation using 7 bar pressures in operation on SF6 gas has a 90 kV / mm bar capabilities. If the CF_3CHCl_2 gas is used in the distended state with a 7 bar chamber pressure such as the treatment of SF6 gas, the gas capacity of CF_3CHCl_2 as a gas isolation can withstand a penetrating voltage of 167.6 kV / mm Bar multiplied by the amount of gas in chamber (7 Bar) obtaining a penetrating voltage value of 1173.2 kV whereas the ability of SF_6 gas to withstand a breakdown voltage of 90 kV / mm Bar multiplied by the amount of gas entered into the chamber (7 Bar) gets a breakdown voltage of 630 kV. The percentage of incremental voltage gain that can be achieved by CF_3CHCl_2gas in SF_6 gas is 86%.

By using the binding energy method, SF_6 gas binding energy of 1962 kJoule / mol, while the bonding energy for CF_3CHCl_2 gas is 2524 kJoule / mol. In accordance with the description above the size of the binding energy affects the ability of the gas to withstand breakdown voltage. The greater the binding energy possessed by a compound the greater the compound's ability to withstand breakdown voltage or to become more insulating. If the binding energy of 1962 kJoule / mol, an SF6 gas capable of withstanding voltage of 630 kV at 7 Bar chamber pressure, the ability of CF_3CHCl_2 gas breakdown voltage which has 2524 kJoule / mol binding energy with the same code in the chamber with 7 Bar pressure is obtained a breakdown voltage of 810 kV.

In the above description and explanation of those two methods (HV test method and bonding energy calculation method) can be performed with the comparative value of breakdown voltage on CF_3CHCl_2 gas. By testing the high voltage gas CF_3CHCl_2 has a total penetration value at 7 bar pressures of 1173.2 kV, while the value of breakdown voltage on CF_3CHCl_2 gas by using the binding energy method obtained the value of 8-10 kV penetrating voltage. Utilizing the HV test the breakdown voltage obtained at CF_3CHCl_2 gas has 363.2 kV differences greater than the binding energy method.

IOP Conf. Series: Materials Science and Engineering **336** (2018) 012004 doi:10.1088/1757-899X/336/1/012004

Comparison of CF_3CHCl_2 gas through voltage test value by HV test method compared with penetrating voltage value of binding energy method obtained CF_3CHCl_2 gas capability can withstand 1.45 times penetration voltage greater than penetrating voltage with binding energy method.

4. Conclusion

Based on the description, results and discussion on the comparison of CF_3CHCl_2 gas breakdown voltage using HV test and binding energy method can be concluded:

- Gas CF_3CHCl_2 is very feasible to be used as a gas substitute for gas insulation on SF6 gas in substations because it has a similar specification with SF6 gas.
- Gas CF_3CHCl_2 has the ability as an electric arc extinguisher when there is switching on high voltage equipment because it contains compound F (Fluor) and Cl (Chlor) which is halogen class and electronegative is the ability of the gas to extinguish the electric arc.
- The result of high voltage test of gas CF_3CHCl_2 able to withstand breakdown voltage above SF6 gas capability (SF6 gas through voltage 630 kV and gas CF_3CHCl_2 equal to 1173.2 kV).
- Comparison of CF_3CHCl_2 gas breakdown voltage by using high voltage testing method and bonding energy method obtained a higher yield from SF6 gas breakdown voltage (HV test method 1173.2 kV and using bonding energy method of 810 kV).
- Gas penetration value of CF_3CHCl_2 with HV test method compared to the bonding energy method obtained by the difference of greater penetration voltage 363.2 kV and CF_3CHCl_2 gas breakdown ratio of 1.45 times the binding energy method.

References

[1] S. Meijer, E. Gulski, J. J. Smit, and A. J. L. M. Kanters, "Risk assessment of fixed defects in GIS under different voltage wave shapes," *Electr. Distrib. 2005. CIRED 2005. 18th Int. Conf. Exhib.*, no. June, pp. 1–4, 2005.

[2] S. Okabe, "Insulation properties and degradation mechanism of insulating spacers in Gas Insulated Switchgear (GIS) for repeated/long voltage application," *IEEE Trans. Dielectric. Electr. Insul.*, vol. 14, no. 1, pp. 101–110, 2007.

[3] S. Okabe, S. Kaneko, T. Minagawa, and C. Nishida, "Detecting characteristics of SF6 decomposed gas sensor for insulation diagnosis on gas insulated switchgear," *IEEE Trans. Dielectric. Electr. Insul.*, vol. 15, no. 1, pp. 251–258, 2008.

[4] J. H. Lee, J. S. Ryu, H. W. Joo, S. J. Tak, S. W. Park, and J. H. Lee, "Insulation characteristics of various conditions for eco-friendly high voltage switchgear," in *The XIX International Conference on Electrical Machines - ICEM 2010*, 2010, pp. 1–4.

[5] K. P. Shine *et al.*, "Perfluorodecalin: Global warming potential and first detection in the atmosphere," *Atmos. Environ.*, vol. 39, no. 9, pp. 1759–1763, 2005.

[6] J. Wang, F. Wei, and J. Feng, "Impact of sea surface temperature anomaly on two global warming periods in the 20th century over the arid central Asia," *Int. Geosci. Remote Sens. Symp.*, pp. 3249–3252, 2011.

[7] S. Lee, J.-W. Choi, and S.-H. Lee, "Separation of greenhouse gases (SF6, CF4, and CO2) in an industrial flue gas using apilot-scale membrane," *Sep. Purify. Technol.*, vol. 148, pp. 15–24, 2015.

[8] L. Zhao, W. Zeng, and Z. Yuan, "Reduction of potential greenhouse gas emissions of room air-conditioner refrigerants: a life cycle carbon footprint analysis," *J. Clean. Prod.*, vol. 100, pp. 262–268, Aug. 2015.

[9] Y. C. Ma *et al.*, "Net global warming potential and greenhouse gas intensity of annual rice-wheat rotations with integrated soil-crop system management," *Agric. Ecosystem. Environ.*, vol. 164, pp. 209–219, 2013.

[10] A. S. Khan and S. A. Chowdhury, "GHG Emission Reduction and Global Warming Adaptation Initiatives by UNFCCC," *Dev. Renew. Energy Technol. (ICDRET), 2012 2nd Int. Conf.*, pp. 1–6, 2012.

[11] S. Moharami and M. Jalali, “Effect of acid rain on the fractionation of heavy metals and major elements in contaminated soils,” *Chem. Ecol.*, vol. 31, no. 2, pp. 160–172, 2015.

[12] D. A. Burns, J. Aherne, D. A. Gay, and C. M. B. Lehmann, “Acid rain and its environmental effects: Recent scientific advances,” *Atmos. Environ.*, vol. 146, pp. 1–4, 2016.

[13] Y. H. Wu, T. M. Liu, C. Sun, J. Xu, and C. K. Yu, “Effects of simulated acid rain on corrosion behavior of Q235 steel in acidic soil,” *Corros. Eng. Sci. Technol.*, vol. 45, no. 2, pp. 136–141, 2010.

[14] Y. J. Du, M. L. Wei, K. R. Reddy, Z. P. Liu, and F. Jin, “Effect of acid rain pH on leaching behavior of cement stabilized lead-contaminated soil,” *J. Hazard. Mater.*, vol. 271, pp. 131–140, 2014.

[15] K. S. Kumar and K. Rajagopal, “Computational and experimental investigation of low ODP and low GWP HCFC-123 and HC-290 refrigerant mixture alternate to CFC-12,” *Energy Convers. Manag.*, vol. 48, no. 12, pp. 3053–3062, Dec. 2007.

[16] B. Dai, C. Dang, M. Li, H. Tian, and Y. Ma, “Thermodynamic performance assessment of carbon dioxide blends with low-global warming potential (GWP) working fluids for a heat pump water heater,” *Int. J. Refrig.*, vol. 56, pp. 1–14, Aug. 2015.

[17] H. Zhang, J. Wu, and P. Lu, “A study of the radiative forcing and global warming potentials of hydrofluorocarbons,” *J. Quant. Spectrosc. Radiat. Transf.*, vol. 112, no. 2, pp. 220–229, Jan. 2011.

[18] Tedy Juliandhy, T.Haryono, Suharyanto, and Indra Perdana, “Utilization Of Dichlorotrifluoroethane Gas As An Alternative To Sf6 Gas Insulated Substation,” *Dissertation Propos.*, 2015.

[19] G. Ueta, J. Wada, and S. Okabe, “Evaluation of breakdown characteristics of CO2 gas for non-standard lightning impulse waveforms -Method for converting non-standard lightning impulse waveforms into standard lightning impulse waveforms -Method f,” *IEEE Trans. Dielectric. Electr. Insul.*, vol. 18, no. 5, pp. 1724–1733, 2011.

[20] J. Wada, G. Ueta, and S. Okabe, “Evaluation of breakdown characteristics of N 2 gas for non-standard lightning impulse waveforms under non-uniform electric field - Breakdown characteristics for single-frequency oscillation waveforms,” *IEEE Trans. Dielectric. Electr. Insul.*, vol. 19, no. 1, pp. 263–273, 2012.

[21] Z. Rajovic, M. Vujicic, K. Stankovic, and P. Osmokrovic, “Influence of SF6-N2 Gas Mixture Parameters on the Effective Breakdown Temperature of the Free Electron Gas,” *IEEE Trans. Plasma Sci.*, vol. 41, no. 12, pp. 3659–3665, 2013.

[22] Y. K. Deng and D. M. Xiao, “Analysis of the insulation characteristics of c-C 4 F 8 and N 2 gas mixtures by Boltzmann equation method,” *Eur. Phys. J. Appl. Phys.*, vol. 57, no. 2, p. 20801, Feb. 2012.

IOP Conf. Series: Materials Science and Engineering **336** (2018) 012005 doi:10.1088/1757-899X/336/1/012005

OTEC Potential of East Nusa Tenggara Province in Indonesia

M Widyartono*, **R Rahmadian**

Departmen of Electrical Engineering t, Universitas Negeri Surabaya, Ketintang Street, Surabaya, Indonesia

*mahendrawidyartono@unesa.ac.id

Abstract. Indonesia is the largest archipelago country in the world, located between Indian Ocean and Pacific Ocean. Indonesia has more than 17000 islands with 70 per cent of the region is ocean. The Growth of the economic and population in Indonesia increasing the demand of the electricity annually, in 2015 alone electricity consumption in Indonesia reaching 200 TWh and will continue increasing every year. However, East Nusa Tenggara Province electrification ratio only around 58.64%, this is the second lowest ratio in Indonesia. This electrification ratio describes the level of availability of electrical energy for the community. Power Plant with renewable source placement in East Nusa Tenggara Province or smaller district need to be prioritise to cope with the low electrification ratio. Renewable sources for power plant have a good potential to work with, in example wind power, solar power, geothermal, or biomass. In addition, another renewable source that not yet known is from the ocean itself. Ocean Thermal Energy Conversion (OTEC) is one of the renewable source method from ocean. This paper will uncover the potential of OTEC in East Nusa Tenggara province so it will bring possibility to build an OTEC power plant in the future.

1. Introduction

Astronomically, Indonesia is located between 6° 04' 30" North latitude and 11° 00' 36" South latitude, and between 94° 54' 21" and 141° 01' 10" East longitude and lies on equator line located at 0° latitude line. In terms of geographic location, Indonesia is located between Asian Continent and Australian Continent, and between Indian Ocean and Pacific Ocean. As an archipelago country, Indonesia consist of thousands of islands interconnected by strait and seas. Indonesia has 34 provinces spreading over five main islands and four archipelagoes. Each province has its own legislature and governor which are subdivided into regencies and cities and then further subdivided into districts and again into administrative villages [1].

East Nusa Tenggara (NTT) is one of the province in Indonesia. Land area of NTT is ± 47,931.54 km^2 or 2.49% of Indonesia's land area and water area ± 200,000 km^2 outside the water of Indonesian Exclusive Economic Zone (ZEEI).NTT Province consists of 21 districts and 1 city, which is spread on three large islands and some small islands, among others: Timor Island (Kupang City, Kupang District, South Central Timor, North Central Timor, Belu and Malacca), Sumba Island (East Sumba, Central Sumba, West Sumba and Southwest Sumba), Flores (West Manggarai, East Manggarai, Ngada, Nagekeo, Ende, Sikka and East Flores); Small Islands (Alor, Lembata, Rote Ndao, and Sabu Raijua).The foremost islands in NTT are disadvantaged areas due to their remote and isolated locations that are less touched by development.

Likewise, the electrical conditions in the province of NTT, there has been no distribution of electricity development in all NTT areas [2]. Based on Statistik Ketenagalistrikan 2015, electrification ratio of NTT is 58,64% [3]. That is the second lowest in Indonesia. This electrification ratio describes the level of availability of electrical energy for the community. The total of PLN electricity users in NTT are 628,427 customers, with the largest percentage in Kota Kupang of 102,743 customers. Meanwhile, the lowest PLN electricity users are in the regencies of West Sumba and Central Sumba with 13,143 customers. The only solution to those problems is a good diversification of power resource on both its variety and amount through alternative power resource, especially the renewable one. Ocean Thermal Energy Conversion (OTEC) is one of the renewable source method from ocean.

2. Ocean Thermal Energy Conversion

2.1. OTEC System

OTEC is a technology of electric power generation that utilizes the temperature difference between water at sea level and deeper seawater. OTEC uses heat energy stored in warmer sea levels to produce steam from ammonia fluid to rotate steam turbine. Meanwhile cold water from within the ocean is pumped to the surface to cool down steam out from turbine.

Temperature difference between the sea surface and deep sea, with depth not more than 1000m (3280 ft), is around 20 degrees Celsius, or about 68 degrees Fahrenheit. The greater the temperature difference will generate bigger electricity as well [4]. OTEC system uses circulate fluids in cycle and with help of heat exchanger, the heat from the surface water of the will be absorbed by fluids which then will evaporate it. The steam generated will turn the turbine that coupled mechanically with electricity generator. The steam the exits from the turbine will be condensed with seawater taken from deep-sea level, so it will complete the operation cycle. The diagram of the system can be seen below.

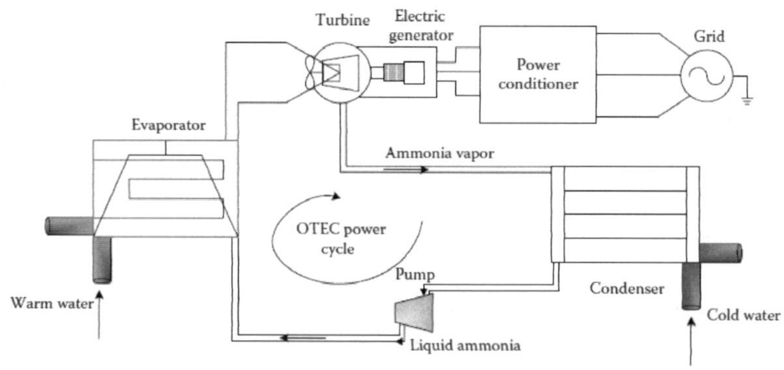

Figure 1. OTEC system [5]

OTES system can be applied offshore or onshore near the beach. Offshore OTEC usually uses cold seawater taken from depth around 600 – 1000m below sea level, and warm water taken from sea level up to 30m below. Offshore OTEC is a floating construction which anchored to the seabed, this construction could be a lighthouse or a buoy. Underwater cable use to transmit electricity generated to the main building on the shore. Onshore OTEC system needs longer pipe for the seawater intake because of the distance between the construction and the sea. However, onshore OTEC system have the benefits because the system will discharge a desalination water that can be used for other purposes.

2.2. Classification of OTECs

2.2.1 Closed-Cycle OTEC Systems. A closed-cycle OTEC systems uses low boiling-point fluids like ammonia, that later will be boiled until changed into steam. This steam will turn the turbines that coupled with turbo-generator. Warm seawater pumped through the heat exchanger which where the boiling happens, the second heat-exchanger served as the cooler, with cold deep seawater serve as a cooler. The steam then will be reverted back to fluids through condensation process, and the cycle continues. Figure 2 shows closed-cycle OTEC systems.

2.2.2 Open-Cycle OTEC systems. In the Open-Cycle OTEC system, uses seawater as the main components (figure 2b). The boiling-point of water can be changed by changing the pressure around it. In this system warm seawater pumped into a vacuum chamber with low pressure around 2% that lowers the seawater boiling point. The seawater then will be boiled and change into steam that will turn the low-pressure turbine. The steam and then will be cooled down with cold deep seawater. The result of this process is desalination water that can be used as irrigation and/or other purposes. This system benefits for the area that have rare access to clean water.

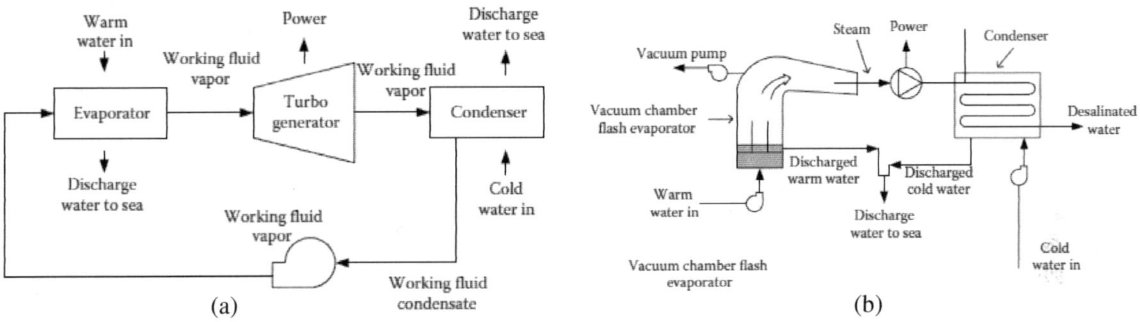

Figure 2. Operational system diagram of OTEC: (a) close cycle (b) open cycle

2.2.3 Hybrid Cycle OTEC systems. This system called hybrid cycle because it combines both from open-cycle and closed-cycle method. In this system both ammonia and warm seawater servers as the main components. The warm seawater will be divided into 2 purposes, first one used as evaporator for the ammonia and the second one will be evaporated just like the ammonia. Each steam (seawater and ammonia) will be separated and turns the turbine just like closed cycle for ammonia and open cycle for seawater. Then, ammonia will be condensed and reuse as main components, but seawater will be distilled and discharged for other purposes.

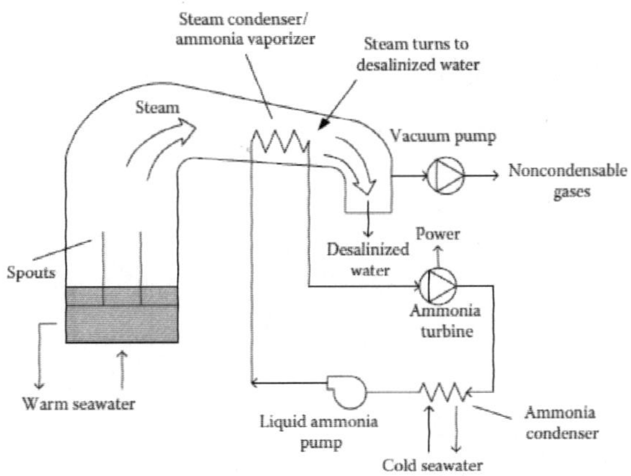

Figure 3. Hybrid Cycle OTEC systems

IOP Conf. Series: Materials Science and Engineering **336** (2018) 012005 doi:10.1088/1757-899X/336/1/012005

2.3. Thermodynamic Efficiency

According to the thermodynamics first law, the energy balance for the fluid served as the main component is stated by:

$$W = Q_H - Q_C \tag{1}$$

Where Q_H is the value of the heat energy absorbed from the heat reservoir and Q_C is the value of the heat energy absorbed form the cold reservoir, which has negative value. According to the thermodynamics second law, the system efficiency restricted by the Carnot efficiency cycle. The maximum possible efficiency would be

$$\eta = \frac{W}{Q_H} = \frac{Q_H - Q_C}{Q_H} = \frac{T_H - T_C}{T_H} = 1 - \frac{T_C}{T_H} \tag{2}$$

Equation 2 define the principal from Carnot's efficiency. Based on this equation, can be said that maximum efficiency from the heated engine depends on both hot and cold reservoir. Because of this, the seawater temperature may affect the power generated by OTEC system, in addition temperature of the seawater are varied throughout the year.

3. Potential OTEC in East Nusa Tenggara (NTT)

OTEC is one of the renewable resource system that can be used in Nusa Tenggara Timur (NTT). There are 2 suburbs and one city that can OTEC system could be applied, which is Waingapu, Ende and Kupang respectively. These three locations have a high potential because of several reasons. The First one is because the location is near the sea. The second is because the seawater temperature near these location is relatively stable throughout the year. The last one is because the distance from the deep seawater intake is not far. Figure 5 show the distance between shore and the minimum depth (1000m) for deep seawater intake, while table 1 show the average temperature for the seawater on the surface and the temperature at 1000m depth. Figure 6 taken from Ocean Data View [6], show yearly temperature from different depth form three different location, Waingapu, Ende, and Kupang respectively.

Table 1. Annual surface and seabed (1000 m) temperatures on several location in NTT.

Location	T_{max} (°C)	T_{min} (°C)	Efisiensi (η)
Waingapu	27,97	4,88	0,826
Ende	27,76	4,87	0,825
Kupang	28,06	4,63	0,835

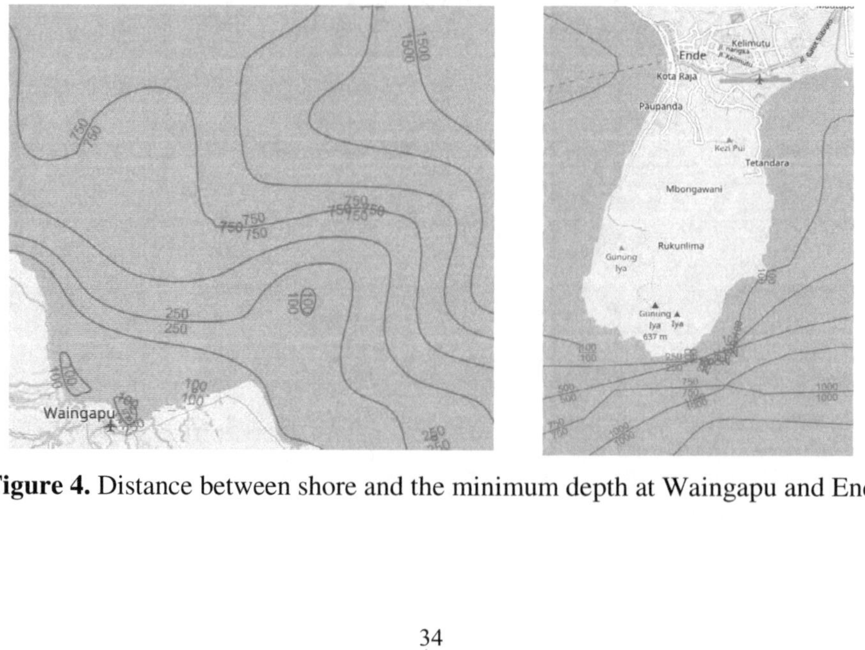

Figure 4. Distance between shore and the minimum depth at Waingapu and Ende

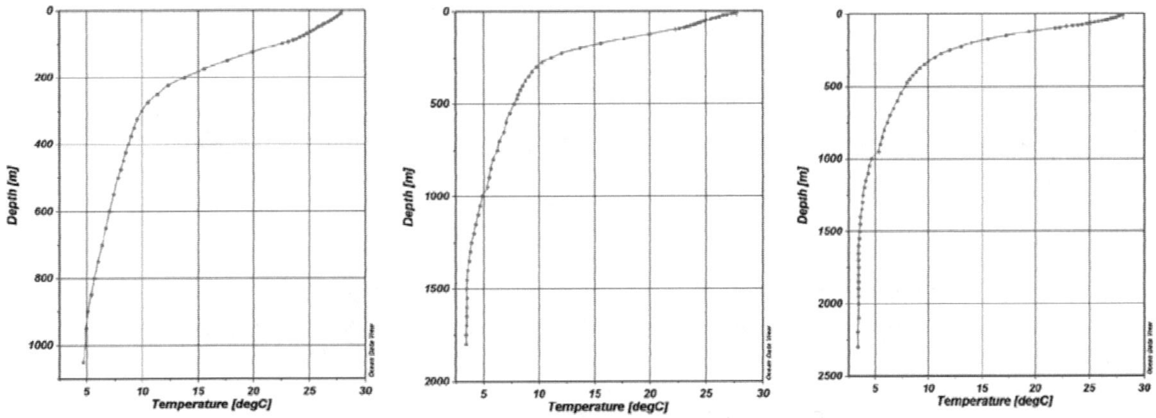

(a) (b) (c)

Figure 5. Profil temperature at different water depths in (a) Waingapu (b) Ende dan (c) Kupang

4. Conclusion

This journal discussed OTEC potential in three different location in NTT, Waingapu, Ende and Kupang. From this journal can be concluded that if applied, OTEC system in Waingapu have an efficiency of 0,826, and OTEC system in Ende have 0,825 efficiency, lastly in OTEC system in Kupang will have efficiency of 0,825. OTEC system have a large potential in Indonesia because of its sea temperature is ideal for OTEC system. These results are expected to help optimize the use of OTEC in Indonesia in the future.

Reference

[1] Badan Pusat Statistik 2016 *Statistical Yearbook of Indonesia 2015*. BPS Statistic Indonesia

[2] Seri Analisis Pembangunan Wilayah Provinsi Nusa Tenggara Timur 2015

[3] Agoes Triboesono 2016 *Statistik Ketenagalistrikan 2015* Kementerian Energi dan Sumber Daya Mineral

[4] Khaligh A and Onar C 2010 *Energy, Power Electronics and Machines Series*CRC Press Taylor & Francis Group

[5] Koto J 2016 *Potential of Ocean Thermal Energy Conversion in Indonesia* Vol.4 No.1 International Journal of Environmental Research & Clean Energy

[6] World Ocean Atlas 2013 [Cited 13 Septermber 2017] http://odv.awi.de/en/data/ocean/world-ocean-atlas-2013/

[7] Syamsudin M L and friend 2015*OTEC Potential in The Indonesian Seas*Energy Procedia 65 (2015) 215 – 222 Conference and Exhibition Indonesia – New, Renewable Energy and Energy Conservation

[8] Avery W H and Wu C 1994 *Renewable Energy from the Ocean : A Guide to OTEC* Oxford University Press

[9] Sinuhaji A R 2013 Potential Ocean Thermal Energy Conversion (OTEC) in Bali Vol. 1 (2015) 5-12 Renewable Energy and Energy Conversion Conference and Exhibition

IOP Conf. Series: Materials Science and Engineering **336** (2018) 012006 doi:10.1088/1757-899X/336/1/012006

Implementation of Fuzzy Decision to Control Patient Room Facilities using Eye Blink

Ilham A.E. Zaeni [1,*], Aji P.Wibawa[1], Aripriharta[1], Siti Sendari[1]

[1]Department of Electrical Engineering, State University of Malang, Indonesia

* ilham.ari.ft@um.ac.id

Abstract. This study proposed the implementation of Fuzzy decision to control patient's room facilities. In this study, four icons were sequentially displayed on the computer screen. The icons representing four option that can be selected by the patient is including switch the light on/off, switch the fan on/off, moving the bed's backrest downward, and moving the bed's backrest upward. The eye blink was extracted from subject's electroencephalograph (EEG) signals which acquired from the FP1 region. The attention was also extracted from subject's EEG signals to ensure that subject concentrate to the task. The eye blink and attention level were used for Fuzzy decision inputs, while the output is a decision that states the selection is valid or not. The selected option is the command that appears on the screen when the selection is valid. In this study, subjects were asked to choose each command several times and the accuracy was computed based on the number of correct selection.

1. Introduction

Some people cannot perform some daily activities including walking, grasping etc. because of some condition such as Amyotrophic Lateral Sclerosis (ALS) diseases. ALS is defined as diseases that attach body muscle which cause the subject cannot conduct any voluntary movement [1], [2]. In some cases, the patient cannot perform any activities and only lying on the bed. Because of the condition, these people need some assistive device to help them conducting some daily activities.

Some researcher uses the subject brain signal as an input to control some assistive device. The technology that utilizes subject electroencephalograph (EEG) signal to control some device is defined as Brain-computer interface (BCI) devices [3]. Some gaming devices have been developed by using brain signal to control game object movement such as air swimmers [4]. BCI-based feeding robotic has been developed using arm robot controlled by brain signal [5], [6]. BCI-based wheelchair has been designed to assist patients to move from one to another place [7], [8].

In the ALS diseases, the eye muscle still retains its function while the other voluntary muscle has been paralyzed [9]. Because of this condition, the motion generated by eye muscle such as eye blinking can be used as input to control assistive device for ALS patients. There are several researches that used eye blinking signal as input to control some devices such as text typing systems [10], [11], and home lighting equipment [12]. The eye blinking can be detected from the EEG signal by filtering the signal with frequency less than 5 Hz and the amplitude is greater than 4 μV [12]. The Fp1 channel can be set as the location to acquire EEG signal and obtain the eye blink signal [13].

The ALS patient usually spent their time by staying in their room and lying on the bed. Sometimes they want to move their bed backrest upward or downward, switch the light on/off, and switch the fan on/off to make them feel comfortable. Based on the description, a system is developed to provide patient room facilities that can be controlled by subject eye blinking gestures. The room equipment

IOP Conf. Series: Materials Science and Engineering **336** (2018) 012006 doi:10.1088/1757-899X/336/1/012006

that can be controlled including switch the light on/off, switch the fan on/off, moving the bed's backrest downward, and moving the bed's backrest upward.

2. Experimental Details

2.1. System Design

In this system, a computer screen was placed in front of the subject to display four icons representing the control command. The control command that is provided is including switch the light on/off, switch the fan on/off, moving the bed's backrest downward, and moving the bed's backrest upward. Only one icon appeared on the computer screen, while the other icons were hidden. Each icon appears in the computer screen for the same period. The sequence of icons that displayed on the computer screen is shown in Figure 1. The command can be selected by blinking the subject eye when the suitable icon is displayed on the computer screen. The subject should also keep his/her attention at a specific level in order to obtain a valid selection.

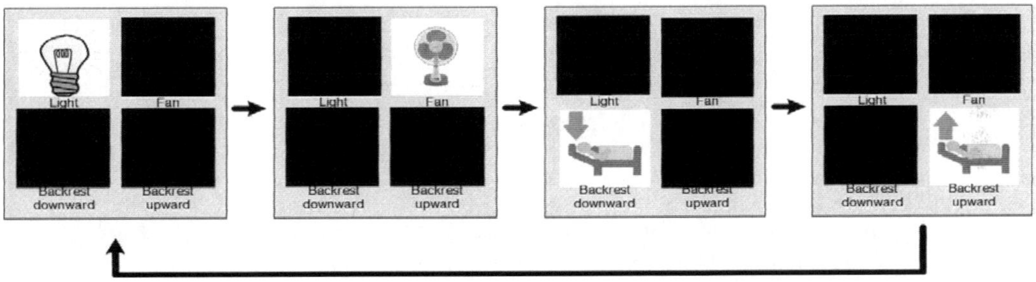

Figure 1. The sequence of icons that displayed on the computer screen

During the experiment is conducted, the subject EEG signals were acquired. The EEG acquisition is conducted using Mindwave Mobile headset from Neurosky Company. The electrode was placed at FP1 channel while the ground and reference are located at the left earlobe. The raw EEG signal is sampled at 512 samples/second. The raw EEG signal is digitized using 10-bit analog to digital converter by ThinkGear ASIC Module (TGAM) in the headset. The digitized raw EEG signal is sent through Bluetooth communication to the personal computer (PC) and used to detect the blink strength.

The eye blink strength was obtained from raw EEG signal after several processing steps. Because the eye blink has low frequency, the raw EEG signal is filtered using fourth order Butterworth low pass filter. The filter's cut off frequency is set to 5 Hz. Based on the filtered EEG signal, the maximum value in one time frame was computed. The maximum value is used as the eye blink strength. The attention level was also measured and sent to the PC for further analysis. The attention level is used to ensure that the subject is focus on the task. If the attention level is higher, it is mean that the subject more focuses on the task. When the subject is rest, the attention level will be low.

The blink strength and attention level are used as Fuzzy input, which the Fuzzy output is decision expressing that the selection has been made or not. When the selection has been made based on the Fuzzy output, the selected option is the icon that appears on the computer screen. The decision of the selected option is used to control the room facilities. The overall system design is shown in Figure 2.

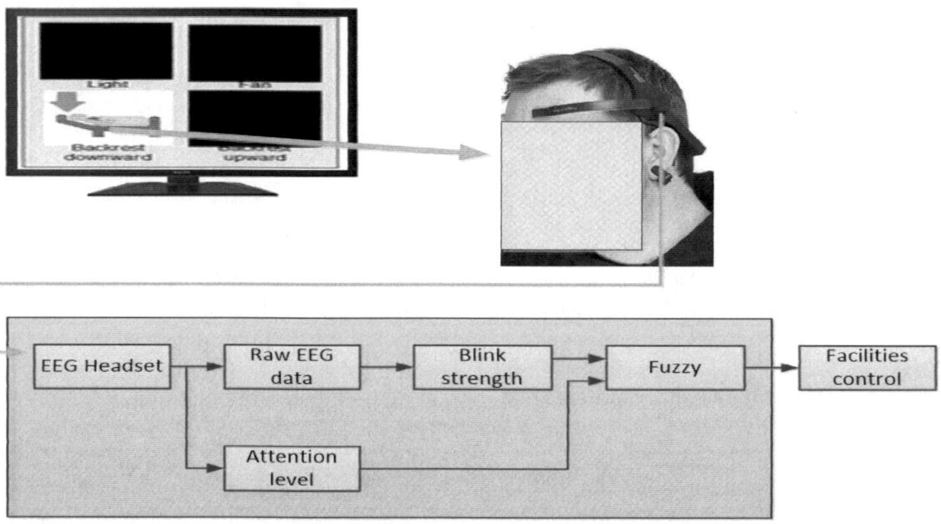

Figure 2. System Design

2.2. Fuzzy Decision Model

Fuzzy inference was used as decision model in this study. The structure of the Fuzzy decision model is shown as part of the system in Figure 2. The blink strength (BS) and attention level (AT) were used as fuzzy inputs. The attention and blink strength value are normalized into range 0 to 100 before used as fuzzy inputs. The fuzzy input sets are divided into 5 linguistic value, including very low (VL), low (L), normal (N), high (H), and very high (VH) for both inputs. While the fuzzy output sets are defined into 5 linguistic value, including very not valid (VNV), not valid (NV), normal (N), valid (V), and very valid (VV). The membership function of the fuzzy that used in this study is shown in Figure 3.

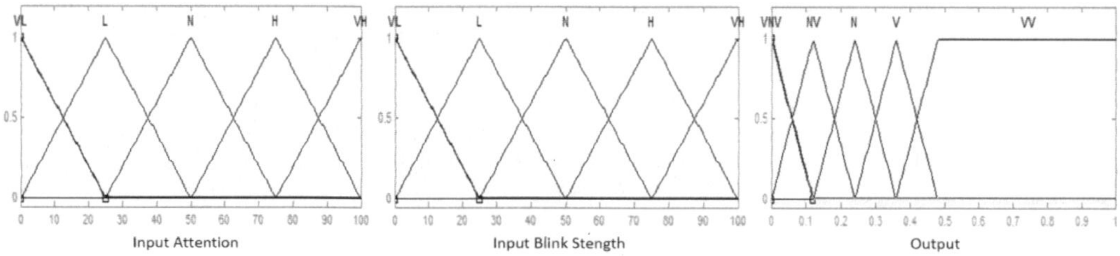

Figure 3. Fuzzy Membership

The inference rule that is used in this study is defined as

$$\text{if AT is Ai and BS is Bj, then y is Cij,} \quad (1)$$

where i and j are the indexes of rules. Ai and Bj are the membership functions of attention (AT) and blink strength (BS), respectively. Cij is the fuzzy decision output based on the rules that shown in Table 1.

Table 1. The rules of the fuzzy decision

Bj \ Ai	VL	L	N	H	VH
VL	VNV	VNV	NV	N	N
L	VNV	NV	NV	N	N

N	NV	NV	N	N	V
H	N	N	N	V	VV
VH	N	N	V	VV	VV

The fuzzy decision output variable, F'D is the result of the defuzzification process and computed using center of gravity, as

$$F'D = \frac{\sum_{i=1}^{n} S_i B_i}{\sum_{i=1}^{n} S_i} \qquad (2)$$

where the variable Si is the grade of the fuzzy membership of the i-th premise based on the inference rule, and variable Bi represents the center value of the i-th conclusion.

The decision output (FD) is computed by rounding the fuzzy output variable (F'D) up. The decision output can be defined as:

$$FD = \text{round}\,(F'D) \qquad (3)$$

If the decision output (FD) is 1, then it can be concluded that the selection is valid and the selected command is the command that appears on the screen when the decision is made.

3. Results and Discussion

The experiment was conducted by asking the subject to sit in front of computer monitor. The subject is asked to use NeuroskyMindwave. The subject brain wave signals were acquired and recorded from Fp1 channel using NeuroskyMindwave. The acquired EEG signal is processed and analyzed to obtain the decision. Based on the decision, the command selection is made by choosing the icon that appears when the decision is made.

The first experiment is conducted by changing the time interval between icons to appear on the computer screen. Several time intervals were used in this experiment, including 1, 2, and 3 seconds. In this experiment, subject was asked to make 20 correct selection. Based on the number of correct and wrong selection, the accuracy was computed. The experiment result is shown in Table 2.

Table 2. The Experiment Result using different time interval

Subject	**Time interval**		
	1 second	**2 second**	**3 second**
A	83%	87%	91%
B	83%	91%	91%
C	80%	83%	83%

The experiment result is shown that time interval 1 second yields the worst result. The subjects said that the time interval 1 second is not enough for them to make a decision and choose the command by blinking their eye. Sometimes when they make the decision and choose the command, the icon has changed and yielded a wrong selection. The time interval 3 seconds yields the best result, while time interval 2 seconds yields the result that almost same as time interval 3 seconds. Because time interval 2 seconds give the result that almost same as the best result, therefore the time interval 2 seconds is used in the next step. The time interval 3 seconds is not used in next step because it will take time to wait the option changed from one icon to another icon.

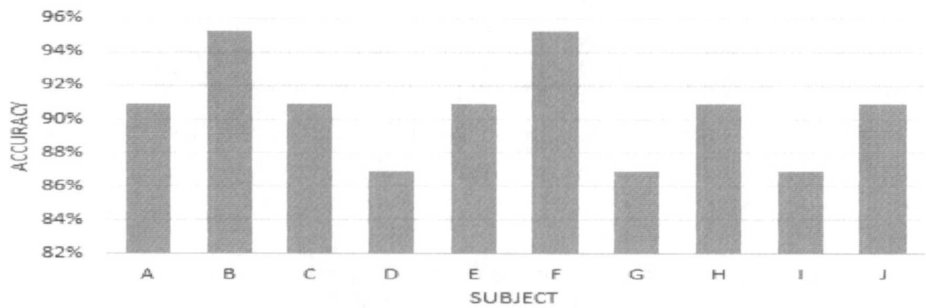

Figure 4. Experiment Result

The second experiment is conducted by asking 10 subjects to participate in the experiment. This experiment is conducted to test the performance of the proposed approach using time interval 2 seconds. In this experiment, subject is asked choose every option for 5 times, therefore the subject should make 20 correct selections to finish the experiment. Based on the number of correct and wrong selection, the accuracy was computed. The experiment result is shown in Figure 4. The result shows that the average accuracy was 90.59%, while the maximum and minimum accuracy is 95.24% and 86.96%, respectively.

4. Conclusion

All subjects in this experiment yield the accuracy more than 80%, so the accuracy of the proposed approach is high enough. Therefore, the proposed approach can be implemented as decision model to provide patient's room facilities control using eye blink. Using the eye blink and attention level as input of Fuzzy system, the system can be used to select the command to switch the light on/off, switch the fan on/off, moving the bed's backrest downward, and moving the bed's backrest upward.

References

[1] ALS association, "About ALS." [Online]. Available: http://www.alsa.org/about-als/. [Accessed: 31-May-2017.

[2] National Institute of Neurological Disorders and Stroke (NINDS), "Amyotrophic Lateral Sclerosis (ALS) Fact Sheet | National Institute of Neurological Disorders and Stroke." [Online]. Available: https://www.ninds.nih.gov/Disorders/Patient-Caregiver-Education/Fact-Sheets/Amyotrophic-Lateral-Sclerosis-ALS-Fact-Sheet. [Accessed: 31-May-2017].

[3] J. R. Wolpaw, N. Birbaumer, D. J. McFarland, G. Pfurtscheller, and T. M. Vaughan, "Brain–computer interfaces for communication and control," Clin. Neurophysiol., vol. 113, no. 6, pp. 767–791, 2002.

[4] Y.-J. Chen, S.-C. Chen, I. A. E. Zaeni, and C.-M. Wu, "Fuzzy Tracking and Control Algorithm for an SSVEP-Based BCI System," Appl. Sci., vol. 6, no. 10, 2016.

[5] S. C. Chen, C. H. Hsu, H. C. Kuo, and I. A. E. Zaeni, "The BCI Control Applied to the Interactive Autonomous Robot with the Function of Meal Assistance," in Proceedings of the 3rd International Conference on Intelligent Technologies and Engineering Systems, vol. 345, J. Juang, Ed. New York: Springer, 2016, pp. 475–483.

[6] S.-C. Chen, C.-M. Wu, I. A. E. Zaeni, and Y.-J. Chen, "Applying fuzzy decision for a single channel SSVEP-based BCI on automatic feeding robot," Microsyst. Technol., pp. 1–9, 2016.

[7] T. F. Bastos, S. M. T. Muller, A. B. Benevides, and M. Sarcinelli-Filho, "Robotic wheelchair commanded by SSVEP, motor imagery and word generation," in Engineering in Medicine and Biology Society, EMBC, 2011 Annual International Conference of the IEEE, 2011, pp. 4753–4756.

[8] A. Widyotriatmo, Suprijanto, and S. Andronicus, "A collaborative control of brain computer interface and robotic wheelchair," in Control Conference (ASCC), 2015 10th Asian, 2015, pp. 1–6.

IOP Conf. Series: Materials Science and Engineering **336** (2018) 012006 doi:10.1088/1757-899X/336/1/012006

[9] Umeå universitet, "Eye muscles are resilient to ALS -- ScienceDaily." [Online]. Available: https://www.sciencedaily.com/releases/2017/01/170126093252.htm. [Accessed: 31-May-2017].
[10] B. Ashtiani and I. S. MacKenzie, "BlinkWrite2: An Improved Text Entry Method Using Eye Blinks," in Proceedings of the 2010 Symposium on Eye-Tracking Research & Applications, 2010, pp. 339–345.
[11] I. Scott MacKenzie and B. Ashtiani, "BlinkWrite: efficient text entry using eye blinks," Univers. Access Inf. Soc., vol. 10, no. 1, pp. 69–80, 2011.
[12] M. S. b A. Rani and W. bt. Mansor, "Detection of eye blinks from EEG signals for home lighting system activation," 2009 6th International Symposium on Mechatronics and its Applications. pp. 1–4, 2009.
[13] M. Abo-Zahhad, S. M. Ahmed, and S. N. Abbas, "A new multi-level approach to EEG based human authentication using eye blinking," Pattern Recognit. Lett., vol. 82, Part 2, pp. 216–225, Oct. 2016.

IOP Conf. Series: Materials Science and Engineering **336** (2018) 012007 doi:10.1088/1757-899X/336/1/012007

Monitoring System and Temperature Controlling on PID Based Poultry Hatching Incubator

S Shafiudin*, N Kholis

Electrical Engineering Department, Faculty of Engineering, Universitas Negeri Surabaya

*Sofyanshafiudin@gmail.com

Abstract. Poultry hatching cultivation is essential to be observed in terms of temperature stability by using artificial penetration incubator which applies On/Off control. The On/Off control produces relatively long response time to reach steady-state conditions. Moreover, how the system works makes the component worn out because the lamp is on-off periodically. Besides, the cultivation in the market is less suitable to be used in an environment which has fluctuating temperature because it may influence plant's temperature stability. The study aims to design automatic poultry hatching cultivation that can repair the temperature's response of plant incubator to keep stable and in line with the intended set-point temperature value by using PID controller. The method used in PID controlling is designed to identify plant using ARX (Auto Regressive eXogenous) MATLAB which is dynamic/non-linear to obtain mathematical model and PID constants value that is appropriate to system. The hardware design for PID-based egg incubator uses Arduino Uno R3, as the main controller that includes PID source, and PWM, to keep plant temperature stability, which is integrated with incandescent light actuators and sensors where DHTI 1 sensor as the reader as temperature condition and plant humidity. The result of the study showed that PID constants value of each plant is different. For parallel 15 Watt plant, Kp = 3.9956, Ki = 0.361, Kd = 0, while for parallel 25 Watt plant, the value of Kp = 5.714, Ki = 0.351, Kd = 0. The PID constants value were capable to produce stable system response which is based on set-point with steady state error's value is around 5%, that is 2.7%. With hatching percentage of 70-80%, the hatching process is successful in air-conditioned environment (changeable).

1. Introduction

In daily life, human being's activities require stable temperature both for their comfort in their activities and for their work as in poultry farms. Poultry hatching cultivation has to be heeded in terms of successful process, even more for local chicken eggs. Hatching eggs has two ways; 1) incubated by the chicken hen directly and 2) through egg hatching machine with a system to control the temperature of the radiation of heating lamp. According to the field survey on egg hatcher, the result showed that: 1) the suitable temperature of incubation is at 37 ° C; 2) the span of the egg-laying period is approximately 21 days; 3) and the characteristics of incubator's design and egg. Stability of hatchery room temperature should always be maintained in order to obtain maximum results (Angga, 2016)

Fluctuation in environmental temperatures can influence the temperature of the hatchery (*forced air*). Moreover, the working system is still On / Off with the outages of the heating lamp which cause wear-worn components. Besides, while the system is still not precise, there is a big error which has a relatively long settling time to reach the next set point due to the instability of the radiation beam temperature. Therefore, an idea to control the temperature of the incubator appears in order to reduce error level, wear out on components and keep the system working so that the temperature is more stable (steady state) based on the set point with PID control method. This method can be used to determine each parameter which is done by plant identification process to get mathematical model through ARX (Auto Regressive eXogenous).

IOP Conf. Series: Materials Science and Engineering **336** (2018) 012007 doi:10.1088/1757-899X/336/1/012007

This study used Arduino Uno R3, DHT11 sensor and lamp driver as PWM driver (Pulse Width Modulation) incandescent lamp 15 Watt / 25 Watt Parallel and fan driver for DC 12 Volt fan switch. In addition, it is to monitoring the temperature graph in the system using GUI MATLAB. This egg incubator uses a manual egg-activator every 6 hour which aims to prevent embryo hatching when attached to the eggshell. The research problems include the realization of automatic egg incubator with PID control method so that the temperature in the system can be stable (steady state) in fluctuating environmental temperature for optimal quality and quantity of eggs.

2. Research Methodology

2.1. Research Approach
This study uses Arduino IDE and Matlab 2015a software for simulating and operating PID system that would be used and designing plant build which would produce temperature response in egg hatching incubator.

2.2. Research Setting and Time
The research was carried out in the Laboratory of Engineering Physics of the Department of Electrical Engineering of the State University of Surabaya and its execution time was done in the even semester of 2016/2017.

2.3. Research Design
The design of this study aims to answer the problem in order to formulate the conclusion, as described in Figure 1. Flow diagram as follows:

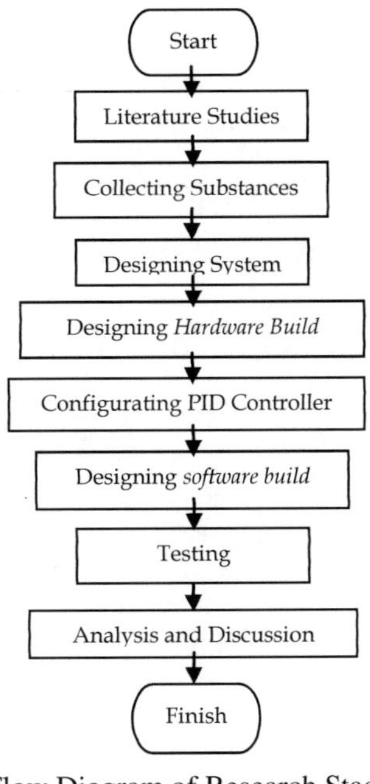

Figure 1. Flow Diagram of Research Stages

2.4. Designing Hardware Build
The design of hardware build includes several block designs namely; power supply block, sensor block, lamp driver block, fan driver block, LCD block, and egg incubator plant block.

2.5. Power Supply Block
Power Supply (Power Supply) on this system has an important role as a source of DC voltage on the system. With PLN input voltage mesh as input, Power Supply has two outputs which each of them has 9 Volt and 12

IOP Conf. Series: Materials Science and Engineering **336** (2018) 012007 doi:10.1088/1757-899X/336/1/012007

Volt output. Voltage of 9 Volt is used for Arduino Uno R3 while the voltage of 12 Volt is for fan driver supply needs. Figure 2 is schematic of power supply circuit.

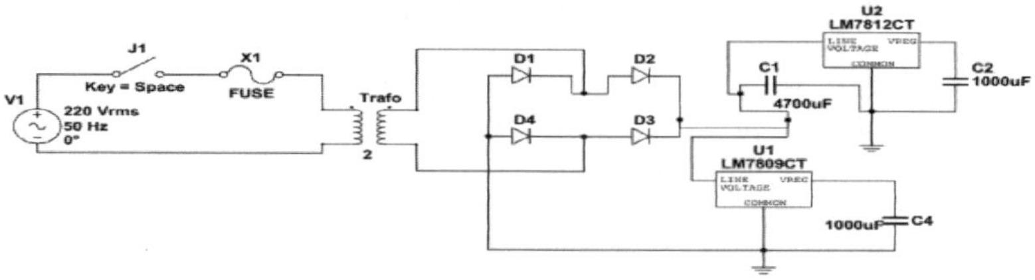

Figure 2. Power Supply

2.6. DHT11 Sensor Block

The DHT 11 sensor is a calibrated sensor with an accuracy level of ± 2 ° C and a humidity level of ± 5% RH which is connected to pin 2 of Arduino Uno with a 40 cm cable to be treated and processed as data input. Figure 3 is DHT11 sensor scheme.

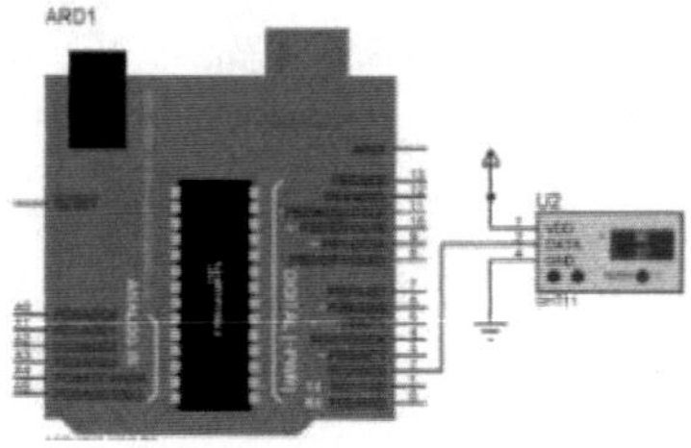

Figure 3. DHT11 Sensor

2.7. Lamps Drivers Block

The lamp driver has a principle to use a 220 Volt AC voltage input rectified through several *Wheatstone*-generated diodes that produce DC voltages with great power and current. There is an Optocoupler component as a switch interface with MOSFET. The lamp driver has a role as a PWM actuator that gets signal from Arduino. The following figure is a schematic of the lamp driver circuit.

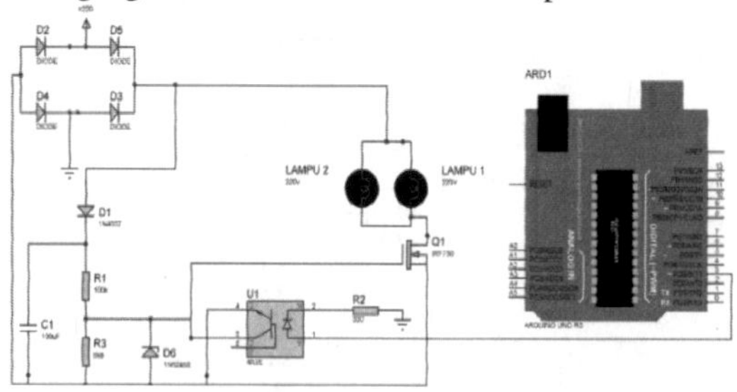

Figure 4. Lamps Driver

2.8. Fan Driver Block

This fan driver has a role to run the fan during the hatching process so that the temperature inside the incubator evenly and throw hot air when the temperature exceeds the *setpoint* value. Figure 5 is fan driver scheme.

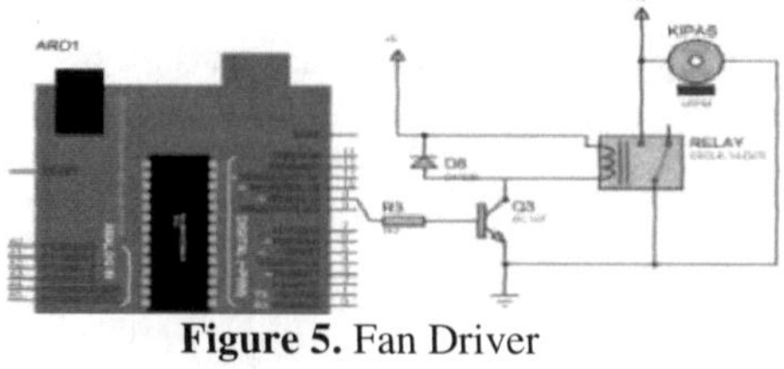

Figure 5. Fan Driver

2.9. 16X2 LCD Block

The LCD display is enabled to display the temperature and humidity values in the egg incubator. It uses I2C LCD module to minimize the number of cables connected to the Arduino pin. Figure 6 is 16X2 LCD Scheme.

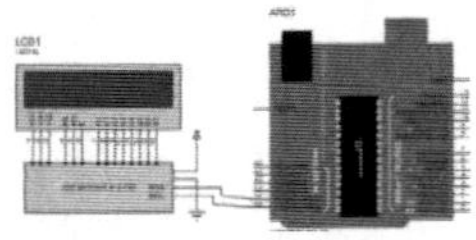

Figure 6. LCD Scheme 16X2

2.10. Egg Incubator Plant Block

This study used a plant made of wood, namely *Triplek*. This type of wood is ideally used as a raw material for incubator egg hatchery. The dimension of incubator is 30 x 40 x 32 cm with a height of 30 cm from the ground. Figure 7 is the design of an egg incubator plant incubator.

Figure 7. Egg Incubator Plant Design

3. Configuration of the PID Control

Before designing the PID control configuration which is appropriate to the plant, the first step is to design the whole system of the egg hatching incubator and then identify the system with ARX (AR-Auto Regressive X-eXogenous) method associated with the PWM stage input and output in the form of temperature that is measured to obtain a mathematical model of a dynamic system. The data object was created using Matlab system identification toolbox with Simulink ARX. Figure 8 is Simulink used for system identification for each plant.

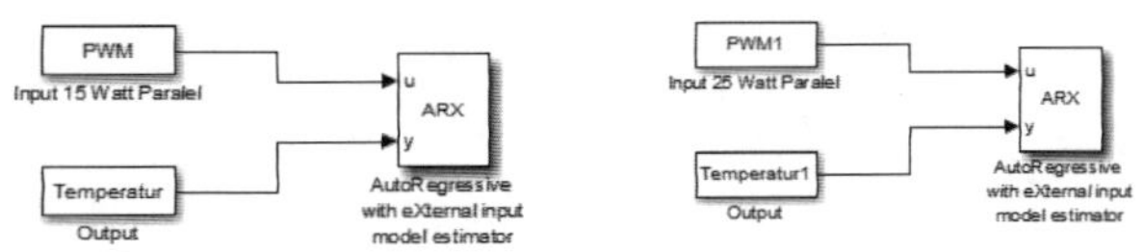

(a) **(b)**

Figure 8. Simulink System Identification of (a)15 Watt Parallel Watts (b) 25 Watt Parallel

The result of transfer function obtained from Simulink is still a discrete transfer function. To get transfer function continuous (Laplace Transformation), it used sysc = d2c (sysd) syntax. Below is each Transfer Function continuous plant 15 Watt Parallel and 25 Watt Parallel:

$$\text{Plant 15 Watts} = \frac{0.496s+1.658}{s^2+37.6s+3.313}$$

$$\text{Plant 25 Watt} = \frac{1.711s+0.4965}{s^2+31.02s+1.419}$$

The testing of open loop response model of mathematical plant used step test signal with step 37 value according to the targeted temperature. Figure 14 and Figure 15 is open loop response result of each plant:

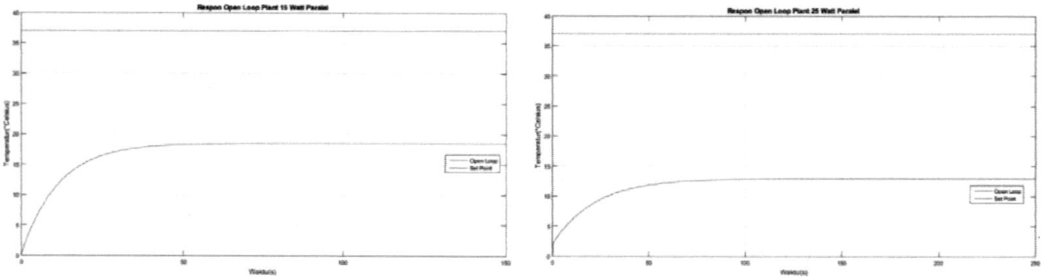

(a) (b)

Figure 9. Open Loop Plant Response of (a) 15 Watt Parallel (b) 25 Watt Parallel

The results of Fig. 9 are characteristic of first order system response graphs. The steady state output value of each system still has not reached the 37 setpoint value of 18.52 ° C for the 15 Watt and 12.95 ° C for 25 Watt plants. Therefore, a PID control is required with the first order method to achieve the targeted steady state value. With the method of first order system approach, it is determined of the value of $\tau = \tau i$ to determine the value of Kp, Ki, Kd.

4. Designing Software Build

Software in this study used IDE Arduino 1.6.12 and Matlab 2015a. The design of the program used for the PID eggs hatching system begins by creating a flowchart program. The whole flowchart program can be seen in Figure 10.

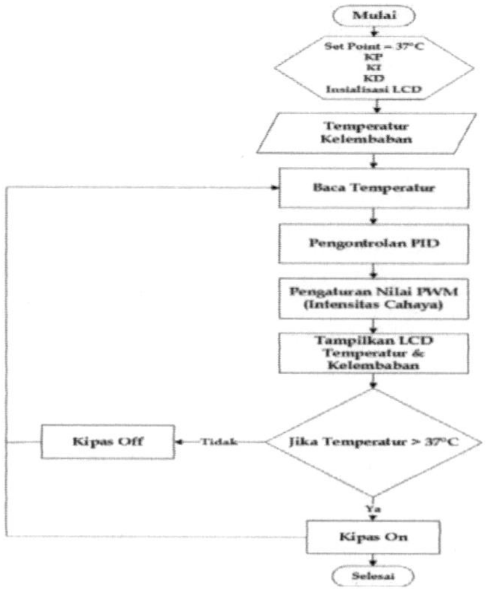

Figure 10. Flowchart Program System

5. Results and Discussion

Testing and research analysis include; 1) analysis and testing of plant control system circuit design for PID, 2) application of PID control configuration on plant, 3) PID control test result.

5.1. Testing of circuit design

The test circuit design includes power supply, DHT11 sensor, light driver, and fan driver. Below is each block of circuit testing:

6.1.1 Power supply. Power Supply (Power Supply) on this system has an important role as a source of DC voltage on the system. Based on Power Supply input measurement of 224.6 Volt AC, for output 1, the voltage is 9.06 Volts as the Arduino source, while the output voltage 2 is 12.20 Volts as the DC fan source. DHT11 Sensor

In this research, this thesis compares DHT 11 temperature reading data with other temperature instrumentation which has ± 3% temperature accuracy level with 1 ° C resolution, that are Clamp Meter and Digital Thermometer with accuracy of ± 1 ° C. Testing by comparing the readings of other types of temperature sensors aims to see the accuracy of DHT11 sensors. The following table is Comparison of sensor accuracy:

Table 1. DHT Sensor Test 11 and Clamp Meter

No.	Time	DHT 11 Sensor		Clamp Meter	Difference
		°C	%RH	°C	°C
1.	Morning (07.00-08.00)	22	61	24	2
2.	Afternoon (13.00-14.00)	24	54	26	2
3.	Night (22.00-23.00)	21	62	22	1
				Average	1.7

Table 2. DHT Sensor 11 Testing and Digital Thermometer

No.	Time	DHT 11 Sensor		Digital Thermometer	Difference
		°C	%RH	°C	°C
1.	Morning (07.00-08.00)	22	61	25.3	3.3
2.	Afternoon (13.00-14.00)	24	54	27.9	3.9
3.	Night (22.00-23.00)	21	62	24.7	3.7
				Average	3.6

6.1.2 Lights Driver. This lights driver has a principle of PWM to supply voltage to the incandescent lamp. With the PWM 0-255 value indicator of the Arduino control that will provide the light level response. Calculation analysis of lamp driver component specifications:

Zener Diode component used is 15 Volt 1 Watt.

$$\mathrm{P} = \mathrm{Vzener.\,I} \tag{1}$$

$$\mathrm{I} = \frac{P}{Vzener} \tag{2}$$

$$\mathrm{I} = \frac{1}{15} = 0.07\ \mathrm{A}$$

Voltage after the wheatstone bridge rectifier :

$$Vm = VAC\,x\,\sqrt{2} \tag{3}$$

$$Vm = 224.6\ \mathrm{x}\ \sqrt{2} = 317.63\ \mathrm{Volt}$$

$$VDC = \frac{2Vm}{\pi} \tag{4}$$

$$VDC = \frac{2\ X\ 317.63}{3.14} = 202.31\ \mathrm{Volt}$$

The current flowing in R1

$$VR1 = Vm - Vzener \tag{5}$$

$$VR1 = 317.63 - 15 = 312.63\ \mathrm{Volt}$$

$$IR1 = \frac{VR1}{R1}$$

$$IR1 = \frac{312.63}{100K} = 3.1263\ \mathrm{mA}$$

The power produced is

$$P = V\,x\,I \tag{6}$$

$$P = 312.63\,x\,3.126\,x\,10^{-3} = 0.977\ \mathrm{Watt} \approx 1\ \mathrm{Watt}$$

So, for the use of R1 component is 100KΩ with 2 Watt power in order to be more secure. Specification diode zener Vzener 15 Volt, to get the value of R3:

$$R3 = \frac{Vzener}{IR1}$$

$$R3 = \frac{15}{3.1263} = 4.798 \approx\approx 4k7\ \Omega$$

So for the use of R3 component is 6K8 Ω with 1 Watt power in order to be more secure.

6.1.3 Fan Driver. This circuit works when there is an input signal with logic 1 of the Arduino that will activate the On condition transistor which makes the active relay from normally open to normally close. With the active relay normally close, the fan will be activated to a predetermined temperature limit. This fan driver will be Off when receiving input with logic 0 when the temperature is met which makes the transistor will be on Off position. Besides, the relay will be Off to the normally open position.

5.2. Application of PID Control Configuration

Based on the results of the open loop response, each plant indicates the response type including first order system characteristics because there is no Overshoot. So, the control method uses PI (Proportional Integral) controller with Kd = 0, τ = τi for optimum output response value suitable to input value.

For Parallel 15 Watt Plant:

$$K = \frac{Yss}{Xss} \tag{7}$$

$$K = \frac{18.52}{37} = 0.50054$$

$$\tau = 11{,}046 \text{ s accelerated 2x}$$

$$\tau^* = 5{,}523 \text{ s}$$

To get the value of Kp refers to equation (4).

$$Kp = \frac{\tau i}{K\tau *} = \frac{11.046}{0.50054x5.523} = 3.9956$$

$$K_i = \frac{K_p}{\tau_i} = \frac{3.9956}{11.046} = 0.361$$

Validation control of 15 Watt PI to know the response after controlled.

$$PI = Kp\ (1 + \frac{1}{\tau is})$$

$$PI = 3.9956\ (1 + \frac{1}{11.046}) = 4.357$$

For Parallel Plant 25 Watt:

$$K = \frac{12.95}{37} = 0.35$$

$$\tau = 18.1140 \text{ s accelerated 2x}$$

$$\tau^* = 9{,}057 \text{ s}$$

To obtain the value of Kp refers to equation before.

$$Kp = \frac{\tau i}{K\tau *} = \frac{18.1140}{0.35x9.057} = 5.714$$

$$K_i = \frac{K_p}{\tau_i} = \frac{5.714}{18.1140} = 0.315$$

Validation of 25 Watt PI control to determine response after controlled.

$$PI = Kp\ (1 + \frac{1}{\tau is})$$

$$PI = 5.714\ (1 + \frac{1}{18.1140}) = 6.0294$$

Figure 17 and Figure 18 simulate of each plant using a PI controller.

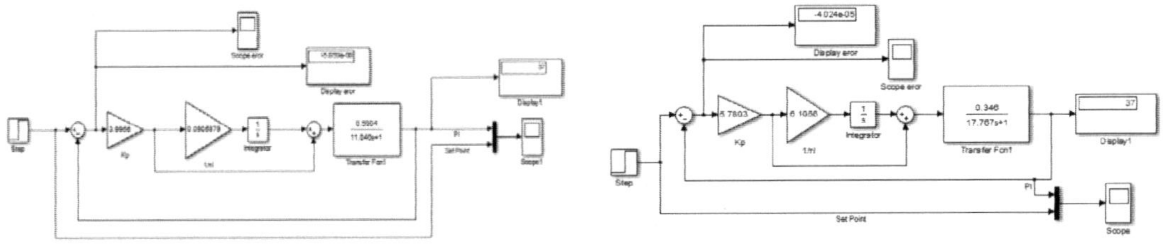

(a) (b)

Figure 11. Simulation of PI Controller at (a)15 Watts Plant (b) 25 Watt Plant

Each plant has different value of PID constant to achieve a stable value. In the plant with 15 Watt parallel lamps have been identified the response system in accordance with first order system which it generally uses PID controller constant value Kp = 3.9956, Ki = 0.361, Kd = 0. While the plant with 25 Watt parallel lamps has been identified the response system constant value Kp = 5.714, Ki = 0.315, Kd = 0. After that, the PID value is implemented into the Arduino program. Testing of each plant with a PI controller results a temperature response that matches with the step setpoint input value of 37 ° C. Here are Figure 12 about Plant Response with PI controller.

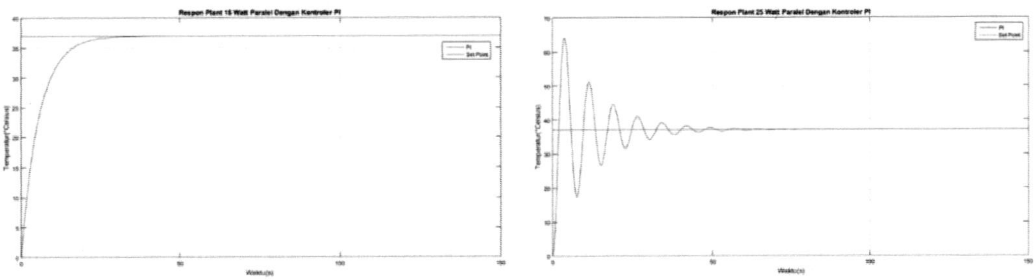

Figure 12. Plant Response with PI Controller of (a) 15 Watts (b) 25 Watt

5.3. PID Control Test Result

6.3.1 Monitoring Plant Train Graph. In the process of monitoring, PID control train chart for each plant has different response according to the result of PID constant value which has been analyzed. The test is carried out with the initial temperature adjusting the ambient temperature to the system between 25-26 ° C by providing a *setpoint* value of 37 ° C.

For 15 Watt parallel plant in the egg incubator has a constant value of Kp = 3.9956, Ki = 0.361, Kd = 0 yields a rise time response of 2178 seconds to achieve the first steady state (settling time) and delay time of 1089 seconds. In this 15 Watt plant PID control test, it has a steady state error value of 0% which indicates successful control from 5% tolerance. In this test, the result showed that the result value of Overshoot Maximum 2.7% with peak value of 38 ° C. The time to reach the peak value of the response (peak time) is 4100 seconds.

The 25-watt parallel plant on the egg incubator has a constant value of Kp = 5.714, Ki = 0.351, Kd = 0 yielding a rise time of 1244 seconds to reach the first *setpoint* value and a delay time of 622 seconds. In this test, the result showed that the result value of Overshoot Maximum 2.7% with peak value of 38 ° C. The time to reach the peak value of the response (peak time) is 1740 seconds. To reach the steady-state value after the peak value has a settling time of 5010 seconds. In this case, 25 Watt plant PID control test has a steady state error value of 2.7% indicating successful control of 5% tolerance.

Figure 13 is the response graph analysis on a 15 Watt GUI plant display using PID controller.

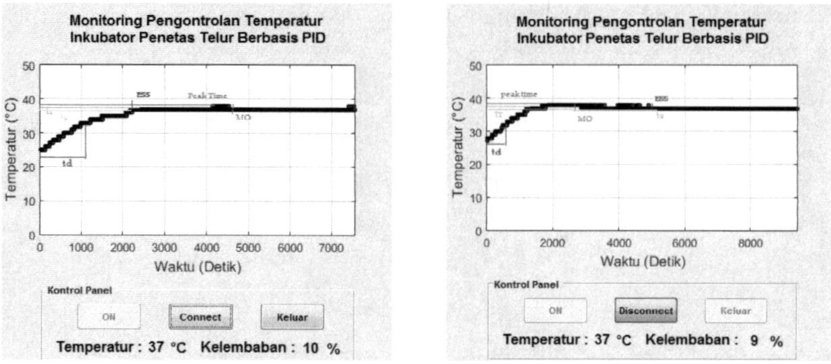

(a) (b)

Figure 13. Graphic Response Analysis On (a) 15-Watt and (b) 25 watt GUI Display Plant Using PID Controller

Testing a 15 watts plant of interference is given at 2368 seconds. When the disruption is given by opening the incubator door for 20 seconds, the response graph indicates that the system is experiencing instability, the temperature drop 3°C from the *setpoint*. It takes a settling time of 215 seconds to reach the *setpoint* withstand steady-state conditions. This indicates good enough response system in responding to interference. Figure 14 is the graphical analysis of 15 Watt plant response to disturbance.

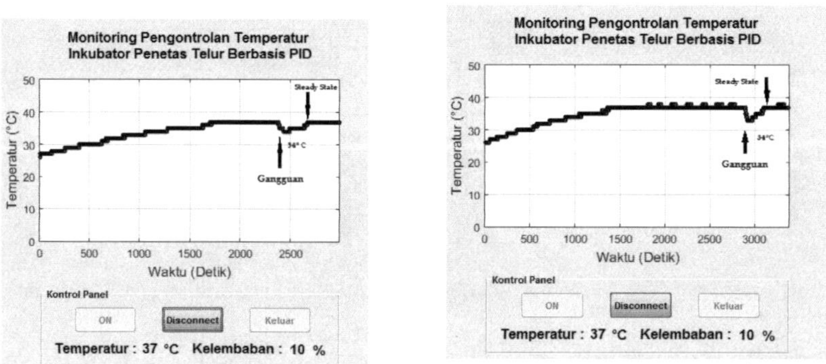

(a) (b)

Figure14. Graph Analysis of (a) 15 Watt and (b) 25 Watt Response Plant to Interference

While for the 25-watt plant is given interference at 2891 seconds. When disturbed by opening the incubator door for 20 seconds, the response graph is instability, the temperature drop to 3°C reaching 34°C. It takes 150 seconds of settling time to reach the *setpoint* value in the steady state. This indicates that the response of a 25-watt system plant reaches steady state faster than the response of a 15-watt plant system in responding to interference. Figure 24 is the analysis of 25 Watt plant response to the disturbance chart.

6.3.2 Hatching Results. Based on research that has been conducted on April 7 to May 2, 2017, the temperature stability treatment significantly affects the hatch results. Through stable temperature control, it has a faster temperature time level in terms of hatching. The result of a 15 Watt lamp plant with PID controller that has a stable temperature at the *setpoint* is able to incubate eggs with a time span of 19 days. While the results of a 25 Watt plant that has temperature stability up and down from the *setpoint* is able to incubate eggs with a time span of 21 days. The following are Table 2 and Table 3 of hatchery testing results of each plant.

Table 3. Test Results of 15 Watt Plant Hatching

Plant 15 Watt Parallel (10 eggs)			
Egg Hatching			Eggs Not Hatching
70%			3 Eggs (30%)
Normal	Defect	Dead	
7	0	0	

Table 4. Test Results of 25 Watt Plant Hatching

Plant 25 Watt Parallel (10 eggs)			
Egg Hatching			Eggs Not Hatching
80%			2 Eggs (20%)
Normal	Defect	Dead	
7	0	1	

6. CLOSING

6.1. Conclusion

Based on the research, it is concluded that the monitoring and temperature control system in egg incubator for PID controller has been successfully integrated automatically with PWM method in incandescent lamp. To apply the PID control on the egg incubator plant using the egg incubator method is by identification method of ARX system with the constant value of Kp = 3.9956, Ki = 0.361, Kd = 0 for the parallel 15 Watt plant and the constant value of Kp = 5.714, Ki = 0.351, Kd = 0 for the plant 25 Watt parallel. Using these constants, the response of the control system is able to stabilize the egg incubator's temperature. In PID control plant test, the 15 Watt plant produces slower response for the rise time of τr = 2178 seconds, delay time τd = 1089 seconds. While the 25 Watt plant response rise time of τr = 1244 seconds, delay time τd = 622 seconds. But for a steady-state response (stable) the 15 Watt plant is capable to produce faster response of τs = 2178 seconds than the steady-state plant 25 Watt of τs = 5010 seconds. The steady state error value of 2.7% indicates that the temperature control runs well in ambient conditions with 70-80% hatching rate of eggs.

6.2. Suggestion

Based on the conclusion above, there are some suggestions that can be done for the development of the system to get maximum results that is to use a more accurate and precise sensor such as SHT 11 type, to apply the air humidity control system on the plant, to apply rotary shelf system automatically, to apply the UPS as a backup power Electricity if the electricity goes off, to implement an online monitoring system or Internet Of Thing (IoT), and to change the controller of adaptive, fuzzy logic and neural network (ANN) controls to reduce the steady state error value .

REFERENCES

[1] Gunterus, Frans.1994. Falsafah Dasar: Sistem Pengendalian Proses. Jakarta: PT. Elex Media Komputindo
[2] Hastono, Djoko Tri. 2009. *Sistem Pengendali dan Pengukur Suhu pada Mesin Penetas Telur Berbasis Mikrokontroler AT89S51.* Semarang: Universitas Negeri Semarang.
[3] Jaya, Mitra. 2016. *Buku Panduan Praktis C-500 & C 1000.* Malang: Mitra Jaya Company
[4] Ljung L., 2011. *System Identification Toolbox™ User's Guide.* Natick: The MathWorks, Inc.
[5] Ogata, Katsuhiko. 1985. *Teknik Kontrol Automatik jilid 1.* Terjemahan Edi Laksono. Jakarta: Erlangga.
[6] Wardana, I Nyoman Kusuma. 2015. *Teknik Antarmuka MATLAB dan Arduino.* Denpasar: Vaikutha International Publication.
[7] Web.https://www.arduino.cc_Diakses 24 Januari 2017.

IOP Conf. Series: Materials Science and Engineering **336** (2018) 012008 doi:10.1088/1757-899X/336/1/012008

Effect OF NaOH Treatment on Bending Strength Of The Polyester Composite Reinforce By Sugar Palm Fibers

Mochamad Arif Irfai[1]*, Diah Wulandari [1], Sutriyono[2], Eko Marsyahyo[2]

[1]Department of Mechanical Engineering, Universitas Negeri Surabaya, Kampus Unesa Ketintang Surabaya, East Java P.O. 60231, Indonesia

[2]Department of Mechanical Engineering, National Institute of Technology, Jalan Sigura-gura Malang, East Java P.O. 65145, Indonesia

*arifirfai@unesa.ac.id

Abstract. The objective of this research is to investigate the effect of NaOH treatment on bending strength of lamina composite reinforced by sugar palm fiber. To know of mechanism fracture can be done with visual inspection of the fracture surface. The Materials used are random sugar palm fibers that have been in the treatment of NaOH, polyester resin and hardener. Sugar palm fibers after washed and dried then soaked NaOH with a long time soaking 0, 2, 4, 6 and 8 hours. The bending test specimens were produced according to ASTM D 790. All specimens were post cured at 62°C for 4 hours. The Bending test was carried out on a universal testing machine. The SEM analysis has conducted to provide the analysis on interface adhesion between the surfaces of fiber with the matrix. The result shows that polyester composite reinforced by sugar palm fiber has highest bending stress 176.77 N/mm^2 for 2 hours of a long time soaking NaOH, the highest flexural strain 0.27 mm for 2 hours of a long time soaking NaOH, elongation 24.05% for 2 hours of a long time soaking NaOH and the highest bending modulus 1.267 GPa for 2 hours of a long time soaking NaOH. Based on the results, it can be concluded that the polyester composite reinforced by sugar palm fiber has the optimum bending properties for a long time soaking 2 hours. The fracture surface shows that the polyester composite reinforced by sugar palm fiber pull out that indicate weakens the bond between fiber and matrix.

1. Introduction

The development of composite materials is increasingly widespread in various areas that effort innovation continuously. Composite materials have been applicated in various areas such as sport, transportation in the land, water, and air so construction areas and the world of space because it had been many advantages application composite materials i.e ratio strength and specify density is high, cheap and simple production [2].

Composite with reinforced fiber more applicated and development than composite with reinforced metal or other reinforced composite. This is due to the fiber has high tensile stress so it is suitable for reinforcing the material. Moreover, another factor that encourages for using natural fiber is abundant availability, light, corrosion resistant, water resistant, interesting performance and no machining

IOP Conf. Series: Materials Science and Engineering **336** (2018) 012008 doi:10.1088/1757-899X/336/1/012008

process. Because the characteristic composite is lightweight, then expenses load as a result of the construction can be eliminated.

Natural fiber is trying to shift synthetic fiber, as E-Glass, Kevlar-49, Carbon/Graphite, Silicone carbide, aluminum oxide, and boron. Although not fully shift, but the use of natural fiber replace synthetic fiber is a wise step in saving the environment from waste made and limited natural resources which are not renewable. However, the strength of natural fiber composites still lower than synthetic fiber composites, as E-glass fiber, S-glass as well as metal fiber such as aluminum. Treatment with an alkaline solution can improve the strength of the composite so that the strength of natural fiber composites able to approach synthetic fiber. The research with treatment soaked jute fiber in solution NaOH, it was able to in crease the strength of jute fiber up to 79% and toughness up to 46% [9].

The use of wood is continuously to raw materials for shipbuilding trigger illegal logging which resulted in the forest become disappear. Therefore, it is necessary to find an alternative solution substitute wood which is used raw material for making boats. One of the wise step is the use of polyester composite reinforcement with sugar palm fiber. In addition, the potential of sugar palm fibers that overflow needs to be optimized also keep environmental sustainability by not exploiting continuously the forest which can cause to environmental degradation. In order for the design of safe use, it is necessary to bending test adapted to real conditions where the real conditions, composite panels receive the load bending/press of waves.

2. Experimental

Sugar Palm fibers were washed with water then proceed drying in a closed room for 3 x 24 hours, as shown in Figure 1. Sugar Palm fibers that have been dried and then soaked in the alkaline solution NaOH 5% for 0, 2, 4, 6 and 8 hours, as shown in Figure 2. The materials used in this study is the sugar palm fibers that have been treated NaOH for 0, 2, 4, 6 and 8 hours, resin unsaturated polyester (UPRs) type 157 BQTN and hardener metil ethyl ketone peroxide (MEKPO).

The tools used in this study as mold, digital scales, Vernier calipers and a hydraulic jack. Free water content contained in the fibers removed by inserting it in the oven for 15 minutes. Manufacture of composite made with press mold. Manufacture of the composite is done by placing the fibers that have been soaked in NaOH for 0, 2, 4, 6 and 8 hours respectively then the surface of the fiber wet unsaturated polyester resin. Manufacture of composite made one process for each variable a long time soaking solution of NaOH (0, 2, 4, 6 and 8 hours).

Bending test specimen is prepared by cutting the polyester composite reinforced sugar palm fiber which has been processed using a grinder. Dimensions of specimen and test method designed according to ASTM D 790-02. Long span bending test is 60 mm. Bending test was conducted using TORSEE-universal testing machine. Photo fracture cross-section did with a macro photo and SEM (Scanning Electron Microscopy) to determine the mechanism and the type of fracture.

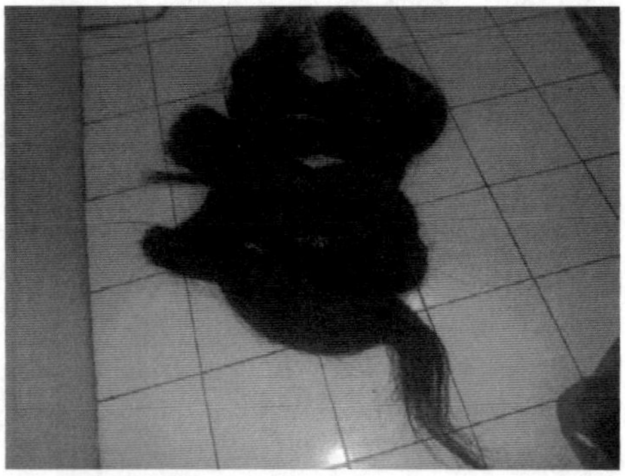

(a).

(b)

Figure 1 a. raw materials (sugar palm fiber) b. Sugar palm fiber are washed with mineral

(a)

(b)

Figure 2 (a.) Sugar palm fiber immersion in a solution of NaOH; (b). pouring resin on the sugar palm fiber

(a).

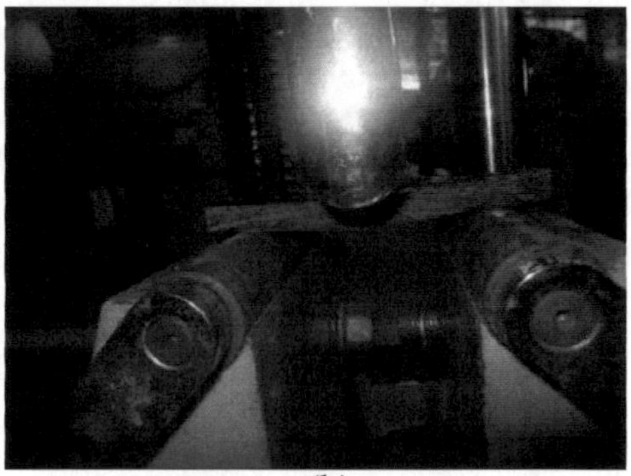

(b)

Figure 3 (a). Specimens were ready to be tested; (b). Bending test process

IOP Conf. Series: Materials Science and Engineering **336** (2018) 012008 doi:10.1088/1757-899X/336/1/012008

3. Result

3.1. Bending Tests

The data obtained in the bending test then entry and processed into the equation 1, 2 and 3 than shown in table 1.

$$\sigma_f = \frac{3PL}{2bd^2} \tag{1}$$

$$\varepsilon_f = \frac{6Dd}{L^2} \tag{2}$$

$$E_B = \frac{L^3 m}{4bd^3} \tag{3}$$

Table 1. Relationship the long time soaking NaOH with bending strength (bending stress, flexural strain, elongation and modulus of elasticity)

A Long time soaking (hours)	Bending stress (N/mm^2)	*Flexural strain* (mm)	Modulus of elasticity(GPa)	Elongation (%)
0	95.225	0.18	0.461	17.15
2	176.771	0.27	1.267	24.05
4	157.402	0.25	0.790	23.27
6	152.727	0.19	0.680	21.81
8	152.531	0.17	0.532	18.58

From table 1 above is then processed and presented in the form of the curve relationship between the long time soaking NaOH with bending stress, flexural strain, elongation and modulus of elasticity as shown in figures 4.

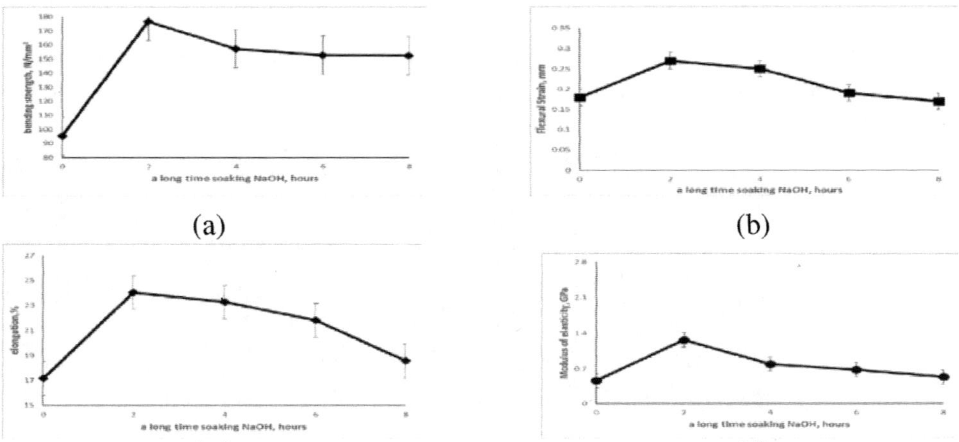

(a) (b) (c) (d)

Figure 4 The curve relationship a long time soaking NaOH (a) with bending stress; (b) with flexural strain; (c) with elongation; (d) with modulus of clasticity

3.2. Photos of cross-sectional fracture

(a) (b) (c)

Figure 5 Fracture test samples NaOH 2 hours soaking time (a). Fracture surface appears on the front (b). Fracture surface appears on the top (c). Fracture surface with 60X magnification.

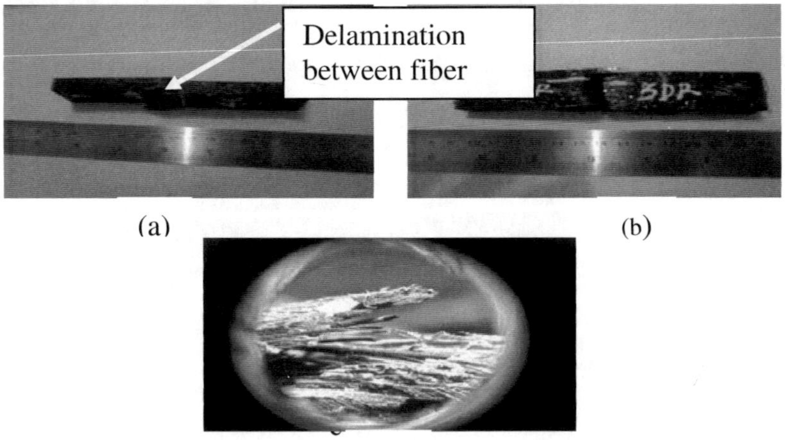

(a) (b) (c)

Figure 6 Fracture test samples NaOH 0 hours soaking time (a). Fracture surface appears on the front (b). Fracture surface appears on the top (c). Fracture surface with 60X magnification.

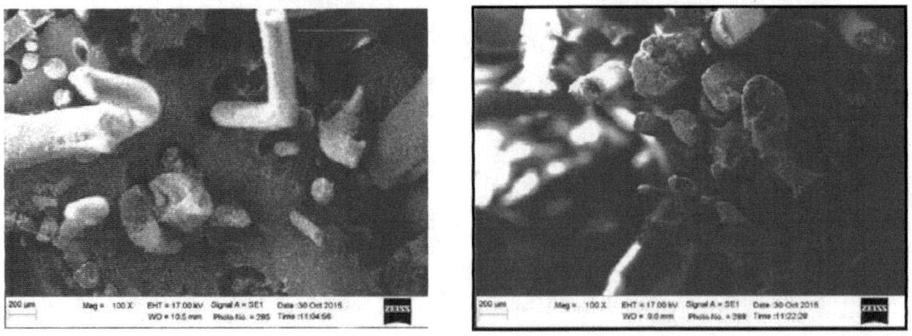

(a) (b)

Figure 7 The result of the SEM photograph sugar palm fiber composite was soaked NaOH for (a) 2 hours. (b) 0 hours

4. Discussion

A long time soaking NaOH affects sugar palm fiber strength. This is evidenced by the longer time soaking NaOH, it also increases the bending stress. However, this condition only applies when NaOH long time soaking reaches 2 hours. If the longtime soaking NaOH followed by increasingly longer, then it will cause a decrease in fiber strength. This is evidenced by the length of time of immersion NaOH that the longer (more than 2 hours) then the trend-bending stress also decrease, as shown in figure 1.

Bending stress increases with the length of time immersion NaOH. It is strengthened from the long immersion time curve with flexural strain. As shown figure2 that the longer of immersion time then also increase the flexural strain. The flexural strain is defined as the composite ability to break up a composite failed so the higher flexural strain of a material's ability to withstand the loads is also higher. At 2 hours soaking time NaOH, strength composite has the best performance in weight holding. Then the length of time immersion longer so sugar palm fiber properties decrease.

The ability to elongation of the composite material is also increasing with the length of time immersion NaOH. The most excellent elongation occurs in a long time soaking NaOH 2 hours, as shown in figure 3. At the time of immersion of NaOH with a longer time, it causes a decrease in sugar palm fiber properties. It happens on the properties of other sugar palm fiber composite (modulus of elasticity). Modulus of elasticity of composite polyester reinforced sugar palm fiber has the highest modulus of elasticity with long time soaking NaOH for 2 hours is equal to 1.267 GPa.

Fibers that have been soaked in a solution of NaOH 5% produce an effect on the properties and strength of the fiber. At the fracture test sample with fibers have been soaked in NaOH 2 hours presence fibers pull out from the matrix. It occurs when the bending load received test samples not fully passed on to the fiber due to the weak part (mechanical bonding fiber and matrix) no ability to hold bending load so that load should be forwarded to the fibers are not fully accepted by the fiber as a reinforce but a failure on the part between the fiber and the matrix. Fiber and matrix bond strength lower than the strength of the fiber itself, as shown in figure 4.

On test samples without soaking NaOH shows fiber strength much weaker. Dirt on the surface of the fiber suspected as the cause of the low strength of the fiber. Dirt on the surface of the fiber covers the bonding of the fiber and the matrix so that mechanical bonding between the fibers and the matrix becomes less strong, as shown in figure 5. Type of failure in the specimen without soaking NaOH is delamination fiber and matrix as a result weakens the bond between the fiber and the matrix.

In the test result SEM-ijuk fiber composite polyester soaked NaOH for 2 hours commonly found fiber pull out. This shows that the NaOH treatment provides effect strength fiber. Load received forwarded to the fiber composite through the matrix. If the load received over the fiber strength then it causes part failure is on the part of the fiber, as shown in figure 6. On the contrary, if the strength of the bond between the fiber matrix is greater than the strength of the fiber itself then the first failure is sugar palm fiber, as shown in figure 7.

5. Conclusion

Sugar palm fiber properties that are obtained in the treatment of NaOH for 2 hours with bending stress, flexural strain, elongation and modulus of elasticity respectively 176.771 N/mm^2, 0.27 mm, 24.05% and 1.267 GPa. Characteristics of fracture on sugar palm fiber in NaOH treatment for 2 hours is commonly found fiber pull out (withdraw of fiber from the matrix) whereas the fibers without soaking NaOH found any kind of failure of the laminate.

6. Acknowledgement

This work was supported by the Ministry of Research, Technology and Higher Education, Indonesia, through the Pekerti Research Program with Contract No. 0002.51/UN38.11-P/LT/2016 dated March 14 2016

IOP Conf. Series: Materials Science and Engineering **336** (2018) 012008 doi:10.1088/1757-899X/336/1/012008

References

[1] Irfai M.A., Wulandari D., Sutriyono and Marsyahyo E., 2015, Effect of Fiber Volume Fraction to bending strength of the Polyester Composite Reinforced by Sugar Palm Fiber, Proceedings of the National Seminar on Research and Community Services LPPM Unesa 237-245

[2] Ishak M.R., Sapuan S.M., Leman Z., Rahman M.Z.A, Anwar U.M.K and Siregar J.P., 2013, Sugar Palm (Arenga pinnata): Its fibres, polymers and composites, Carbohydrate Polymers Journal **91** 2, 699-710

[3] Kaw, Autar K., 2006, Mechanics of Composite materials, ed. 2 nd, CRC Press

[4] Khan R. A., 2011, Fabrication and Characterization of Jute Fiber Reinforced Low Density Polyethylene Based Composites: Effects of Chemical Treatment, Dhaka, Journal of Scientific Reseach **3** 2.

[5] Marsyahyo Eko, 2008, Identification of Ramie Single Fiber Surface Topography Influenced by Solvent Based Treatment, Journal of Industrial Textiles, Vol.38, no.2

[6] Marsyahyo Eko, 2008, Surface Characteristic and Direct Interaction Measurement on Ramie(Boehmeria nivea) Single Fiber - Polymer Matrix, International Research Journal of Material Sciences **5** 1 15-24

[7] Marsyahyo E., Suryanto H. and Santhiarsa I.G.N, 2013, Review on Engineering Green-Materials and Applications from Tropical Plants Sources for Sustainable Future in Indonesia, Malang, Presented for the 4 th International Conference GT-2013 UIN Maliki Malang.

[8] Marsyahyo E., Soekrisno, Rochardjo H.S.B. and Jamasri, 2008, Investigation of Chemical Surface Treatment of Ramie Fiber (Boehmeria nivea) on Surface Morphology, Tensile Strength and Single Fiber Fracture Modes, Yogyakarta, International Conference Product Design and Development, Universitas Gadjah Mada

[9] Ray D., Sarkar B. K, Rana A. K and Rose N. R, 2001, Effect of alkali treated jute fibres on composite properties, Indian Academy of Sciences, Bull. Master Sci, **24** 2 129-135

[10] Santhiarsa I.G.N, Pratikto, Sonief A and Marsyahyo E., 2012, Potensi Serat Ijuk (Arenga Pinata Fiber) sebagai Penguat Komposit Poliester, Bali, Prosiding Konferensi Nasional Engineering Hotel III, Universitas Udayana

IOP Conf. Series: Materials Science and Engineering **336** (2018) 012009 doi:10.1088/1757-899X/336/1/012009

Patients' Heart Monitoring System Based on Wireless Sensor Network

T S Sollu[1], Alamsyah[1*], M Bachtiar[1] and A G Sooai[2]

[1]Department of Electrical Engineering, Faculty of Engineering, Universitas Tadulako, Palu, Indonesia

[2]Department of Informatics Engineering, Faculty of Engineering, UNWIRA, Kupang, Indonesia

* alamsyah.zakaria@untad.ac.id

Abstract. Wireless sensor network (WSN) has been utilized to support the health field such as monitoring the patient's heartbeat. Heart health monitoring is essential in maintaining health, especially in the elderly. Such an arrangement is needed to understand the patient's heart characteristics. The increasing number of patients certainly will enhance the burdens of doctors or nurses in dealing with the condition of the patients. Therefore, required a solution that could help doctors or nurses in monitoring the progress of patients' health at a real time. This research proposes a design and application of a patient heart monitoring system based on WSN. This system with using electrocardiograph (ECG) mounted on the patients' body and sent to the server through the ZigBee. The results indicated that the retrieval of data for 15 seconds in male patients, with the age of 25 years was 17 times rate or equal to 68 bpm. For 884 data packets sent for 15 minutes using ZigBee produce a data as much as 4488 bytes, throughput of 2.39 Kbps, and 0.24486 seconds of average delay. The measurement of the communication coverage based on the open space conditions within 15 seconds through ZigBee resulting throughput value of 4.19 Kbps, packet loss of 0 %, and 6.667 seconds of average delay. While, the measurement of communication range based on closed space condition through ZigBee resulting throughput of 4.27 Kbps, packet loss of 0 %, and 6.55 seconds of average delay.

1. Introduction

The WSN-based communication system in the health field has shown a relatively rapid and positive development [1] and has applied to monitoring system applications. Implementation health control system, among others: supervision of the patient's heartbeat [2], the human health control system and the detection of the spread of toxic chemicals [3], and the monitoring of patients' rehabilitation process based on distance [4]. The system aims to provide ease in monitoring the patients' conditions, operational cost efficiencies, reducing diagnostic errors, and converting conventional systems into computer-based digital systems, as well as supporting professional practice of healthcare services involving multiple disciplines [5].

One of the problems that occurred at the moment is the increasing number of patients and limitations of medical personnel, the workload of doctors will increase in dealing with the patient's health condition. Furthermore, the equipment used in monitoring the patient's heartbeat is still using a cable connected to the computer. Therefore required a solution that can assist medical personnel in the follow-up to the patient's health real time.

One solution that can apply to a patient's heart health monitoring system is the design and implementation of a WSN-based heartbeat system using the ZigBee protocol as a transmission medium. This designed system will provide the following benefits: 1) reduce the workload of medical personnel, 2) reduce medical errors, because patient data can monitor in real time.

2. Review of Literature

2.1. Health Monitoring

Health monitoring is the monitoring of a person to identify changes in health status due to certain substances. Health monitoring may include: consultation, answering questions related to medical history or lifestyle such as smoking, alcoholic beverages, and related that affect health [6].

In creating the architecture design of a patient's health, there are at least three layers to perform its function [7]. The layers are: 1) Perception Layer, this layer is the bottom layer of e-health monitoring architecture. Perception this layer consists of sensors that take individual data, related to the patient's health, 2) Middleware and API's Layer, this layer is where the data processing place, and 3) E-Health Application and Service Layer, it is the top layer of e-health monitoring architecture. This layer is an interface to display supervision of the patient.

2.2. Wireless Sensor Network (WSN)

WSN is a collection of nodes organized into a partner network [8]. Each node has processing capabilities, has several types of memory (programs, data and flash memories), has RF transceivers (Radio Frequency), has resources (e.g., Batteries and solar cells), and accommodates various sensors, and the system can revolutionize the way work to live again [9].

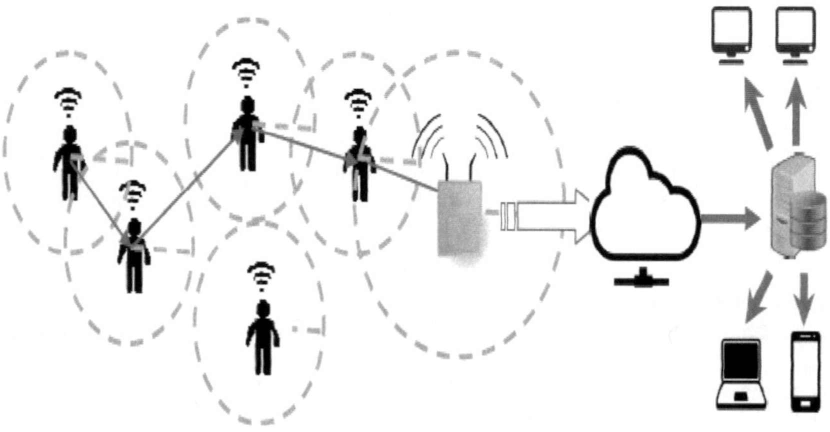

Figure 1. Topology wireless sensor network

2.3. Sensor Network

A sensor is a device used to detect or measure the value of a specified physical quantity. Sensors are a type of transducer used to convert mechanical, magnetic, heat, rays and chemical variations in voltage and electric current. Sensors play a critical role in the control of modern fabrication processes [10]. Meanwhile, according to [3], the sensor can be defined as a device that receives and responds to signals or stimulus.

The software that implemented on the network sensor is TinyOS. TinyOS is an open source operating system designed specifically for wireless sensor networks, which is responsive to sensory input and event-oriented [11], and TinyOS provides a relatively simple method of developing and

IOP Conf. Series: Materials Science and Engineering **336** (2018) 012009 doi:10.1088/1757-899X/336/1/012009

quickly implementing WSN [12]. TinyOS has a component-based architecture that supports the innovation and implementation of WSN by minimizing the required code size, as well as characteristics of sensor networks that have less memory.

2.4. Heart
The heart is the center of the human circulatory system [13]. The heart is a muscular organ that can push blood to various parts of the body. The human heart shape is like a cone and sized as an adult's fist, that's located in the left chest cavity and wrapped by a membrane called the pericardium. The heart is formed by muscular organs, apex, right and left atria, right and left ventricle. Heart size is about 12 cm long, 8 to 9 cm, wide as well as about 6 cm thick, heart weight is approximately 200 to 425 grams.

2.5. Electrocardiograph (ECG)
The electrocardiograph is a graph that depicts the electrical recording of the heart. The electrical activity of the heart in the body can be recorded and written by electrodes mounted on the body surface. An ECG sensor can detect signals at a frequency of 0.01 s / d 250 Hz. The main principle of the ECG is to identify the anatomy of the heart physiology and nerves. Thus it can be used to detect everything that occurs related to the performance of the heart. The signals issued by the ECG are analog signals of 0.5 mV up to 4 mV. The use of an optocoupler components needed in filtering ECG signals from sensor [14]. The physical form can see in Figure 2.

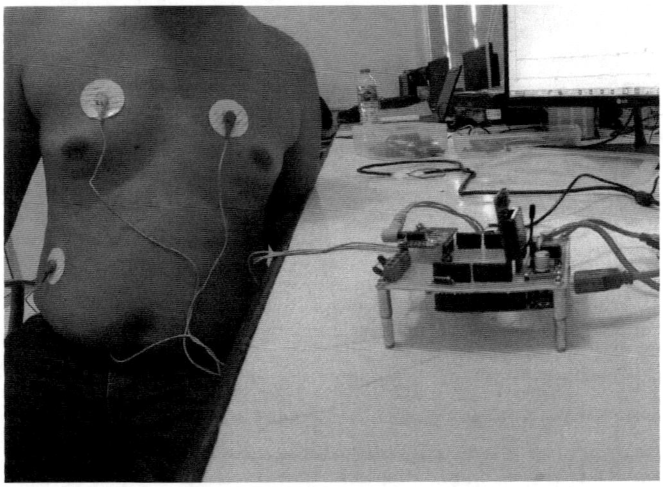

Figure 2. Electrode, lead cable and ECG sensor

The analog signals generated by the electrocardiogram are signals that interfered with other signals present in the body. Therefore, a sign is needed to separate it into two parts, i.e., a sign of the instrument (Signal Instrumentation Amplifier) and filter (low-pass filter and band reject filter).

From the results, the signal selected will be only the heartbeat signal. This weak signal will then be amplified by Instrumentation Amplifier and will scale. Once the signal processed at the signal processing and scaling level has performed, the data will be inserted into the ADC (Analog to Digital Conversion), or utilize the internal ADC in the Arduino module.

3. Research Methods
Figure 3 shows the stages of research that have done, including stages literature studies, tool design, and making heartbeat signal monitoring device, hardware testing, software preparation of heartbeat monitoring system, and analysis of test results.

IOP Conf. Series: Materials Science and Engineering **336** (2018) 012009 doi:10.1088/1757-899X/336/1/012009

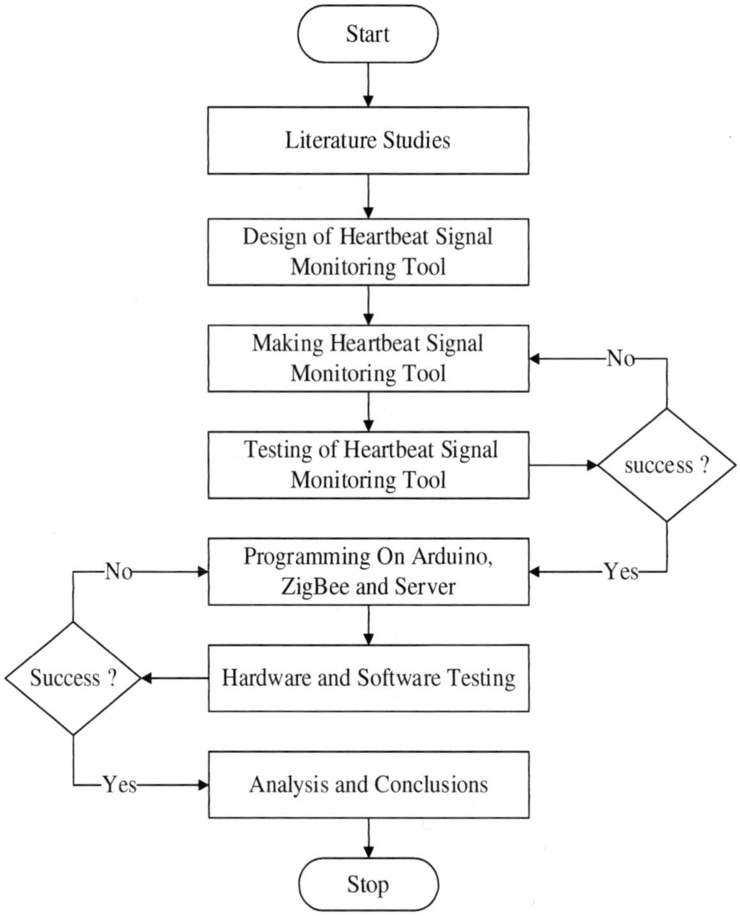

Figure 3. Flowchart of research

3.1. Literature Studies

In the literature study phase, a reference search will be made by the problem to be considered, especially about the heartbeat signal monitoring system. The target at this stage is to support the hypothesis in the design of a patient heartbeat monitoring system based on WSN.

3.2. Designing and Making Tools

Stage of the design and making of the heartbeat signal control system using hardware and software devices. The device consists of: sensor-board AD8232 is an ECG sensor, Arduino is the processing of ECG sensor data, XBee is a transmission medium, and the power supply is a voltage source. While the software consists of Integrated Development Environment (IDE) application program for Arduino, a XCTU application program for XBee, and node.ijs application program for the server.

IOP Conf. Series: Materials Science and Engineering **336** (2018) 012009 doi:10.1088/1757-899X/336/1/012009

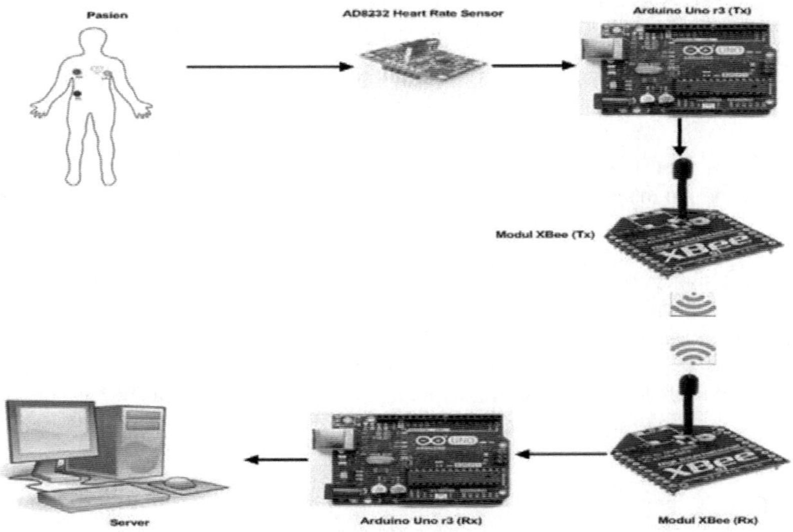

Figure 4. Block Diagram of the Patient's Heartbeat Pulse Monitoring System Based WSN

Figure 4 shows the block diagram of a patient's heartbeat pulse monitoring system. On this policy will be attached an AD8232 sensor-sensor (ECG sensor) on the patient's body. treatedECG sensor data results will be processed and treated by Arduino. Further, the data handled by the Arduino module will send to the server computer via the XBee module, and the patient ECG data will be visible on the server.

3.2.1 Design Hardware. Figure 5 shows the block diagram hardware, where the heartbeat signal is a weak analog signal. Therefore, the heartbeat signal to be connected to the AD8232 sensor-board pin will be amplified and filtered. Connect the ECG sensor, the signal conditioned according to the AD8232 sensor board specification. The sensor board AD8232 and Arduino are configurations from a sensor node. The sensor nodes are tasked to report analog to digital converter (ADC) readings to the server via XBee transmission media. At a later stage, the system test will be done entering the heartbeat signal into the Arduino and the result processed heartbeat data by Arduino will be forwarded to the server via XBee.

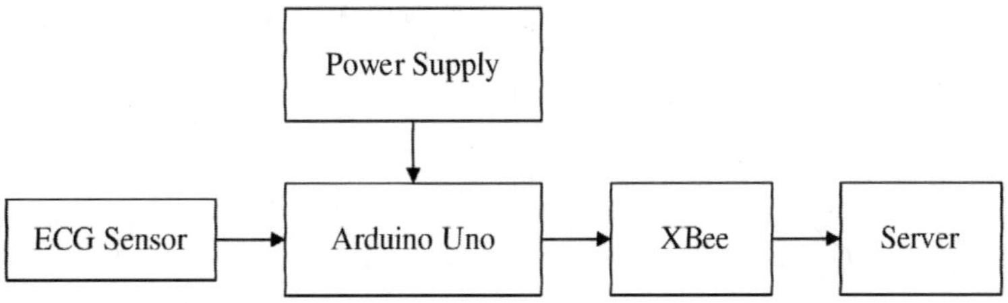

Figure 5. Block Diagram Hardware

3.2.2 Design Software. The application program used in software design consists of the IDE application program for Arduino by using C programming language, a XCTU application program which will be used to manage XBee module, and node.ijs application program used to run server

IOP Conf. Series: Materials Science and Engineering **336** (2018) 012009 doi:10.1088/1757-899X/336/1/012009

computer. The XCTU program implemented on a patient's heartbeat monitoring system using the JavaScript language.

4. Results and Discussion

4.1. Sensor Reading Validation

The voltage test results on the AD8232 board sensor shows that the observed value of the AD8232 sensor is 3.3 volts and the measurement value is 3,286 volts.

4.2. The Measurement of Heartbeat Signals Using Serial Plotter

Testing of heartbeat signals originating from the AD8232 sensor board output is not transmitted using XBee, but the output of the AD8232 sensor board connected to the Arduino will be linked directly to the server. In testing of heart rate test of male patients with age 25 years for 15 seconds resulted in 68 bpm heart signal period, the average delay of 0.245 seconds, and received data of 4,888 bytes. For a maximum voltage of 3.25 volts and a minimum voltage of 0 volts. The results of measurement of male heart signals using serial plotter could see in Figure 6.

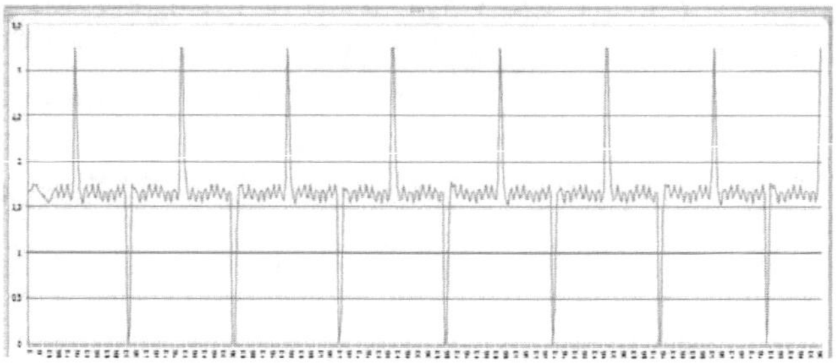

Figure 6. Heartbeat Signals Using Serial Plotter

4.3. Heart Signal Measurement with XBee

Testing the patient's heartbeat signal transmitted using XBee in males with age of 25 years, obtained a maximum voltage value of 3.25 volts and a minimum voltage value of 0 volts. While the delivery of 884 data packets conducted for 15 minutes through XBee able to receive data as much as 4488 bytes, a throughput of 2.39 Kbps and an average delay of 0.24486 seconds. The result of packet transmission of the heartbeat signal received through XBee can see in Figures 7 and 8.

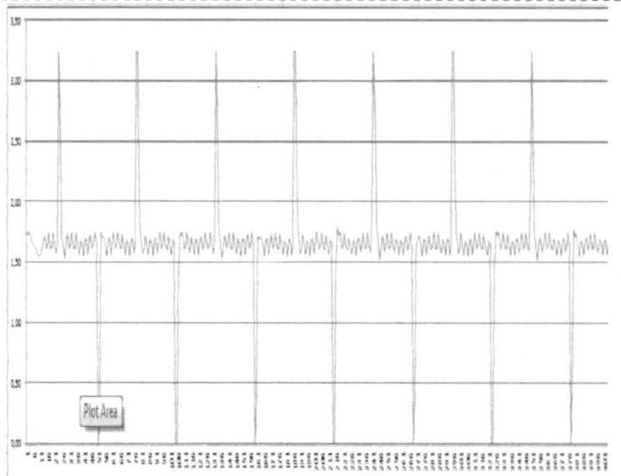

Figure 7. Results of heartbeat using XBee

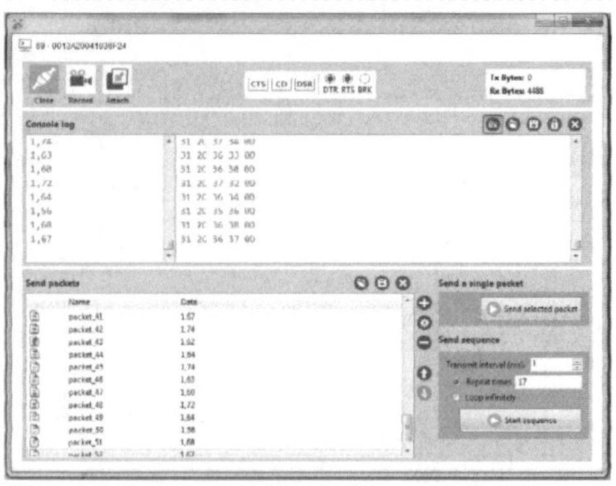

Figure 8. Display of Console XBee results

4.4. Communication Coverage Measurement
The measurement of XBee signal coverage done in two scenarios, namely the condition of enclosed space and open space conditions. In closed space conditions, the position between the first XBee and the second XBee placed along 23 meters with a height of 50 to 200 cm. While the status of open space, a position between the first XBee and XBee second positioned in the 38 meters with a height of 50 to 200 cm.

4.4.1 Closed Space Scenario. Table 1 shows the testing of the coverage of the communication signal using XBee in a closed space scenario with a height of 50 cm to 200 cm, send data of 80 bytes for 60 seconds.

Table 1. Closed space communication test results

XBee Position (cm)	Tx Data (byte)	Tx Packet	Rx Packet	Time (s)
50	19280	249	239	60
100	20080	251	251	60
125	30880	386	386	60
150	31200	390	390	60
200	31440	393	393	60

The test results show, the XBee module placement based on the height, very influential on the throughput generated. The result of throughput can see in Figure 9.

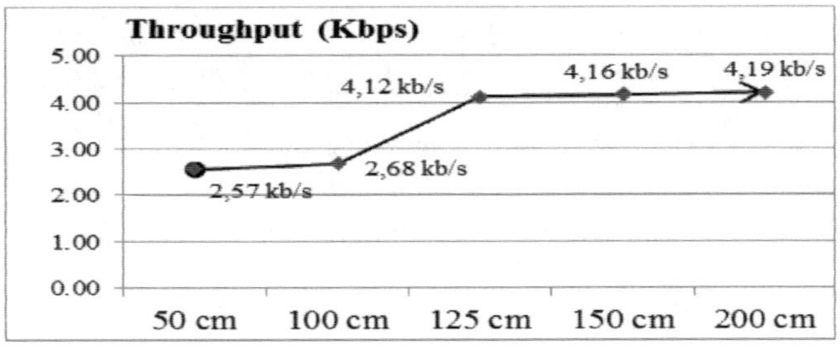

Figure 9. Throughput results in closed space scenarios

Figure 9 shows, the placement of XBee modules from 50 cm to 125 cm by throughput value increased from 2.57 Kbps to 4.12 Kbps. The location of the XBee module from 125 cm to 200 cm, throughput value increased by 0.07 Kbps. However, the placement of XBee modules 200 cm, the throughput value will experience a constant position of 4.19 Kbps.

For the test results, the amount of packet loss during the two XBee modules to communicate can see in Figure 10.

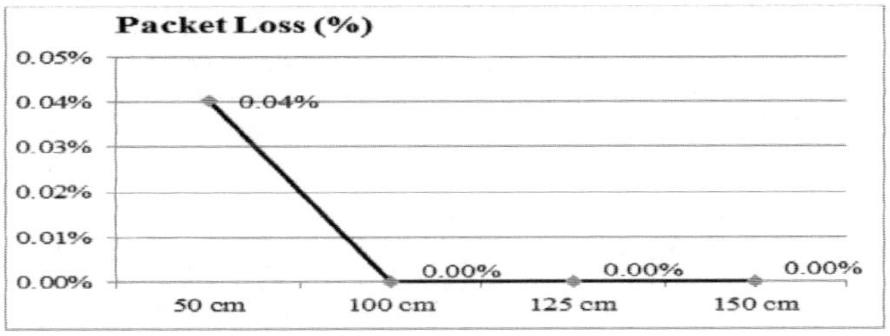

Figure 10. Packet loss occurs in closed space scenarios

IOP Conf. Series: Materials Science and Engineering **336** (2018) 012009 doi:10.1088/1757-899X/336/1/012009

Figure 10 shows, the placement of the XBee module at an altitude of 0 to 50 cm increased packet loss by 0.04%. However, the location of the XBee module at the height of 50 cm to 100 cm a packet loss decrease of 0%. Packet loss experiences a constant position at the height of 100 cm to 200 cm, with a packet loss value of 0%.

The average delay obtained when the two XBee modules communicate, can be seen in Figure 11.

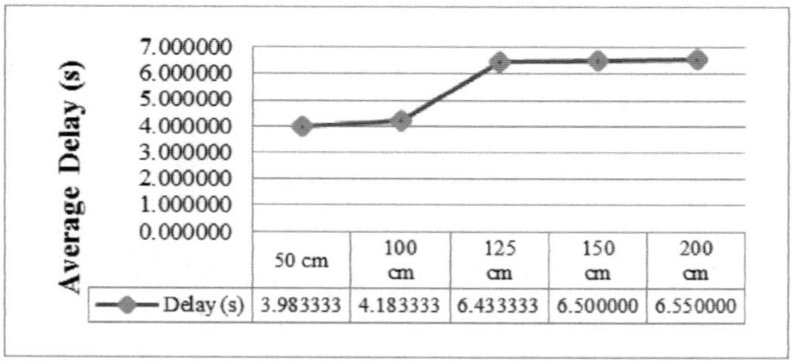

Figure 11. Average delay results in closed space scenarios

Figure 11 shows, the placement of the XBee module at an altitude of 50 cm to 200 cm has an average delay increase of 3.983 seconds to 6.55 seconds. The average delay experiences a constant position at the height of at 200 cm height, with a mean delay value of 6.55 seconds.

4.4.2 Open Space Scenario. Table 2 shows, the testing of the coverage of the communication signal using XBee in an open space scenario with a height of 50 cm to 200 cm, send data of 80 bytes for 15 seconds.

Table 2. Open space communication test results

ZigBee Position (cm)	Tx Data (byte)	Tx Packet	Rx Packet	Time (s)
50	6160	100	77	15
100	8000	100	100	15
125	8000	100	100	15
150	8000	100	100	15
200	8000	100	100	15

The test results show, the XBee module placement based on the height, very influential on the throughput generated. The result of throughput can see in Figure 12.

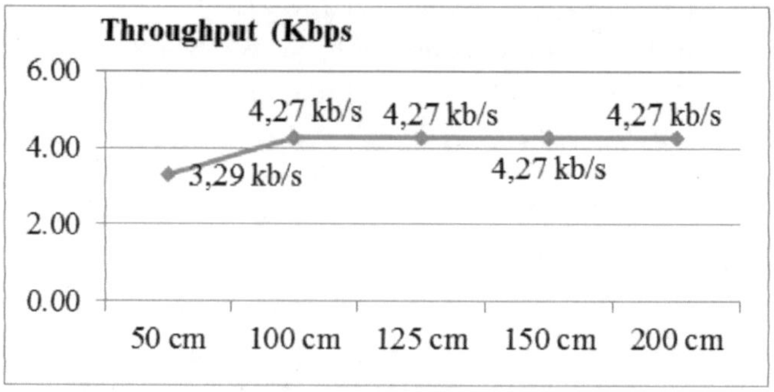

Figure 12. Throughput results in open space scenarios

Figure 12 shows, the placement of the XBee module at an altitude of 50 cm to 100 cm, the throughput value increased from 3.29 Kbps to 4.27 Kbps. The throughput value will be constant at a height of 100 cm to 200 cm, with a throughput value of 4.27 Kbps.

The test results in the amount of packet loss during the two XBee modules to communicate can see in Figure 13.

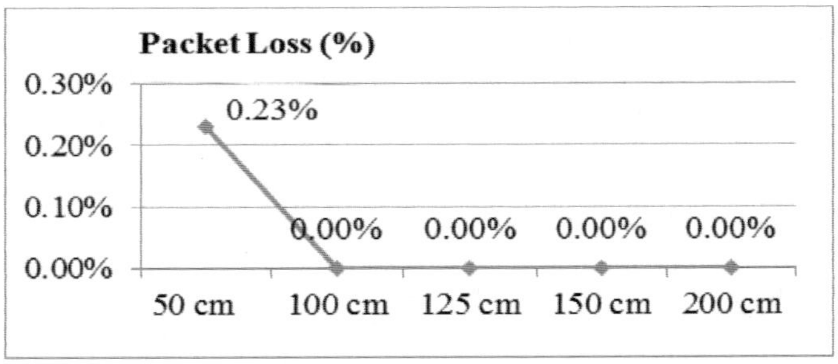

Figure 13. Packet loss results in open space scenarios

Figure 13 shows, the placement of the XBee module at an altitude of 0 to 50 cm increased packet loss of 0.23%. However, the location of the XBee module at the height of 50 cm to 100 cm a packet loss decrease of 0%. Packet loss experiences a constant position at the height of 100 cm to 200 cm, with a packet loss value of 0%.

The average delay obtained when the two XBee modules communicate, can be seen in Figure 14.

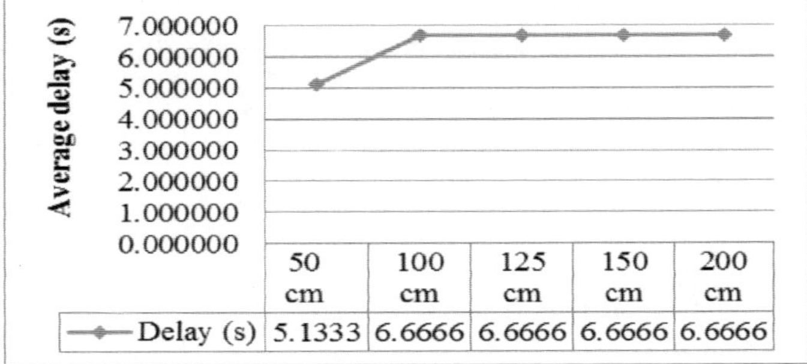

Figure 14. Average delay results in open space scenarios

Figure 14 shows, the placement of the XBee module at an altitude of 50 cm to 200 cm has an average delay increase of 5.133 seconds to 6.666 seconds. The average delay experiences a constant position at the height of at 200 cm height, with a mean delay value of 6.666 seconds.

5. Conclusion

Testing of heart signal sensor data which is done using a serial plotter for 15 seconds in men with the age of 25 years, resulted in a heart signal period of 68 bpm, received data of 4888 bytes, a throughput of 2.39 kbps and an average delay of 0.24486 seconds. The maximum voltage value is 3.25 volts, and the minimum voltage value is 0 volts.

The results of tests conducted on the XBee module based on the height of 50 cm - 200 cm and the distance of 23 meters within 60 seconds in the closed space, obtained the data transmission will be stable at an altitude of 100 to 200 cm.The higher the position of the module placed, it will result in a significant throughput value, and the resulting packet loss is decreasing. The test conducted on the XBee module based on the height of 50 to 200 cm and the distance of 38 meters within 15 seconds in the open space obtained that the data transmission will be stable at an altitude of 100 to 200 cm. The higher the position of the module placed, it will produce a significant throughput value and the resulting packet loss is decreasing.

References

[1] Santosh S and Anup Kale S., 2010. Remote patient monitoring network cluster. *Proceedings of the 2nd Makassar International Conference on Electrical Engineering and Informatics (MICEEI)*, Makassar Golden Hotel (MGH) Makassar, South Sulawesi, Indonesia, pp. 55-58
[2] Elyas P, Muhammad B and Andani., 2012. A 2.5 GHz wireless ECG system for remotely monitoring heart pulses, "accepted for presentation at IEEE *Antenna and Propagation Society (APS) Symposium/URSI/USNC Meeting*. Chicago, IL, USA
[3] Kazem S, Daniel M and Taieb Z., 2007. Wireless sensor network: technology, protocol, and applications, Wiley-Interscience
[4] Paolo B., 2005. Advances in wearable technology and applications in physical medicine and rehabilitation. *Journal of neuroengineering and rehabilitation, vol. 2 no.*
[5] Rhoanda C., 2003. Home care technology and telehealth. *Home healthcare nurse, volume 21*
[6] Safe Work Australia, 2013. Health monitoring for exposure to hazardous chemicals guide for workers. *Australian Industry Group*
[7] Srijani M, Koustabh D and Soumya K D., 2014. Patient health management system using e-health monitoring architecture. *IEEE International Advance Computing Conference (IACC)*, ITM University Gurgaon, India
[8] Jason H, Robert S and Kristofer P., 2000. System architecture directions for networked sensors. ASPLOS, Cambridge, MA, USA
[9] John A S., 2006. Wireless sensor networks. *Department of Computer Science University of Virginia Charlottesville, Virginia*
[10] Petruzella, F.D. 1996. Elektronika Industri. *Andi Offset, Yogyakarta*
[11] Jonathan H., 2004. TinyOS network programming. *Found in Jun. 2017 at http://www.cs.berkeley.edu/ jwhui/research/deluge/deluge-manual.pdf*
[12] Chris M., 2009. A tutorial for programming in TinyOS. *WCNG, University of Rochester*
[13] Evelyn P., 2007. Anatomy and physiology for paramedics. *Gramedia, Jakarta*
[14] Predrag C and Anda A., 2007. Processing of medical signals (ECG) in wireless sensor networks.

IOP Conf. Series: Materials Science and Engineering **336** (2018) 012010 doi:10.1088/1757-899X/336/1/012010

Indonesian Sign Language Number Recognition using SIFT Algorithm

Isa Mahfudi*, Moechammad Sarosa, Rosa Andrie Asmara, M. Azrino Gustalika

Electrical Engineering Department, State Polytechnic of Malang, Malang, Indonesia

*isa_mahfudi@polinema.ac.id

Abstract. Indonesian sign language (ISL) is generally used for deaf individuals and poor people communication in communicating. They use sign language as their primary language which consists of 2 types of action: sign and finger spelling. However, not all people understand their sign language so that this becomes a problem for them to communicate with normal people. this problem also becomes a factor they are isolated feel from the social life. It needs a solution that can help them to be able to interacting with normal people. Many research that offers a variety of methods in solving the problem of sign language recognition based on image processing. SIFT (Scale Invariant Feature Transform) algorithm is one of the methods that can be used to identify an object. SIFT is claimed very resistant to scaling, rotation, illumination and noise. Using SIFT algorithm for Indonesian sign language recognition number result rate recognition to 82% with the use of a total of 100 samples image dataset consisting 50 sample for training data and 50 sample images for testing data. Change threshold value get affect the result of the recognition. The best value threshold is 0.45 with rate recognition of 94%.

1. Introduction

Indonesian sign language (ISL) is generally used for deaf individuals and poor people communication in communicating. They use sign language as their primary language which consists of 2 types of action: sign and finger spelling. The system Indonesian sign language (SISL) dictionary as a formal standard in communicating their daily. The SISL dictionary is shown in figure 1. However, not all people understand their sign language so that this becomes a problem for them to communicate with normal people. This problem also becomes a factor they are isolated feel from the social life. There is a solution for them to be able to communicate with normal people. One of them is using the services of a sign language interpreter. this requires additional costs because of the need to pay the translator. Other solutions use a sign language recognition technique.

Techniques for sign language Recognition can be implemented with 3 methods[1]. Utilize the image processing technology, using glove data [1][2] and using sensors EMG (electromyography)[1]. Utilization technology of image processing to recognize sign language has its own challenges.Many research that offers a variety of methods in solving the problem of sign language recognition based on image processing [3][4][5][6][7][8][9]. Until now these existing methods still have shortcomings and will continue to do the repairs so that generated the best sign language recognition. SIFT (Scale Invariant Feature Transform) algorithm is one of the methods that can be used to identify an object.

SIFT is claimed very resistant to scaling, rotation, illumination, and noise [10][11]. In this paper, researchers proposed the SIFT method to use on the recognition of Indonesian sign language number.

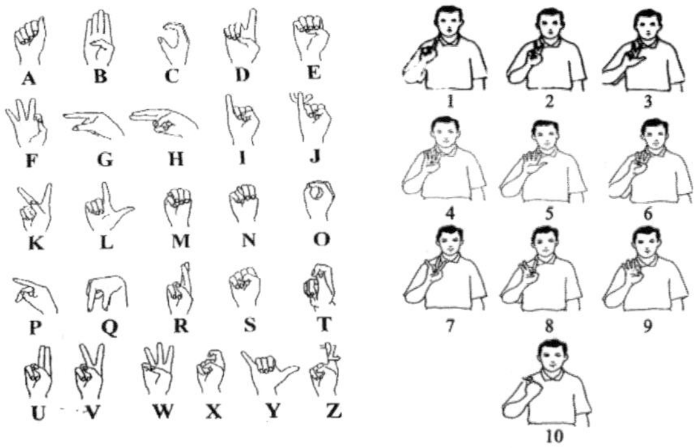

Figure 1. The system Indonesian sign language (SISL) Dictionary

2. Related Work

Sign language recognition has been widely performed by the researcher, some of the methods proposed for produce the high accuracy. A method of HMM [9][12] has been used for sign language recognition. HMM used for classification of techniques. The using of HMM method resulted in accuracy 85%. The using of Edge Orientation Histogram for recognition of the 26 ASL characters use 2600 samples training datasets result in accuracy to 88.26% [8]. Researchers [6][7] used a Principle Component Analysis (PCA) algorithm as a technique of classification for sign language recognition of Vietnamese and evaluate its Euclidean and Mahalanobis distances in classification. This resulted in the success of the introduction to 90.4% and 91.5%. The use of the Kinect Sensor is also applied [5], this Kinect sensor used to detect the numbers and letters of the United Kingdom and integrated with the SVM method for classification process. This resulted in 100% accuracy for numbers and 70.59% for letters with a distance of 300 mm – Kinect sensor from 1000mm.

SIFT algorithm has been applied as an extraction and classification method. Researchers [15][16][]7]. SIFT as a feature extraction method for an approach to batik pattern recognition. Using a collection of 120 batik images generated from 20 basic batik patterns, the proposed method shows an improvement over the original SIFT matching method with an equal error rate of 8.47% [15]. Integrating feature extraction between BOF and SIFT and using SVM for classification method for batik image classification results average accuracy of this method reaches 97.67%, 95.47% and 79% in normal image, rotated the image and scaled image, respectively[15]. Researcher [14] applied SIFT on Brain CT Images as Matching Algorithm. SIFT method combines with the gray feature. This experimental results accuracy from 89.83% to 92.78% with the time complexity basically unchanged, at the same time, the number of correct matches are not much reduced.

3. Proposed System

This research scheme refers to previous researchers who have proposed [7][13]. The Block diagram scheme proposed is shown figure 2.The workings of this system started taking pictures from the webcam as input data. the result image capture will do the process of segmentation to separate the object from the background. This segmentation using the HSV color model. This segmentation process results will be processed to minimize image cropping as well as do a filter for removing noise morphology and smooth image. next process will be applied extraction and classification method using SIFT algorithm[17][18].

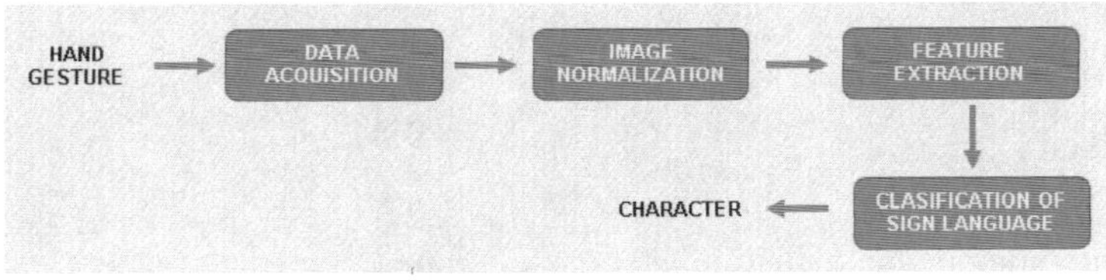

Figure 2. Proposed System.

3.1. Data Acquisition

This step is the stage for collect training dataset or testing dataset. This step started from capturing by webcam to take pictures with 1.3-megapixel resolution. When taking picture, the position of the camera facing straight with the position of the hands. The distance of the camera is set as far as 1 m from the hand. Captured hand gesture images describe Indonesian sign language number standard, starting from character number 1 to number 10. The total of the dataset is 100 sample pictures consists of 50 sample pictures for training data and 50 pictures for testing data. all sample datasets are conditioned with the same background and lighting. Indonesian sign language numbers dataset shown in Figure 3.

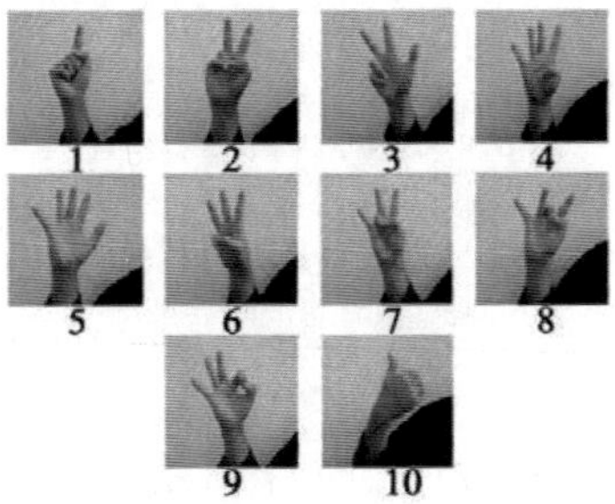

Figure 3. List of Indonesian Sign Language Number

3.2. Preprocessing

Preprocessing consist image acquisition, segmentation and morphological filtering method. The first step preprocessing is segmentation, started by filtering skin color for getting the skin color. The input RGB Image is transformed into HSV color space. After the skin object has been separate from the background, applied morphology filters for reducing noise and smooth the object. morphology operations used are dilation, erosion, and opening. The block diagram preprocessing is shown in Figure 4.

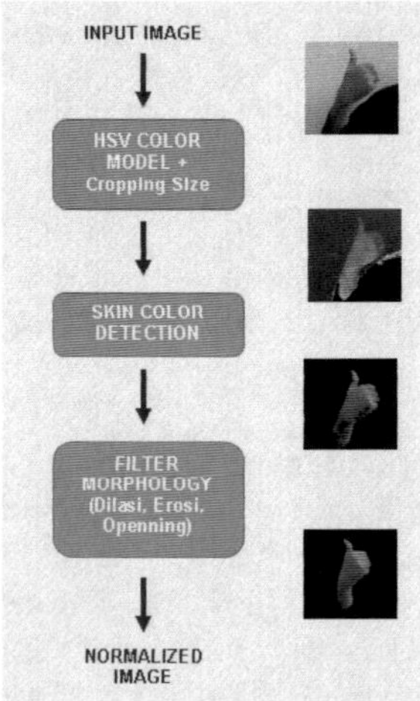

Figure 4. Block Diagram Preprocessed

3.3. Sift Based Image Matching

The SIFT algorithm has five major stages:

3.3.1. Scale-space construction. Suppose *I*(x,y) refers to the input image. In order to find reliable feature points in scale-space, filters are needed to process the images [17][18]. Gaussian transformation of images in scale-space can be expressed as:

$$L(x, y, \sigma) = G(x, y, \sigma) * I(x, y) \quad (1)$$

In which, $G(x, y, \sigma)$ is a scale variable Gaussian function:

$$G(x, y, \sigma) = \frac{1}{2\pi\sigma^2} e^{\frac{-(x^2+y^2)}{2\sigma^2}} \quad (2)$$

σ is called scale space coordinate and decides image's smoothness. Large-scale corresponds to image's profile feature, and small-scale corresponds to minutiae. I(x,y) represents an image[18]. In order to compute reliable feature points in scale-space effectively, DOG (Difference-of- Gaussian) operator is applied to the approximate Laplacian-Gaussian operator. Suppose the approximate operator is $D(x, y, \sigma)$ after processing the image we will have[17]:

$$\begin{aligned} D(x, y, \sigma) &= \big(G(x, y, k\sigma) - G(x, y, k\sigma)\big) * I(x, y) \\ &= L(x, y, k\sigma) - L(x, y, \sigma) \end{aligned} \quad (3)$$

3.3.2. Scale-space extrema detection. After approximating Laplacian-Gaussian operator with DOG operator, extrema detection is required in DOG scale space. The gray scale value of the point to be detected is compared with 18 pixels at the corresponding position of the front and rear images in the scale space, as well as the surrounding ones of the point itself. Altogether, a single point requires 26 comparisons. When an extremum (maximum or minimum) is detected, it is kept for further processing[17].

3.3.3. Orientation assignment. Local extrema detected in DOG scale-space are called keypoints after the operations of improving positioning accuracy and eliminating low-contrast points. To determine the keypoint orientation, a gradient orientation histogram is computed in the neighborhood of the

IOP Conf. Series: Materials Science and Engineering **336** (2018) 012010 doi:10.1088/1757-899X/336/1/012010

keypoint. The contribution of each neighboring pixel is weighted by the gradient magnitude and a Gaussian window with a value of δ that is 1.5 times the scale of the keypoint. Peaks in the histogram correspond to dominant orientations. A separate keypoint is created for the direction corresponding to the histogram maximum and any other direction within 80% of the maximum value. All the properties of the keypoint are measured relative to the keypoint orientation, which provides invariance to rotation[17].

3.3.4. Keypoint descriptor. The descriptor is formed from a vector containing the values of all the orientation histogram entries. The best results are achieved with a 4x4 array of histograms with 8 orientation bins in each histogram[17]. this paper uses a 4x4x8 = 128 element feature vector for each keypoint.

3.3.5. Feature vector matching. As each feature point corresponds to a 128 element feature vector, searching for matching points from two images becomes searching for matching feature vectors. Euclidean distance is used for measuring the similarity degree between two feature vectors, the ratio between the closest Euclidean distance and that of the second closest are computed and if the ratio is lower than a certain threshold [17]. When the ratio of two distances is less than a given threshold, accept the match points. The Euclidian distance of two N dimension functions f1(x) and f2(x) is as follows[19]:

$$d((fx), g(fx)) = \sqrt{(x_1 - g(x_1))^2 + f(x_1 - g(x_1))^2 + \cdots f(x_n\, g(x_n))^2} \quad (4)$$

$$= \sqrt{\sum_{i=1}^{n} f(x_i\, g(x_i))^2}$$

In this paper using 0.65 the vectors are considered to be matched with each other. Example matching result with Threshold = 0.65 is shown figure 5.

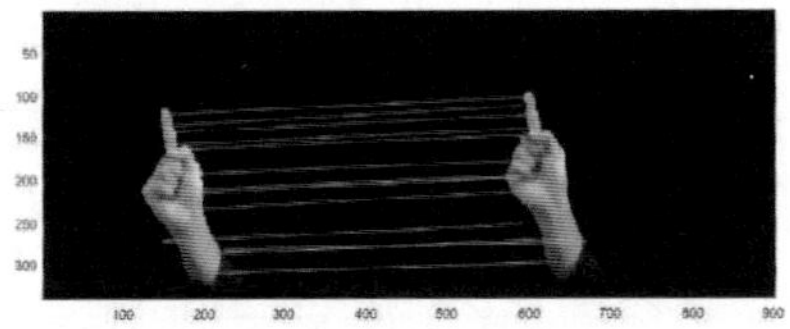

Figure 5. Matching results with T=0.65.

4. Experimental Results and Analysis

This stage will be shown the result of experiments to the recognition of sign language numbers using SIFT algorithm with a variety of characters. Figure 6 shows a matching result with the value Threshold of 0.65.

IOP Conf. Series: Materials Science and Engineering **336** (2018) 012010 doi:10.1088/1757-899X/336/1/012010

Figure 6. Matching results with the value of threshold 0.65.

The next stage will be shown the test results with 100 image dataset that consists 50 sample images testing data and 50 sample images training data. The result of the proposed system is shown in Table 1. there is 10 characters number Indonesian sign language with a total success rate up to 82%. The results show that the success rates of characters number "3", "6", "7", "8" and "9" are low because they are shape-like.Thus the system makes some mistakes while recognition. For the characters such as "1", "2", "4", "5" and "10", the images of hand gesture are good quality after preprocessing stage. Thus the system achieves high success rate in recognition for these characters. Table 2 show test results matching by changing the value of the threshold.

Table 1. The Success Rate of Recognition

Character	Number of test image	Number of correct recognition	Success Rate
1	5	5	100%
2	5	5	100%
3	5	4	80%
4	5	5	100%
5	5	5	100%
6	5	5	80%
7	5	2	40%
8	5	2	40%
9	5	3	80%
10	5	5	100%
	Average		82%

Table 2. Recognition results of sign language number

	Value of Threshold					
	0.45	0.55	0.65	0.75	0.85	0.95
Recognition Rate (%)	94	90	82	78	74	72

Table 2 shows the recognition rate with a different threshold. We obtain the best result with T=0.45 that produces the value recognition rate to 94%. The illustrate of results sign language recognition number shown in Figure 6.

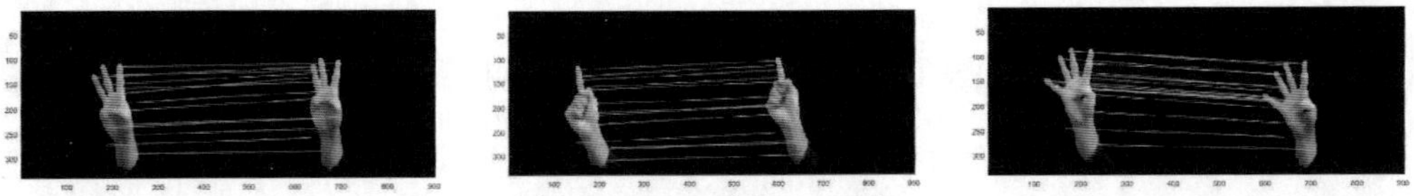

(a) Exact matching results

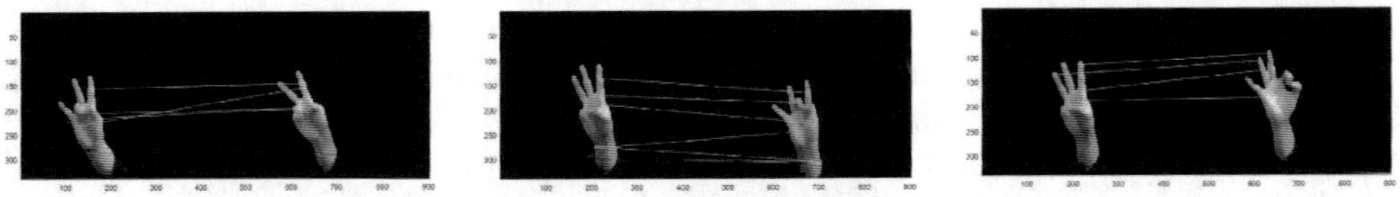

(b) False matching results

Figure 7. Illustrate results Matching performed by SIFT Method.

5. Conclusion

This paper proposes SIFT method for Indonesian sign language recognition number. on the value of the threshold of 0.65, the character numbers can be recognized very well are 1, 2, 4, 5 and 10 with the success rate of 100%. Change the value of the threshold obtained the best recognition rate at T = 0.45 which produces 94%.

6. Acknowledgment

The Authors would like to thank The State Polytechnic of Malang for supporting to attend this conference.

References

[1] M. E. Al-Ahdal and N. Md Tahir 2012 Review in Sign Language Recognition Systems IEEE Symposium on Computers & Informatics (ISCI) 52 – 57

[2] F. Husain, S. Gandhi, T. Nijhawan and Varsha Agarwal 2012 GESTURE RECOGNITION SYSTEM USING MATLAB: A LITERATURE REVIEW International Journal of Scientific Research and Management Studies (IJSRMS) **2** 11 425-432.

[3] A. Izzah and N. Suciati 2014 Translation of Sign Language Using Generic Fourier Descriptor and Nearest Neighbour Int. J. Cybern. Informatics **3** 1 31–41

[4] A. Kuznetsova, L. Leal-Taixe, and B. Rosenhahn 2013 Real-Time Sign Language Recognition Using a Consumer Depth Camera IEEE Int. Conf. Comput. Vis. Work 83–90

[5] W. Yi Yeh, T. Hui Tseng, J. Wei Hsieh and C. Ming Tsai 2016 Sign Language Recognition System Via Kinect: Number And English Alphabet International Conference on Machine Learning and Cybernetics (ICMLC) **2**, 660 - 665

[6] S. N. Sawant 2014 Sign Language Recognition System to aid Deaf-dumb People Using PCA, Int. J. Comput. Sci. Eng. Technol **5** 5 570–574

[7] T. N. T. Huong, T. V. Huu, T. Le Xuan and S. V. Van 2015 Static hand gesture recognition for Vietnamese sign language (VSL) using principle components analysis International Conference on Communications, Management and Telecommunications (ComManTel) 138-141

[8] J. R. Pansare and M. Ingle 2016 Vision-based approach for American Sign Language recognition using Edge Orientation Histogram IEEE International Conference on Image Vision, and Computing 86–90

[9] P. Kumar, H. Gauba, P. P. Roy and D. P. Dogra 2016 Coupled HMM-based Multi-Sensor Data Fusion for Sign Language Recognition, Pattern Recognition Letters **86** 3–10,

[10] R. Azhar, D. Tuwohingide, D. Kamudi, Sarimuddin and N. Suciati 2015 Batik Image Classification Using SIFT Feature Extraction Bag of Features and Support Vector Machine Elsevier Procedia Computer Science 24-30.

[11] X. Zhou, K. Wang, J. Fu 2016 A Method of SIFT Simplifying and Matching Algorithm Improvement International Conference on Industrial Informatics - Computing Technology, Intelligent Technology, Industrial Information Integration (ICIICII) 3-77
[12] M. Jebali, P. Dalle, and M. Jemni 2013 HMM-based method to overcome spatiotemporal sign language recognition issues IEEE Conference Publications 1-6
[13] K. S. Barai and S. Banoth 2016 Hand Gesture Recognition For Human Computer Interaction International Journal of Current Trends in Engineering & Research (IJCTER) **2** 5, 241-245,.
[14] Z. Xiao, L.Yu, Z. Qin, H. Ren, and Z. Geng 2016 A Point Matching Algorithm for Brain CT Images Based on SIFT and Gray Feature IEEE Conference Publications 1-6
[15] I. Nurhaida, A. Noviyanto, R. Manurung, A. M. Arymurthy 2015 automatic Indonesian's Batik Pattern Recognition sing SIFT Approach, International Conference on Computer Science and Computational Intelligence (ICCSCI) **59** 567-576.
[16] R. Azhar, D. Tuwohingide, D. Kamudi, Sarimuddin and N. Suciati 2015 "Batik Image Classification Using SIFT Feature Extraction, Bag of Features and Support Vector Machine", The Third Information Systems International Conference 24-30
[17] W. Wei, H. Jun, and T. Yiping 2008 Image Matching for Geomorphic Measurement Based on SIFT and RANSAC Methods International Conference on Computer Science and Software Engineering **2** 317-320
[18] X. Zhou, K. Wang and J. Fu 2016 A Method of SIFT Simplifying and Matching Algorithm Improvement International Conference on Industrial Informatics - Computing Technology, Intelligent Technology, Industrial Information Integration 73-77
[19] J. Zhu, S. Wang, and F. Meng 2011 SIFT Method for Paper Detection System, International Conference on Multimedia Technology 711-714

IOP Conf. Series: Materials Science and Engineering **336** (2018) 012011 doi:10.1088/1757-899X/336/1/012011

Study of Allocation Guaranteed Time Slot Wireless Body Area Networks Based on IEEE 802.15.4

E Yundra, G D Harsono

Department of Electrical Engineering, Universitas Negeri Surabaya, Indonesia

*eppyyundra@unesa.ac.id

Abstract. This paper aims to determine the size of the Guaranteed Time Slot (GTS) on the super frame structure required for each sensor as well as to know the performance of the GTS resized system compared to the GTS standard on IEEE 802.15.4. This article proposes a scheme to improve IEEE 802.15.4 medium access control, called allocation Guaranteed Time Slot (ALGATIS). ALGATIS is expected to effectively allocate guaranteed time slot to the requested sensors, it adjusts the length of the slot in super frame duration based on the length of the packet data. This article presents a simulation experiment of IEEE 802.15.4, especially for star network, to predict the throughput of networks and average energy consumption. The simulation experiments show that the performance of ALGATIS is better than that of IEEE 802.15.4 standard in term of the throughput of networks and average energy consumption

1. Introduction

According to the IEEE 802.15.4 standard [1], IEEE 802.15.4 can operate on beacon and non-beacon enabled modes. The network coordinator transmits beacon to synchronize and provide necessary information to the devices in beacon enable-mode, while the unslotted carrier sense multiple access with collision avoidance (CSMA/CA) protocol is used in non-beacon-enable mode. The IEEE 802.15.4 MAC supports both contention-based mechanism in contention access period (CAP) and guaranteed time slot (GTS) in contention free period (CFP) under beacon enable-mode. The limited number of allowable retransmissions and the number of back offs as specified in the standard can reduce energy consumption caused by carrier sensing. However, the performance of CAP and CFP are related each other because the number of request GTS packets successfully received by network coordinator may decrease if the contention level in CAP increases, which will decrease the throughput of CFP and vice versa.

The challenges of the IEEE 802.15.4 beacon-enable mode are how to improve throughput and reduce energy consumption in the networks. The author of [2] present an evaluation of the slotted CSMA/CA of IEEE 802.15.4 based on all of its frequency, which only analyze each frequency and compare with each other but not to propose a method. In [3], the authors analyze the performance of IEEE 802.15.4 MAC by using node state and channel state models that are simple but accurate. The authors also present an analytical model for the slotted CSMA/CA algorithm adopted in the CAP of the beacon-enabled mode in IEEE 802.15.4 MAC, which only considers for the saturated mode but not for acknowledgement (ACK).

IOP Conf. Series: Materials Science and Engineering **336** (2018) 012011 doi:10.1088/1757-899X/336/1/012011

The authors of [4] proposed Fibonacci Backoff (FIB) function to compute the backoff interval. In FIB, each node will wait for an increment backoff periods as they need to access channel. This scheme is expected to minimize possibility two or more nodes choose the same backoff period in the networks.

Several mathematical analysis models have been proposed to analyze the performance of IEEE 802.15.4 based on the Markov chain model without considering packet retransmissions [5-9]. However, all of the abovementioned models only consider for the contention-based transmission.

The authors of [10] provide an analysis for channel access during CAP and CFP. However, the purpose of the CFP transmission is to retransmit the packet that is not successful transmitted in CAP to cope with hidden node collisions. A delay bound analysis for an implicit GTS allocation is proposed to analyze the impact on the bandwidth utilization and delay by using numerical network calculus analysis [11]. The authors of [12] proposed a modified superframe structure for supporting immediate emergency data transmission. If any emergency is reported, then a fraction of inactive period is used opportunistically for three new periods for handling emergency events. The authors of [13] analyze the priorities of devices to determine for GTS allocation, while the authors of [14] further propose an adaptive GTS allocation scheme (AGA) using two phases to assign the priorities of devices and schedule GTS.

This article proposed a allocation Guaranteed Time Slot (ALGATIS) for IEEE 802.15.4 to assign adjustable length of GTS slot based on data sensor. ALGATIS is expected to effectively allocate GTS to sensors in order to improve the networks throughput and reduce energy consumption. The performance of ALGATIS for star network is also analyzed by considering packet retransmission, ACK, and defers transmission by modifying the Markov chain model from [14]. The major contribution of this article is to model the channel access for star network that used 7 data sensors for GTS and analyze the overall performance of IEEE 802.15.4 MAC to obtain the networks throughput and energy consumption.

2. The Description of ALGATIS

ALGATIS scheme have been created by a star network consists of one network coordinator and several sensors, while the network coordinator periodically sends beacon frames to the sensors. The network coordinator allocates the dedicated slots for its sensors, if it receives the requests for GTS packets in the CAP period, otherwise the device nodes shall transmit their packets with contention in CAP. This article proposes an allocation Guaranteed Time Slot (ALGATIS), to analyze both CAP and CFP for IEEE 802.15.4 MAC. ALGATIS aims to accurately decide the values of GTS slot based on the length of packet size on the sensors. ALGATIS mainly improves the network goodput and energy consumption by managing the GTS allocation for the requested sensors. ALGATIS can be expanded by considering the length of data packet, the SO value, and packet arrival rate. According to the IEEE 802.15.4 standard, let us denote Tsd to be the time of SD as shown in Eq. (1), where *aBaseSuperframeDuration* and Rs are the minimum duration of a superframe and data symbol rate with the values of 960 symbols and 62500 symbol/second, respectively. Let us also denote Tslot as the time of one slot duration, which can be obtained by Eq. (2).

$$T_{sd} = \frac{aBaseSuperframeDuration \times 2^{SO}}{Rs} \qquad (1)$$

$$T_{slot} = \frac{T_{sd}}{16} \qquad (2)$$

Each sensor with an allocated GTS ensures that the data transmission time, waiting time for ACK, time to transmit ACK packet and interframe spacing (IFS) duration can be completed before the end of its GTS period. Let us denote T_f be the time to transmit one data packet and receive ACK packet, which can be obtained by Eq. (3), where T_{data}, T_{Lack}, T_{ack} and T_{LIFS} are the time to transmit data packet, time to waiting for ACK packet, time to transmit ACK packet and time duration of IFS, respectively. The length of L_{ack} is equal to 88 bits, whereas the length of L_{IFS} is equal to *macMinLIFSPeriod* (160 bits) if the length of packet is greater than *aMaxSIFSFrameSize* (144 bits), otherwise, it is equal to *macMinSIFSPeriod* (48 bits).

$$T_f = T_{data} + T_{Lack} + T_{ack} + T_{LIFS} \tag{3}$$

In CAP, the packet transmission delay cannot be guaranteed, because packets are transmitted by using the CSMA/CA algorithm. Conversely, the packet transmission delay can be guaranteed in CFP, because packets are transmitted by using the allocated GTS without contention. In this article, we assume that each sensor generates time-critical and non-time-critical packets with probabilities P and (1−P), respectively. According to IEEE 802.15.4 standard, the maximum value to allocate the GTS slot duration is seven. Let us denote *adjslot* be the integer value that will be used as the adjustment for the T_{slot} of IEEE 802.15.4 standard become new smaller adjustment time of one slot duration. The value *adjslot* can be calculated by Eq. (4). Let us denote $T_{algatis}$ be the new time of one slot duration in ALGATIS, which can be calculated by Eq. (5)

$$adjslot = \left\lfloor \frac{T_{slot}}{T_f} \right\rfloor, \tag{4}$$

$$T_{algatis} = \frac{T_{slot}}{adjslot} \tag{5}$$

We consider star topology network having one network coordinator and seven sensors with different value of packet data. By using Eq. (5), we get the value of *adjslot* for each sensor. Based on the IEEE 802.15.4 standard, if each sensor request one slot GTS in IEEE 802.15.4, and its request successfully received by network coordinator. Furthermore, if there are more than seven device nodes in the star network, the device nodes which are not allocated GTS can transmits their data packets more in CAP period because the CAP duration of ALGATIS is increased.

3. Flowchart of Research Stages

In this section, the proposed ALGATIS based on the IEEE 802.15.4 MAC use slotted carrier sense multiple access with collision avoidance (CSMA/CA) for part of CAP and GTS transmission for part of CFP. This article also taking into account the case of acknowledged uplink data transmission is investigated comprehensively via simulation experiment as shown in Fig. 1.

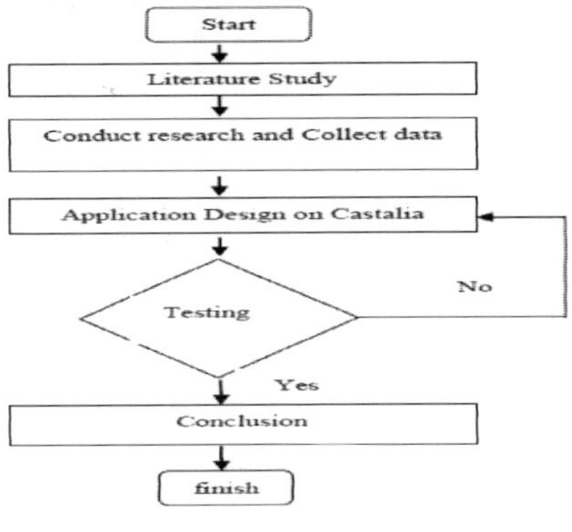

IOP Conf. Series: Materials Science and Engineering **336** (2018) 012011 doi:10.1088/1757-899X/336/1/012011

Figure 1. Flowchart of research stages

Figure 1 shows the flow diagram of the research stages. Beginning with an earlier research study then conducts research and collects data. After that, application design on Castalia. Then is tested by simulation. Finally, the result can be obtained and make conclusions then finish stage. Simulation result data in term of network throughput and energy consumption can be obtained by Castalia simulator.

4. Simulation and Analysis Results

In this section, simulation experiments for ALGATIS are performed by using the extended Castalia simulator to validate the analysis and performance evaluation. In simulation model, we consider a star topology with one PAN coordinator and 21 device nodes, where D_{node} is equal to 10 meters. To simulate the performance of power consumption, we consider the radio parameters of Chipcon's CC2420 2.4 GHz for the IEEE 802.15.4 RF transceiver [15], where the transmitting power PWR_{tx}, the receiving power PWR_{rx}, and the idle power PWR_{idle}, are 31.32 mW, 35.28 mW, and 712 µW, respectively [18]. We compute the network throughput and total network energy consumption, where traffic load varies from 0.1 to 1(full loaded). Table 1 summarizes the simulation parameters.

Table 1. The simulation parameters

Parameter	Value
Physical data rate	250 kbps
UBP	80 bits
NumSuperframeSlots	16
MacPacketOverhead	112 bits
ACK length (L_{ack})	88 bits
D_{node}	10 m
PWR_{tx}	31.32mW
PWR_{rx}	35.28mW
PWR_{idle}	712µW
BE_{min}	3
BE_{max}	5

Figure 2 shows the throughput against the traffic load by simulation in the networks on 3 nodes position. The proposed ALGATIS algorithm can adjust length of data sensor to allocate for GTS, so that ALGATIS algorithm has higher networks throughput than that of IEEE 802.15.4 standard.

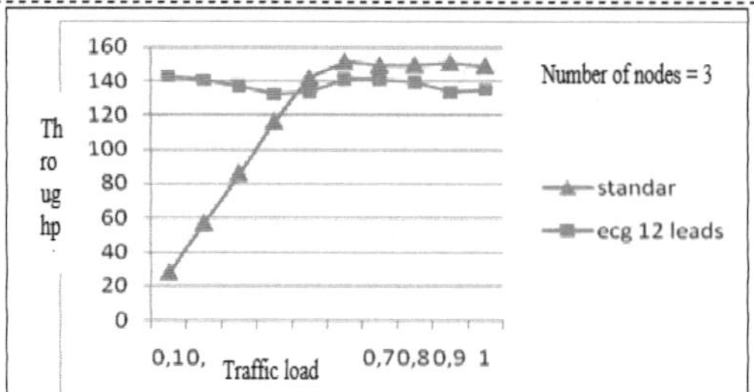

Figure 2. Throughput against traffic load

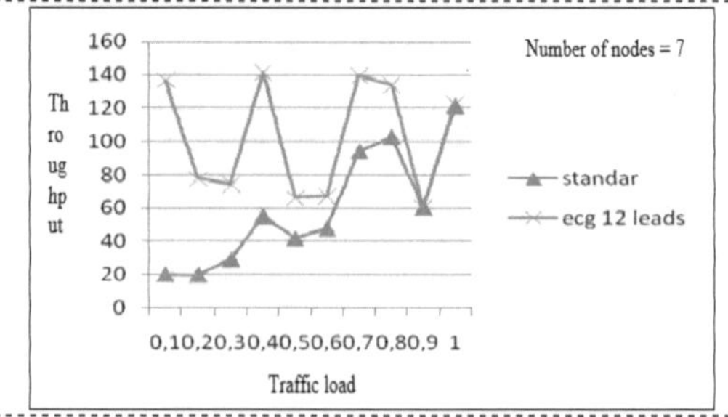

Figure 3. Throughput against traffic load

In networks consist of 7 nodes as shown on figure 3 shows the throughput against the traffic load by simulation in the networks on 7 nodes position. The proposed ALGATIS algorithm still also can adjust length of data sensor to allocate for GTS, so that ALGATIS algorithm has higher networks throughput than that of IEEE 802.15.4 standard.

Figure 4 shows the throughput against the traffic load by simulation in the networks on 13 nodes position. The proposed ALGATIS algorithm still also can adjust length of data sensor to allocate for GTS, so that ALGATIS algorithm has higher networks throughput than that of IEEE 802.15.4 standard.

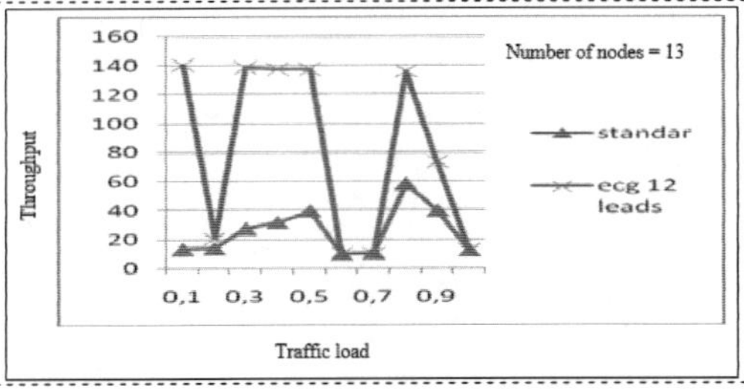

Figure 4. The network throughput against traffic load

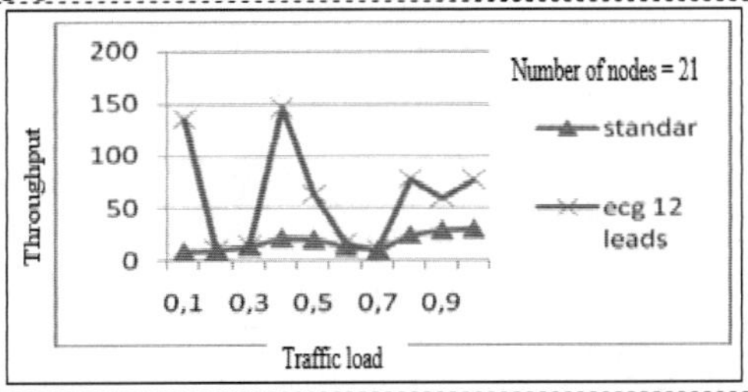

Figure 5. The network throughput against traffic load

Figure 5 shows the network throughput against traffic load. It is obvious the network throughput of ALGATIS is higher than that of IEEE 802.15.4 standard. In the light traffic load (i.e., traffic load is equal to 0.2 – 0.3), the network throughput of ALGATIS is almost the same as that of IEEE standard; however, ALGATIS outperforms to the IEEE 802.15.4 standard as the traffic load increases.

Figure 6 shows the network energy consumption against traffic load. ALGATIS consumes lesser network energy than that of the IEEE 802.15.4 standard, which means that ALGATIS minimizes the energy consumption when retransmitting data packet on number of nodes equal 3.

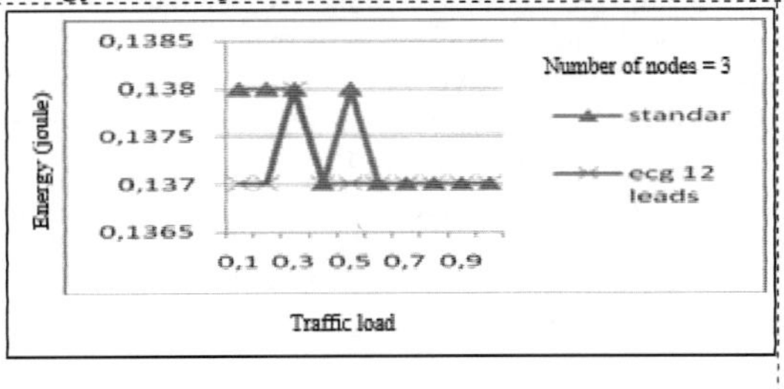

Figure 6. The total of network energy consumption against traffic load

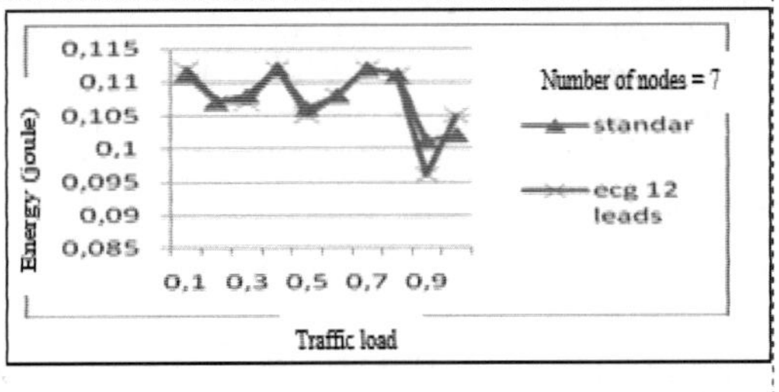

Figure 7. The total of network energy consumption against traffic load

IOP Conf. Series: Materials Science and Engineering **336** (2018) 012011 doi:10.1088/1757-899X/336/1/012011

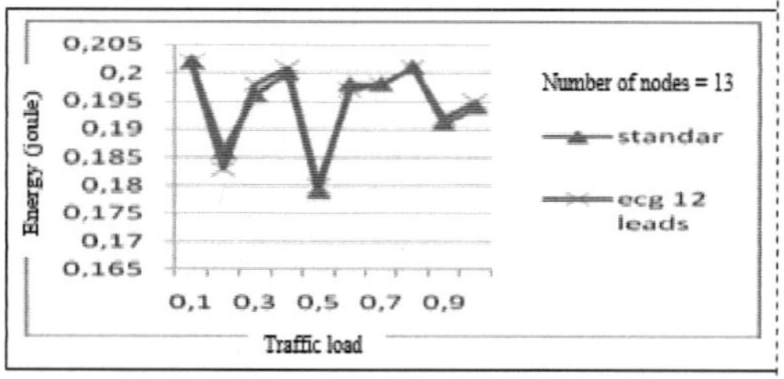

Figure 8. The total of network energy consumption against traffic load

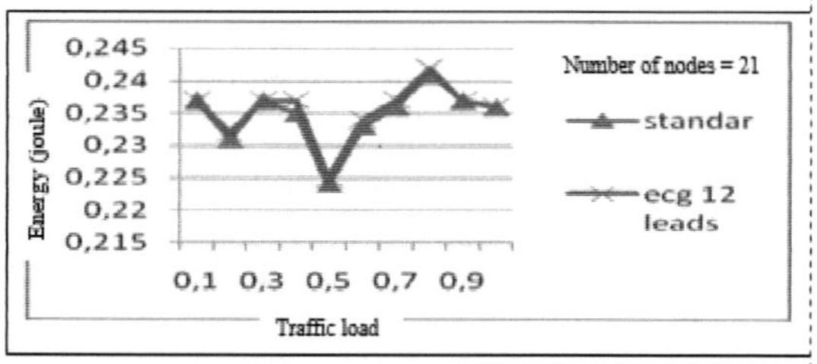

Figure 9. The total of network energy consumption against traffic load

Figure 7, figure 8 and figure 9 respectively, show the network energy consumption against traffic load. ALGATIS consumes lesser network energy than that of the IEEE 802.15.4 standard, because the ALGATIS algorithm can adjust size of data each sensor and allocate to GTS based on its size. Moreover, ALGATIS has greater probability of successful transmission than that of the IEEE 802.15.4 standard, especially in heavy traffic load, which means that ALGATIS minimizes the energy consumption when retransmitting data packet. The energy consumption is obtained by summing the energy consumption of PAN coordinator and all of device nodes in the network.

5. Conclusions

In this article, ALGATIS performs with the adjustable length of the slot in the superframe duration based on the length of sensors, so that it can accurately decide for the starting time, and the GTS length to be allocated for the requested devices to improve networks. ALGATIS is expected to effectively allocate GTS to the requested devices, because the length of CAP on ALGATIS is longer than that of IEEE 802.15.4 standard. This article also presented a comprehensive analysis of IEEE 802.15.4, specifically for star network, to predict the network throughput as well as the network energy consumption. The validity of the analytical model is shown by closely matching its predictions of the simulation results. The results of simulation experiment show that ALGATIS is better than that of IEEE 802.15.4 standard in term of the networks throughput and energy consumption. As a future work, the proposed scheme will be compared by other algorithms

References

[1] IEEE 802.15.4, "Part 15.4: Wireless Medium Access Control (MAC) and Physical Layer (PHY) Specifications for Low-Rate Wireless Personal Area Networks (WPANs)," IEEE standard for information technology, September 2006.

[2] A. N. Alvi, S. S. Naqvi, S. H. Bouk, N. Javaid, U. Qasim, and Z. A. Khan," Evaluation of Slotted CSMA/CA of IEEE 802.15.4," Seventh International Conference on Broadband, Wireless Computing, Communication and Applications, Canada, 2012.

[3] K. Ashrafuzzaman and K. Sup Kwak, "On the Performance Analysis of the Contention Access Period of EEE 802.15.4 MAC," IEEE Communications Letters, Vol. 15, No. 9, September 2011.

[4] T.-R Park, T.-H Kim, J.-Y Choi, S. Choi, W.-H Kwon, "Throughput and energy consumption analysis of IEEE 802.15.4 slotted CSMA/CA," *IEEE Electronics Letters*, vol.41, no.18, pp.

1017-1019, 1 September 2005.

[5] T.-J Lee, H.-R Lee, M.-Y Chung, "MAC throughput limit analysis of slotted CSMA/CA in IEEE 802.15.4 WPAN," *IEEE Communications Letters*, vol.10, no.7, pp. 561- 563, July 2006

[6] S Pollin, M Ergen, S Ergen, B Bougard, L Der Perre, I Moerman, A Bahai, P Varaiya, F Catthoor, "Performance Analysis of Slotted Carrier Sense IEEE 802.15.4 Medium Access Layer," *IEEE Transactions on Wireless Communications*, vol.7, no.9, pp.3359-3371, September 2008.

[7] Y Zhang, F Shu, "Packet Size Optimization for Goodput and Energy Efficiency Enhancement in Slotted IEEE 802.15.4 Networks," *IEEE Wireless Communications and Networking Conference,*pp.1-6, 5-8 April 2009.

[8] J He, Z Tang, HH Chen, Q Zhang, "An accurate and scalable analytical model for IEEE 802.15.4 slotted CSMA/CA networks," *IEEE Transactions on Wireless Communications,* vol.8, no.1, pp.440-448, January 2009.

[9] Z Xiao, C He, L Jiang, "An analytical model for IEEE 802.15.4 with sleep mode based on time-varying queue," *in IEEE International Conference on Communications (ICC)*, Kyoto, Japan, 2011.

[10] M Khanafer, M Guennoun, HT Mouftah, "Adaptive sleeping periods in IEEE 802.15.4 for efficient energy savings: markov-based theoretical analysis," in *IEEE International Conference on Communications (ICC),* Kyoto, Japan, 2011.

[11] B Gao, C He, L Jiang, "Modeling and analysis of IEEE 802.15.4 CSMA/CA with sleep mode enabled," *International Conference on Communication Systems,* pp.6-11, Guangzhou, China, 2008.

[12] C.-Y Jung, H.-Y Hwang, D.-K Sung, G.-U Hwang, "Enhanced Markov Chain Model and Throughput Analysis of the Slotted CSMA/CA for IEEE 802.15.4 Under Unsaturated Traffic Conditions," IEEE Transactions on Vehicular Technology, vol.58, no.1, pp.473-478, January 2009.

[13] B Shrestha, E Hossain, S Camorlinga, "A Markov model for IEEE 802.15.4 MAC with GTS transmissions and heterogeneous traffic in non-saturation mode," *in IEEE International Conference on Communication Systems (ICCS)*, pp. 56–61, Singapore, 2010.

[14] B.-H. Lee, Eppy Yundra, H.-K. Wu and M Udin Harun Al Rasyid, , "Analysis of superframe duration adjustment scheme for IEEE 802.15.4 networks," EURASIP Journal on Wireless Communication and Networking, April, 2015

[15] AS Chipcon SmartRF® CC2420 datasheet (rev 1.2), Chipcon corp., 2004

[16] B Bougard, F Catthoor, DC Daly, "A Chandrakasan, W Dehaene, Energy Efficiency of the IEEE 802.15.4 Standard in Dense Wireless Microsensor Networks: Modeling and Improvement Perspectives". *Proceeding of Design, Automation and Test in Europe Conference and Exhibition. (DATE'05),* pp.196-201, Mar. 2005.

IOP Conf. Series: Materials Science and Engineering **336** (2018) 012012 doi:10.1088/1757-899X/336/1/012012

Brain Tumor Image Segmentation in MRI Image

Hapsari Peni Agustin Tjahyaningtijas

Electrical Engineering Department, Universitas Negeri Surabaya, Indonesia

hapsaripeni@unesa.ac.id

Abstract-Brain tumor segmentation plays an important role in medical image processing. Treatment of patients with brain tumors is highly dependent on early detection of these tumors. Early detection of brain tumors will improve the patient's life chances. Diagnosis of brain tumors by experts usually use a manual segmentation that is difficult and time consuming because of the necessary automatic segmentation. Nowadays automatic segmentation is very populer and can be a solution to the problem of tumor brain segmentation with better performance. The purpose of this paper is to provide a review of MRI-based brain tumor segmentation methods. There are number of existing review papers, focusing on traditional methods for MRI-based brain tumor image segmentation. this paper, we focus on the recent trend of automatic segmentation in this field. First, an introduction to brain tumors and methods for brain tumor segmentation is given. Then, the state-of-the-art algorithms with a focus on recent trend of full automatic segmentaion are discussed. Finally, an assessment of the current state is presented and future developments to standardize MRI-based brain tumor segmentation methods into daily clinical routine are addressed.

1. Introduction

Brain tumor is one a player in the tumor in the sensory system, notwithstanding spinal tumors and fringe nerve tumors. This brain tumor can be an essential tumor or a metastasis from a tumor in another organ. Clinical issues in brain tumors are fairly not quite the same as different tumors on account of their impact, and restricted remedial confinements. Harm to the brain tissue straightforwardly will cause useful disability of the focal sensory system, as engine unsettling influences, tangible, faculties, and even psychological capacity. Furthermore, the impact of mass caused by brain tumor will likewise give a difficult issue considering the tumor is in the cranial hole which in grown-ups is a shut space with settled size.

According to The Central Brain Tumor Registry of the United States (CBTRUS), primary brain tumors are among the top 10 causes of cancer-related deaths. It is estimated that about 13,000 people in the United States die from this tumor every year. Data from Mayo Clinic, based on analysis from 1950 to 1989, it was said that the incidence of primary brain tumors was 19.1 per 100,000 people per year (11.8 per 100,000 for symptomatic tumors and 7.3 per 100,000 for asymptomatic tumors). This data is similar to the data from CBTRUS which gives the figure of 11.47 per 100,000 per year. In Europe the average survival rate of adult malignant brain tumor patients was 18.7%. The prognosis of primary brain tumors varies, in malignant primary malignant brain tumors whose survival is ± 12 months. In another study that measured the survival rate of patients with brain tumor, the survival rate was found in 5 years. The worst brain tumor patients were glioblastoma 3% while the highest was ependimoma, 74%.[1]

Alongside the progress of therapeutic imaging, imaging modalities assume a critical part in the assessment of patients with cerebrum tumors and significantly affect quiet care. Late years, the rising new imaging modalities, for example, XRay, Ultrasonography, Computed Tomography (CT), Magneto Encephalo Graphy (MEG), Electro Encephalo Graphy (EEG), Positron Emission Tomography (PET), Single-Photon Emission Computed Tomography (SPECT), and Magnetic Resonance Imaging (MRI), not just demonstrate the point by point and finish parts of mind tumors, yet additionally enhance clinical specialists to contemplate the component of cerebrum tumors at the point of better treatment. Clinical specialists assume an imperative part in cerebrum tumor evaluation and treatment. Once a mind tumor is clinically suspected, radiologic assessment is required to decide the area, the degree of the tumor, and its relationship to the encompassing structures. This data is imperative and basic in settling on the diverse types of treatment, for example, surgery, radiation, and chemotherapy. Accordingly, the assessment of cerebrum tumors with imaging modalities is presently one of the key issues of radiology offices.

Magnetic Resonance (MR) imaging produces images of the human tissues in a noninvasive manner, revealing the structure, metabolism, and function of tissues and organs. The impact of this image technique in diagnostic radiology is impressive, due to its versatility and flexibility in joining high-quality anatomical images with functional information [2]. MRI is attracting more and more attentions for the brain tumor diagnosis in the clinical.

Four standard MRI modalities used for glioma diagnosis include T1-weighted MRI (T1), T2-weighted MRI (T2), T1-weighted MRI with gadolinium contrast enhancement (T1-Gd) and Fluid Attenuated Inversion Recovery (FLAIR)[3]. Generally, T1 images are used for distinguishing healthy tissues, whereas T2 images are used to delineate the edema region which produces bright signal on the image. In T1-Gd images, the tumor border can easily be distinguished by the bright signal of the accumulated contrast agent (gadolinium ions) in the active cell region of the tumor tissue. Since necrotic cells do not interact with the contrast agent, they can be observed by hypo intense part of the tumor core making it possible to easily segment them from the active cell region on the same sequence. In FLAIR images, signal of water molecules are suppressed which helps in distinguishing edema region from the Cerebrospinal Fluid (CSF).

Lately, restorative imaging and delicate figuring have made huge progressions in the field of brain tumor division. As a rule, the vast majority of strange brain tumor tissues might be effortlessly identified by brain tumor division strategies. In any case, exact and reproducible division results and portrayal of variations from the norm have not been tackled the distance. Since mind tumor division has incredible effect on analysis, checking, treatment making arrangements for patients, and clinical trials, this paper centers around MRI-based cerebrum tumor division and presents a generally nitty gritty review for the current existing techniques for MRI-based cerebrum tumor division.

The rest of this paper is organized as follows: In Section 2, we briefly introduce the segmentation methods of MRI images., in Section 3, an objective assessment is presented and future developments and trends are addressed for MRI based brain tumor segmentation methods.

2. Brain Tumor Segmentation Methods

Nowadays, brain tumor segmentation methods can be organized into different categories based on different principles. In the clinic, brain tumor segmentation methods are usually classified into three main categories including manual, semi-automatic, and fully automatic segmentations based on the degree of required human interaction.

For semi-automatic brain tumor segmentation, it mainly consists of the user, interaction, and software computing. In the semi-automatic brain tumor methods, the user needs to input some parameters and is responsible for analyzing the visual information and providing feedback response for the software computing. The software computing is targeted at the realization of brain tumor segmentation algorithms. The interaction is in charge of adjusting segmentation information between the user and the software computing. The semi-automatic brain tumor segmentation methods were divided into three main processes: initialization, feedback response, and evaluation. Although brain tumor semi-automatic segmentation methods can obtain better results than manual segmentation, it also comes into being different results from different experts or the same user at different times. Hence, fully automatic brain tumor segmentation methods were proposed.

For fully automatic brain tumor segmentation, the computer determines the segmentation of brain tumor without any human interaction. In general, a fully automatic segmentation algorithm combines artificial intelligence and prior knowledge. With the development of machine learning algorithms that can simulate the intelligence of humans to learn effectively, the study of fully automatic brain tumor segmentation has become a popular research issue. The semi-automatic and fully automatic segmentation of tumor brain images are faced with great challenges due to usually exhibiting unclear and irregular boundaries with discontinuities and partial-volume effects for brain tumor images. The segmentation method for various segmentation listed in Table 1.

Table 1. Various Segmentation method for brain tumor image segmentation

Author	Image Segmentation Method	Performance
Xiaoxiao Liu et.al [4]	low rank atlas	accuracy, Geodesic distance, entropy
Mohammad Majid al-[5] et.al	PSO-LVQ	Effisiency
V. Anitha et.al [6]	adaptive pillar K-means clustering algorithm	sensitivity, accuracy, positive predictive value, negative predictive value, false discovery rate,Mathews correlation coefficient
Atiq Islam et.al[7]	Fractal and Fractional Brownian Motion	Similarity coefficients:Jaccard,Dice,
Matthew C. Clark et.al [8]	Adaptive Histogram .	TP,FP,FN, Tumor Ratio, Corr Ratio
Meiyan Huang et.al [9]	local independent projection-based classification	computation time,dice similarity, jaccard similarity
Bjoern H. Menze et.al [10]	expectation-maximization	Dice score, sensitivities and specificities
Elisee Ilunga–Mbuyamba et.al[11]	Localized Active Contour Model with Background Intensity Compensation (LACM-BIC)	Dice Coefficient,housedroff distance
Charutha S. [12]	Modified Texture Based Region Growing, Cellular Automata Based Edge Detection, Modified Texture Based Region Growing + Cellular Automata Edge Detection	Coefficient of similarity, Spatial overlap

3. Conclusion

Automatic segmentation of the brain tumors for cancer diagnosis is a challenging task. Recently, availability of public datasets provided a common medium for the researchers to develop and objectively evaluate their methods with the existing techniques. In this paper, we provided a review segmentation method based automatic segmentation, and a brief overview of traditional techniques in traditional automatic brain tumor segmentation methods, translating prior knowledge into probabilistic maps or selecting highly representative features for classifiers is challenging task. Future improvements and modifications in modified segmentation method architectures and addition of complementary information from other imaging modalities such as Positron Emission Tomography (PET), Magnetic Resonance Spectroscopy (MRS) and Diffusion Tensor Imaging (DTI) may improve the current methods, eventually leading to the development of clinically acceptable automatic brain tumor segmentation methods for better diagnosis.

References

[1] A. Arber, S. Faithfull, M. Plaskota, C. Lucas, and K. de Vries, "A study of patients with a primary malignant brain tumour and their carers: symptoms and access to services," *Int. J. Palliat. Nurs.*, vol. 16, no. 1, pp. 24–30, Feb. 2010.

[2] L. Landini, V. Positano, and M. F. Santarelli, *Advanced image processing in magnetic resonance imaging*. CRC/Taylor & Francis, 2005.

[3] A. Işın, C. Direkoğlu, and M. Şah, "Review of MRI-based Brain Tumor Image Segmentation Using Deep Learning Methods," *Procedia Comput. Sci.*, vol. 102, pp. 317–324, Jan. 2016.

[4] X. Liu, M. Niethammer, R. Kwitt, N. Singh, M. McCormick, and S. Aylward, "Low-Rank Atlas Image Analyses in the Presence of Pathologies," *IEEE Trans. Med. Imaging*, vol. 34, no. 12, pp. 2583–2591, Dec. 2015.

[5] D. J. Hemanth, A. Aber, and M. M. al-Rifaie, "Deploying swarm intelligence in medical imaging identifying metastasis, micro-calcifications and brain image segmentation," *IET Syst. Biol.*, vol. 9, no. 6, pp. 234–244, Dec. 2015.

[6] V. Anitha and S. Murugavalli, "Brain tumour classification using two-tier classifier with adaptive segmentation technique," *IET Comput. Vis.*, vol. 10, no. 1, pp. 9–17, Feb. 2016.

[7] A. Islam, S. M. S. Reza, and K. M. Iftekharuddin, "Multifractal Texture Estimation for Detection and Segmentation of Brain Tumors," *IEEE Trans. Biomed. Eng.*, vol. 60, no. 11, 2013.

[8] M. C. Clark, L. O. Hall, D. B. Goldgof, R. Velthuizen, F. R. Murtagh, and M. S. Silbiger, "Automatic tumor segmentation using knowledge-based techniques," *IEEE Trans. Med. Imaging*, vol. 17, no. 2, pp. 187–201, Apr. 1998.

[9] M. Huang, W. Yang, Y. Wu, J. Jiang, W. Chen, and Q. Feng, "Brain Tumor Segmentation Based on Local Independent Projection-Based Classification," *IEEE Trans. Biomed. Eng.*, vol. 61, no. 10, pp. 2633–2645, Oct. 2014.

[10] B. H. Menze, K. Van Leemput, D. Lashkari, T. Riklin-Raviv, E. Geremia, E. Alberts, P. Gruber, S. Wegener, M.-A. Weber, G. Szekely, N. Ayache, and P. Golland, "A Generative Probabilistic Model and Discriminative Extensions for Brain Lesion Segmentation— With Application to Tumor and Stroke," *IEEE Trans. Med. Imaging*, vol. 35, no. 4, pp. 933–946, Apr. 2016.

[11] E. Ilunga–Mbuyamba, J. G. Avina–Cervantes, J. Cepeda–Negrete, M. A. Ibarra–Manzano, and C. Chalopin, "Automatic selection of localized region-based active contour models using image content analysis applied to brain tumor segmentation," *Comput. Biol. Med.*, vol. 91, pp. 69–79, Dec. 2017.

[12] S. Charutha and M. J. Jayashree, "An efficient brain tumor detection by integrating modified texture based region growing and cellular automata edge detection," *2014 Int. Conf. Control. Instrumentation, Commun. Comput. Technol. ICCICCT 2014*, pp. 1193–1199, 2014.

IOP Conf. Series: Materials Science and Engineering **336** (2018) 012013 doi:10.1088/1757-899X/336/1/012013

Opinion Analysis on Rohingya using Twitter Data

N Rochmawati*, S C Wibawa

Department of Informatics, Universitas Negeri Surabaya, Indonesia

*naim.rochmawati@gmail.com

Abstract. Rohingya is an ethnicity in Myanmar. Recently there was a conflict in the area between the Rakhine population and the Myanmar army. Many opinions are pro and contra in addressing this issue. There is a critic, there is a support and there is a neutral. The purpose of this paper is to analyze the world public opinion about the case of Rohingya. The opinion data to be processed is taken from twitter. the reason for using twitter is because twitter has become one of the popular social media and includes the most frequently visited social media. Therefore, it would be a lot of data that can be taken from twitter to be processed in the process of sentiment analysis. The grouping of opinions will be divided into 3 parts of positive, negative and neutral. the method used in grouping is the naïve Bayes method.

1. Introduction

Rohingya is an ethnic Indo-Aryan from the Rakhine region (also known as Arakan, or Rohang in Rohingya) in Myanmar. Recently there was violence in Rohingya between the Rakhine population and the Myanmar army. Many opinions are pro and contra in addressing this issue. Some are criticized and some are supportive and some are neutral.

Twitter is one of the social media that allows users to send and read text-based messages. this message is called tweet. tweet is limited to 140 characters. twitter was first launched in 2006 (Daniel, Twitter, Edward, Twitter, & Earth, n.d.). people use microblogging like twitter to talk about their daily activities and share information(Finin & Tseng, 2007).

The high popularity of Twitter can be used for various purposes in various aspects, for example for political campaign, learning tool, and as an emergency communication media. Twitter is also faced with various issues and controversies such as user security and privacy issues, lawsuits, and censorship. Twitter can also be used to know the opinions of the world community on news about Rohingya.

The purpose of this paper is to know the opinion analysis of people about Rohingya where the opinion will be divided into 3 groups: positive, negative and neutral. An opinion analysis will be taken from twitter. Grouping data will use naive Bayes.

2. .Literature

2.1. State of the art

Sentiment analysis is widely developed in various fields such as to analyze hotel rating (Elango & Narayanan, 2002; Hargreaves, 2015; Kasper & Vela, nd), politics (Spoelstra, nd), stock (Joshi, nd), health (Database, 2016) and so on.

2.2. Sentiment analysis
Sentiment analysis is one of the fastest growing research areas in computer science (Mäntylä, Graziotin, & Kuutila, 1937). Hence it is not surprising that the sentiment analysis is much in demand as a research material.

Sentiment analysis is a way to retrieve information from opinions or language. Therefore information sentiment analysis is mostly taken from social media because opinions and user information can be extracted from users tweets. Sentiment analysis is divided into two levels: document level (coarse grained sentiment analysis) and fined grained sentiment analysis (Fink, Chou, Kopecky, & Llorens, 2011).

2.3. Text mining
Text mining is the application of data mining concepts and techniques to find patterns in the text. the process of analyzing texts is intended to extract useful information for a particular purpose. Based on the irregularities of text data structures, the text mining process requires several initial steps which in essence is preparing for the text to be changed to be more structured. Text mining has many methods and techniques (Gaikwad, 2014).

The purpose of text mining is to get useful information from a set of documents. Thus, the data source used in text mining is a collection of text that has a unstructured or at least semi-structured format The specific tasks of text mining include categorization of text (text categorization) and text clustering text groupings).

2.4. Naïve bayes
Bayes's theorem is the theorem used in statistics to calculate the odds for a hypothesis. Bayes Optimal Classifier calculates the chances of a class from each of the existing attribute groups, and determines which classes are the most optimal.

Naïve Bayes Classifier is a classification method rooted in Bayes's theorem. The method of classification using the probability and statistical methods proposed by the British scientist Thomas Bayes, predicting future opportunities based on past experience, is known as Bayes's Theorem. The main feature of the Naïve Bayes Classifier is a very strong (naïve) assumption of the independence of each condition or event.

3. Methodology
The steps used to solve the problem are:

3.1. Collecting data
The first step is to collect data from twitter. Because that will be analyzed is the opinion of Rohing so that the data taken using the keyword "Rohingya". The data will be used as a sample of 3500 samples of data.

3.2. Cleaning data
The second step is cleaning the data (Paper, 2013). Before the data is analyzed, the data must be cleaned first because if not cleaned it will disturb the data analysis. What will be removed here are: retweet, @ people, punctuation, numbers, html links, unnecessary spaces, unnecessary symbols. In addition, the removal of punctuation and symbols other than alphabet and change it from uppercase to lowercase.

3.3. Perform sentiment analysis
In this stage the cleared data will be classified using the naïve Bayes algorithm. To classify emotions and positive or negative, use library sentiment.

4. Result
The data in this paper is processed using R in order to do wordcloud. In R there is a facility to create a word cloud. Word cloud (also called text cloud or tag cloud) is one method to display text data visually. These graphs are popular in text mining because they are easy to understand. By using the word cloud, the picture frequency of words can be displayed in an interesting yet informative form. The more often one word is used, the larger the word size is displayed in the word cloud.

The results of the testing that has been done is as follows:

IOP Conf. Series: Materials Science and Engineering **336** (2018) 012013 doi:10.1088/1757-899X/336/1/012013

1. Crawling data. Data

Data taken from twitter as many as 3500 with keyword Rohingya. The result is shown in figure 1:

```
[[1]]
[1] "klinecj: Rohingya - the most persecuted people in the world - Thai PBS English News
https://t.co/jnJNkPu3zB via @twitterapi"

[[2]]
[1] "shuriemfraha: RT @cjwerleman: Thousands of Rohingya Muslims crosssed this river, fl
eeing Burma's death squads today. https://t.co/W1Z9TEeMVd"

[[3]]
[1] "JFCrisp: Very confusing for the public. Homepages of IOM, UNHCR and UNICEF all feat
ure separate appeals for Rohingya refugee… https://t.co/LxO8IRbAsk"

[[4]]
[1] "PanickerUma: RT @TrueIndology: Even Islamic Republic of Bangladesh arrests Rohingya
terrorists posing as refugees, but @ndtv finds \"no evidence of any R…"
```

Figure 1. Crawling data

2. Cleaning Data

One example of the results when cleaned re-tweet then the results obtained can be seen in Figure2:

```
[1] "Rohingya - the most persecuted people in the world - Thai PBS English News https://t.co/jnJNkPu3zB "
```

Figure 2. Cleaning data

There is a difference between the data in Figure 1 before cleaned retweetnya and image 2 after cleaning retweets. Looks via and @ is gone. The purpose of cleaning this data is to remove non-essential data which can interfere with the process of opinion analysis.

3. Sentiment analysis

After the process of analysis, the results obtained in Figure 3

	emotion	polarity
1	sadness	positive
2	unknown	negative
3	unknown	negative
4	unknown	positive
5	sadness	negative
6	unknown	positive
7	unknown	neutral
8	unknown	neutral
9	unknown	positive
10	unknown	positive
11	unknown	negative
12	unknown	positive
13	sadness	negative
14	unknown	positive

Figure 3. Result of sentiment analysis

The result in Fig. 3 if formed in the plot diagram will be like figure 4

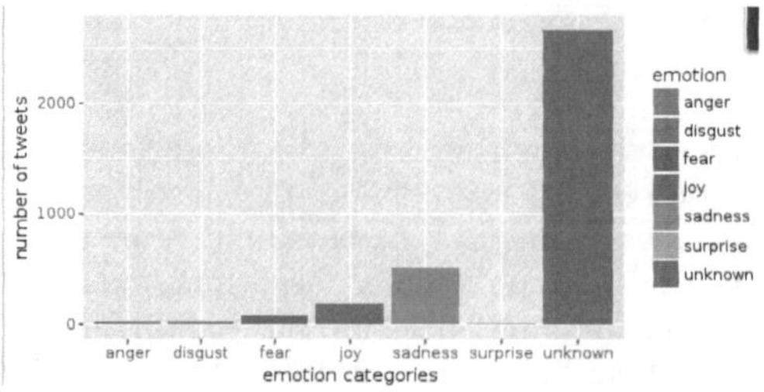

Figure 4. Diagram result of grouping emotion

Unknown results are from groupings that are not included in the group anger, disgust, fear, joy, sadness and surprise. It turns out that the largest group is unknown. The second highest group is sadness. It means that the people of the world turn out to be a lot of sympathizers and feel sorrow for what is experienced by the inhabitants of Rohingya. It is also interesting that the third highest group in the diagram shows some people who actually feel joy with the existence of Rohingya case.

The result of negative, positive and neutral opinion grouping can be seen in figure 5.

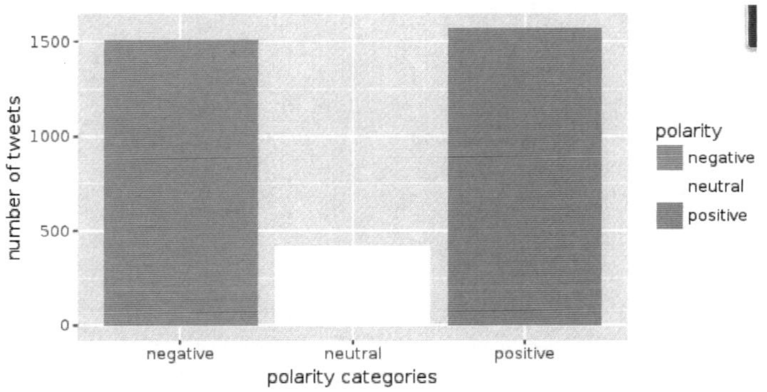

Figure 5. Diagram result of sentient analysis

From the result of figure 5, it can be concluded that between the positive and negative group has the same height. While the group with the lowest number is the neutral group.

The result of plotting wordcloud from this text analysis can be seen at figure 6.

Figure 6. Word cloud

5. Conclusion

Based on the results of data analysis above can be concluded that between the positive vs negative groups balanced. While the neutral groups are fewer. The most number of tweets is unknown. This is due to the difficulty of processing text data so that many text that does not belong to the other 6 groups of emotions. This may be due to the incompleteness of the dictionary or it may be due to something else.

References

[1] Daniel, B., Twitter, T., Edward, W., Twitter, A., & Earth, L. (n.d.). 7 Things You Should Know About Twitter.

[2] Database, R. 2016. An effective model for store and retrieve big health data in cloud computing, *132*, 75–82. http://doi.org/10.1016/j.cmpb.2016.04.016

[3] Elango, V., & Narayanan, G. 2002. Sentiment Analysis for Hotel Reviews.

[4] Finin, T., & Tseng, B. 2007. Why We Twitter : Understanding Microblogging.

IOP Conf. Series: Materials Science and Engineering **336** (2018) 012013 doi:10.1088/1757-899X/336/1/012013

[5] Fink, C. R., Chou, D. S., Kopecky, J. J., & Llorens, A. J. (2011). Coarse- and Fine-Grained Sentiment Analysis of Social Media Text, ***30***(1), 22–30.

[6] Gaikwad, S. V. 2014. Text Mining Methods and Techniques, ***85***(17), 42–45.

[7] Hargreaves, C. A. 2015. Analysis of Hotel Guest Satisfaction Ratings and Reviews : An Application in Singapore, *1*(4), 208–214.

[8] Joshi, K. (n.d.). S TOCK TREND PREDICTION USING NEWS SENTIMENT ANALYSIS.

[9] Kasper, W., & Vela, M. (n.d.). Sentiment Analysis for Hotel Reviews, *231527*, 45–52.

[10] Mäntylä, M. V, Graziotin, D., & Kuutila, M. 1937. The Evolution of Sentiment Analysis - A Review of Research Topics , Venues , and Top Cited Papers.

[11] Paper, D. 2013. An introduction to data cleaning with R.

[12] Spoelstra, J. (n.d.). Predicting US Primary Elections with Twitter, 1–8.

IOP Conf. Series: Materials Science and Engineering **336** (2018) 012014 doi:10.1088/1757-899X/336/1/012014

The Development of Wireless Body Area Network for Motion Sensing Application

P Puspitaningayu*, A Widodo, E Yundra, F Ramadhany, L Arianto, D Habibie

Electrical Engineering Department, Universitas Negeri Surabaya

*pradinip@unesa.ac.id

Abstract. The information era has driven the society into the digitally-controlled lifestyle. Wireless body area networks (WBAN) as the specific scope of wireless sensor networks (WSN) is consistently growing into bigger applications. Currently, people are able to monitor their medical parameters by simply using small electronics devices attached to their body and connected to the authorities. On top of that, this time, smart phones are typically equipped with sensors such as accelerometer, gyroscope, barometric pressure, heart rate monitor, etc. It means that the sensing yet the signal processing can be performed by a single device. Moreover, Android opens lot wider opportunities for new applications as the most popular open-sourced smart phone platform. This paper is intended to show the development of motion sensing application which focused on analysing data from accelerometer and gyroscope. Beside reads the sensors, this application also has the ability to convert the sensors' numerical value into graphs.

1. Introduction

The need of quantity and quality of health service these days grows bigger and bigger. It is caused by the increasing amount of elderly who followed by chronic health problems. Moreover, people now become more aware of health, so they require more service to prevent, monitor, and even cure the diseases. The urban society then demand a simpler way to fulfil those needs, and that is how wireless body area networks (WBANs) take part.

As a branch of wireless sensor networks (WSN) technology, WBANs is specifically designed to be place in, at, or around human body. It consists of sensors which able to read any vital signs, transmit the data to a microcontroller sink, then the wireless module forwards the data to another device connected to the network [1]. This network enables remote health care monitoring, the patient, hospital, and physicians are able to read those vital signs anytime without distance limitation. The IEEE 802.15 has standardised this technology with low-power, low-cost, and low bit rate performance [2].

The current WBAN technology, very close to daily life, is built in a smart phone. Many hand-held devices have sensors, even without the user's knowing. Accelerometer and gyroscope is one of the most included sensors. For example, when a smart phone can read your speed while driving using map application, and when it can adjust the screen when we rotate the device. That is accelerometer and gyroscope respectively. The most interesting part is, smart phone is a *one-go* device for WBAN because it has sensors, microcontroller, and wireless module at the same time.

At the same time, the development smart phones technology and its platform also support the mobility of WBAN technology [3]. Smart phones can act as gateway or sink for the nodes (sensors), and connected to the server or further networks. Moreover, Android development as an open source platform makes it possible to make a smart phone acts as the data processor and also the end system at once.

Accelerometer itself can be used for multiple applications related to positioning. The most popular research in WBAN is about heart disease. Patient with high heart attack risk are usually monitored with ECG sensor also can be combined with accelerometer to examine the influence of movement on the heart rate during daily activities [4]. Therefore, accelerometer can also be used in a more complex sensor networks to give more information for health monitoring [5] [6].

The past works of WBAN applications in Android smart phones is classified into two groups: medical and non-medical purpose [7]. Android platform is chosen by many researchers for its popularity and open-source characteristic, so the developers can freely expand the applications for many purposes.

A motion recognition-based pedestrian navigation discussed at [8] used a smart phone which equipped with accelerometer, magnetometer, gyroscope, and barometric pressure sensors. While the common pedestrian navigation used sensors, which mounted on the shoes or helmet, this application simply used a smart phone with all the required sensors the pedestrian using the smart phone can access the indoor navigation service simply by downloading the application. Furthermore, other positioning methods using the radio frequency (RF) chip, such as GPS or Wi-Fi, could be integrated with the PDR system to complement and improve the positioning accuracy.

This paper aims to show recent work in developing a motion sensing application in an Android smart phone. The application collects data from the three axes of accelerometer and gyroscope value. The numerical data then converted into graphs in order to monitor the pattern of the movements.

2. Wireless Body Area Networks

Wireless sensor networks which has been taking attention these past years has a specific field which only focus on human body. This technology changes the old health care paradigm into a more continuous and reliable gathering and objective analysis of patient's physiological aspects [9].

With the nodes or sensors which are placed on the body or everyday clothing, the ultra-low power connectivity connects physiological parameters data to the computer network. The data can be accessed by the user, hospital, doctor or physician, and any other authorized users. Actually, WBANs is not only can be applied for medical purpose but also for entertainment, sports, military, and other applications [10]. The figure below is an illustration of a wireless body area network.

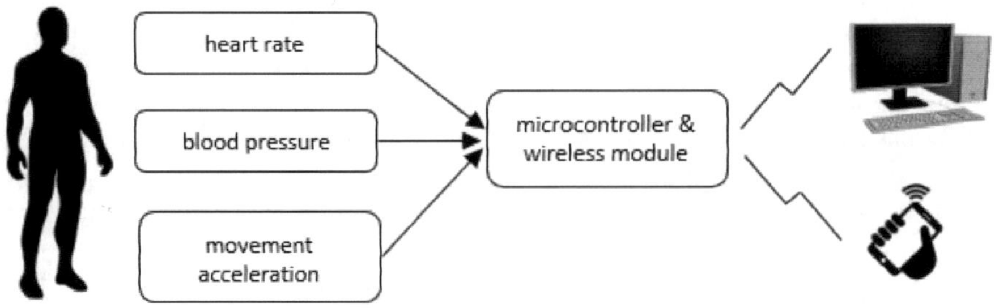

Figure 1. Simplified WBAN network architecture

According to the figure, WBAN architecture has three level of hierarchy: sensor(s), data hub, and medical network. The connectivity of the sensors to the processing module is conducted using radio frequency media. Several wireless standards such as ANT, Bluetooth, Sensium, Zarlink, and Zigbee are the alternatives as written in the IEEE 802.15.6 standard [11].

Furthermore, the data hub handles the data processing, storing, and sending to the next destination or network. The device can be a PDA or smart phone which also equipped with wireless module. The other alternative is, if all the sensors and the hub are built in a single equipment, then between the sensors to the data hub can be connected by wire, and the wireless part is only from the hub to the extended network [12].

At the medical network, the data can be used for further analysis and/or diagnosis to the monitored symptoms of the user. The authorised physicians or doctors can use that data anytime to help them giving the right medication or treatment [13] [14]. As mentioned earlier, WBAN application can be very wide. For medical purpose only, there are various sensors which can be used such as heart rate sensor, ECG, EEG, EMG, blood pressure, blood oxygen saturation, accelerometer, gyroscope, magnetometer, and many more.

Accelerometer and gyroscope are used to monitor the acceleration of the user's movement and the orientation respectively. The data can be continuously changed during the sensors are worn and the user is moving.

3. System Design

The design of the system is actually included in the smart phone. However, the common WBAN architecture is illustrated in figure 2. The development process is conducted by following the Android development process.

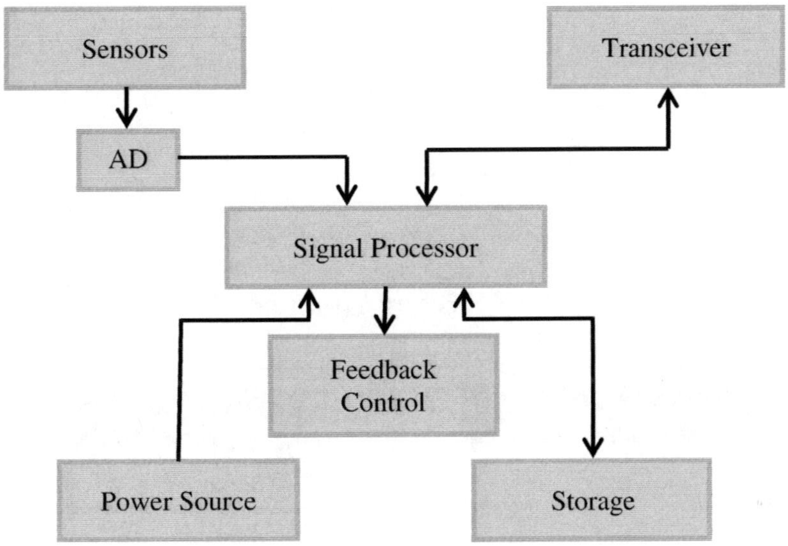

Figure 2. WBAN Architecture

The diagram of the data collection process is shown in the figure 3 where it starts with programming the application, install to the smart phone, move it so the sensors can detect the acceleration and orientation, the application then reads and represent the data into a graph that can be continuously changing as long as the smart phone is being moved.

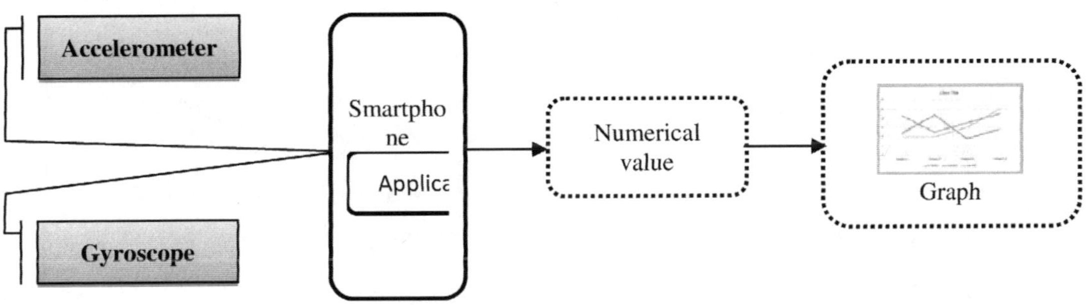

Figure 3. General system design

IOP Conf. Series: Materials Science and Engineering **336** (2018) 012014 doi:10.1088/1757-899X/336/1/012014

4. Result and Discussion

The development of the application first needs to understand how to collect data from the accelerometer and gyroscope of the smart phone. The figure 4 below shows the preview of the application.

Figure 4. Preview of the Application

The application is able to read random movement acceleration and orientation of the user when the smartphone is being held. Then the numerical data can be collected and converted into a graph as shown in the figure 5 and 6 respectively. For example, when the user shakes the smartphone, the movement graph can be shown on the application.

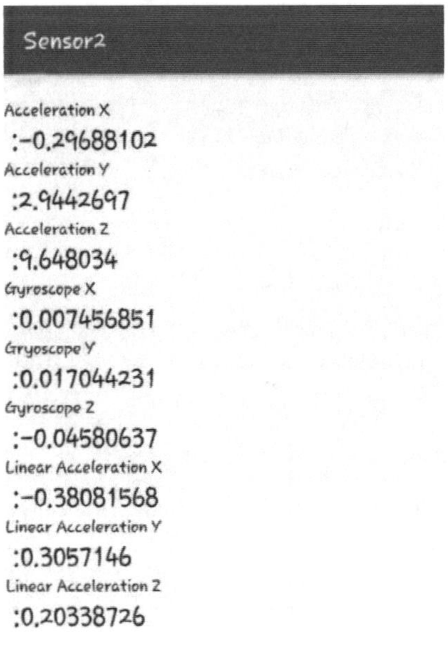

Figure 5. Numerical value

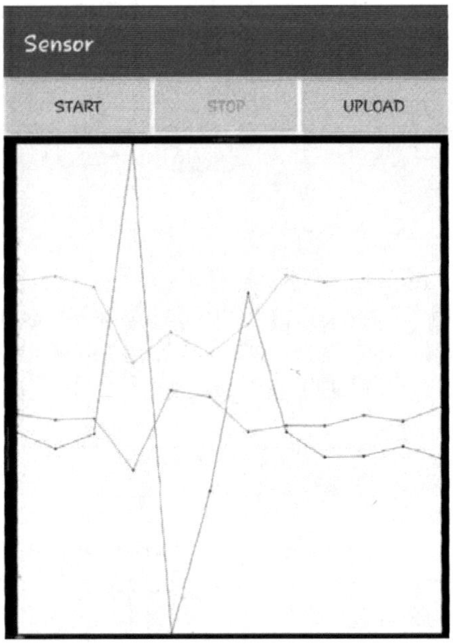

Figure 6. The graph of the movement

5. Conclusion

With the preliminary results, the application has been able to detect the movement and orientation of the user. The numerical data then can also be shown in a graph. The graph can be continuously changing while the user makes any movement.

References

[1] R Cavallari.2014. "A Survey on Wireless Body Area Networks: Technologies and Design Challenges". IEEE Communication Surveys & Tutorials.
[2] Samaneh Movassaghi et al. 2013. "Wireless Body Area Networks: A Survey". IEEE Communications Surveys & Tutorials
[3] G Wogast et al. 2016. "Wireless Body Area Network for Heart Attack Detection". IEEE Antennas & Propagation Magazine October 2016.
[4] Eliasz Kantoch, 2011. "Wireless Body Area Network System based on ECG and Accelerometer Pattern". IEEEComputing in Cardiology
[5] Arsalan Mohsen Nia. 2015."Energy-Efficient Long-term Continuous Personal Health Monitoring". IEEE Transactions on Multi-scale Computing Systems.
[6] H Rezaie, M Ghassemian. 2016. "An Adaptive Algorithm to Improve Energy Efficiency in Wearable Activity Recognition Systems". IEEE Sensors Journal.
[7] W Zang, S Zhang. An Accelerometer-Assisted Transmission Power Control Solution for Energy-Efficient Communications in WBAN, IEEE Journal on Selected Areas in Communications - Series on Green Communications and Networking (Issue 2) 2015
[8] Puspitaningayu et al. 2017. "The Emerging Wireless Body Area Networks: A Review".The 2nd Annual Applied Science and Engineering Conference, Bandung 2017.
[9] B Johny, A Anpalagan. 2014. Body Area Sensor Networks: Requirements, Operations, and Challenges. IEEE Potentials Magazine March/April 2014.
[10] F Hu et al. 2016. A Human Body Posture Recognition Algorithm Based on BP Neural Network for WBANs. China Communications(Volume: 13, Issue: 8, Aug. 2016)
[11] IEEE 802.15.6 standards
[12] A Sangwan, P Battacharya. 2015. "Wireless Body Sensor Networks: A Review". International Journal of Hybrid Information Technology
[13] A Rangarajan. 2016. "Emerging Trends in Healthcare Adoption of Wireless Body Area Networks". AAMI Biomedical Instrumentation & Technology.
[14] M Ghamari et al. 2016. "A Survey on Wireless Body Area Networks for eHealthcare Systems in Residential Environments". Sensors MDPI Journal.

IOP Conf. Series: Materials Science and Engineering **336** (2018) 012015 doi:10.1088/1757-899X/336/1/012015

Multi-Parameter Wireless Monitoring and Telecommand of a Rocket Payload: Design and Implementation

Arga C Pamungkas*, Alma A Putra, Pradini Puspitaningayu, Yulia Fransisca, Arif Widodo

Department of Electrical Engineering, Faculty of Engineering, Universitas Negeri Surabaya

*argapamungkas@mhs.unesa.ac.id

Abstract. A rocket system generally consists of two parts, the rocket motor and the payload. The payload system is built of several sensors such as accelerometer, gyroscope, magnetometer, and also a surveillance camera. These sensors are used to monitor the rocket in a three-dimensional axis which determine its attitude. Additionally, the payload must be able to perform image capturing in a certain distance using telecommand. This article is intended to describe the design and also the implementation of a rocket payload which has attitude monitoring and telecommand ability from the ground control station using a long-range wireless module Digi XBee Pro 900 HP.

1. Introduction

As one of the aerospace vehicles, rocket plays an important role in a various purpose from military to commercial use. A rocket is usually equipped with a set of sensors for monitoring the rocket behaviour or attitude and also surveillance. Many countries develop prototypes of rocket payload to aim a recognition for its aerospace technology [1]. The Indonesian National Institute of Aeronautics and Space (LAPAN) cooperate with the Ministry of Research and Higher Education (Kemenristek Dikti) have been organising a rocket competition called Komurindo since 2009 as an incubator for the next Indonesian satellite system.

A rocket itself usually consist of at least, two parts: the rocket control system and the payload. The payload has several sensors for monitoring the trajectory monitoring and surveillance, then a wireless module to send the monitored data to the ground control station (GCS) [2]. One of the mission carried is the rocket's attitude monitoring which usually performed by using accelerometer and gyroscope sensor to get the 3-dimensional movement data [3]. A compass and a GPS sensor also being used to sense the location of the rocket. Additionally, the payload system also has a camera to capture image in a certain distance.

In [4] the attitude monitoring is proposed using 3 axis gyroscope to measure the angle of rocket payload. However, the use of gyroscope itself has some lacks since it cannot sense the acceleration and forces that occurs to the rocket. Meanwhile, accelerometer itself cannot measure the orientation of the rocket. So that, the use of both sensors, accelerometer and gyroscope are essential to get a more accurate result.

Beside attitude monitoring, a payload can also have surveillance feature. The design of a rocket payload surveillance system has been proposed in [5] by using the uCam-TTL camera to capture 200×200 pixel grey-scale still image. This system has succeeded in capturing an image of a coastline though it suffered some losses.

Furthermore, for the telemetry function, a payload must be equipped with a wireless module as the transmitter of the payload data to the GCS. Prades et al in [6] designed a broadband 3-dimensional biconical antenna which works on 869.4–869.65 MHz and 2.4–2.4835 GHz frequency bands.

Reinaldo et al in [7] has developed a prototype of a rocket dynamics attitude monitoring. It uses n IMU system which able to sense 3-degree movement such as yaw, pitch, and roll of a rocket. Nevertheless, this system uses the 2.4 GHz of frequency performed by the XBee S2.

Meanwhile, this paper come up with a payload system with accelerometer and gyroscope for attitude monitoring, compass and GPS sensors for positioning, also a VC0706 serial camera to capture a 300×300 pixel still image. The organization of this paper as it follows the literature review of the components, the system design, the presented data from the experiments, and conclusion. Additionally, this system is using only a single band 900 MHz Digi XBee Pro HP wireless communication module since the rulebook of Komurindo 2017 only allows a single band communication either 433 or 900 MHz working frequencies.

2. Literature Review

The rocket payload system in this paper is equipped with a GY-80 multi sensor board which has accelerometer and gyroscope sensors, a VC0706 serial camera for surveillance, a compass sensor Devantech CMPS10, and a GPS sensor Ublox Neo M8N.

2.1. Arduino Mega 2560 Board

The *Arduino Mega 2560* is a microcontroller board based on the Atmega2560. It has 54 digital input/output pins (of which 15 can be used as PWM outputs), 16 analogue inputs, 4 UARTs (hardware serial ports), a 16 MHz crystal oscillator, a USB connection, a power jack, an ICSP header, and a reset button. It contains everything needed to support the microcontroller; simply connect it to a computer with USB cable or power it with a AC-to-DC adapter or battery to get started. The Mega 2560 board is compatible with most shields designed for the Uno and the former boards Duemilanove or Diecimila. Figure 1 shows the *Arduino Mega 2560 Board.*

2.2. GY-80 Multi Sensor Board

The GY-80 board combines 5 sensors in a single tiny board which consists of ADXL345 accelerometer, L3G4200D gyroscope, MC5883L magnetometer, and BMP085 as the barometer and temperature sensor. However, this paper only uses two parameters: the linear motion data from the 3-axis accelerometer and the angular rotation from the 3-axis gyroscope [8]. Figure 2 shows the board of GY-08.

Figure 1. Arduino Mega 2560

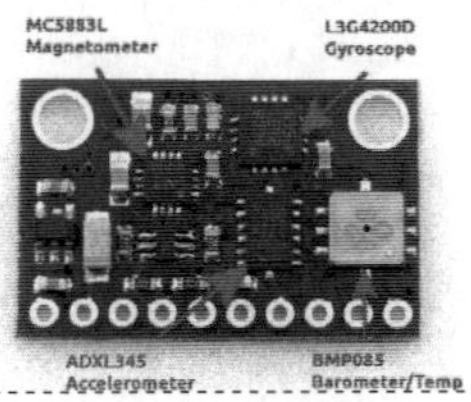

Figure 2. GY-80 multi sensor board

2.3. VC0706 Serial Camera

The VC0706 is an UART VGA camera which has an active pixel array of 649×489. It also has sophisticated functions such as windowing, and mirroring. It uses 1/4 CMOS image sensor MT9V011 and the resulted picture from this camera is in JPEG format [9]. Since this system only requires 300×300 pixel of image, the speed of image transfer is supposed to be faster. The figure 3 below shows the VC0706 camera.

2.4. U-Blox NEO-M8N GPS module

The u-blox NEO-M8N utilizes concurrent reception of the global navigation satellite system (GNSS), simultaneously recognizes multiple constellations, and provides accuracy in positioning. It supports UART, USB, SPI, and DDC interfaces and works in a range of 2.7 – 3.6 V supply. This module can also be used in data logging applications which includes continuous storage of position, velocity, and time information to an onboard SQI flash memory. The information then can be downloaded from the receiver, which is the GCS for location analysis [10]. The figure 4 shows the picture of u-blox NEO-M8.

2.5. Digi XBee Pro 900HP

This wireless module works in 902 – 928 MHz band of frequencies which supports serial data transmission up to 200 kbps of data rate and a maximum line-of-sight range of 15.5 km. It is equipped with ADF7023 transceiver, cortex-M3 EFM32G230 @ 28 MHz processor. The module also needs transmit power up to 24 dBm or 250 mW software selectable [11]. Though it supports several networking topologies, in this paper it will be used in a point-to-point communication between the payload and the ground control station. The following figure shows the XBee Pro 900HP wireless module.

Figure 3. VC0706 camera board

Figure 4. U-blox NEO-M8 GPS

Figure 5. Digi XBee Pro 900HP

3. System Design

The system design for developing a rocket payload dynamics attitude monitoring system divided into two main parts. First, the rocket payload transmitter and the second is ground segment receiver. Figure 6 is a block diagram of the rocket payload and the block diagram of ground control segment shown in Figure 7. And the system was worked on one of three modes, i.e. idle mode, attitude monitoring mode, surveillance mode. As shown in Figure 8.

Idle Mode, Where the payload is powered on and no data being transmitted to the Ground Control Segment (GCS). In addition the GCS could send a command to the payload to start the attitude monitoring mode or surveillance mode.

Attitude Monitoring Mode, Continuously sends the rocket's 3-dimensional movement data to the Ground Control Segment (GCS).

And this mode, the GCS can send a command to stop monitoring the attitude.

Surveillance Mode, The payload sends a 200 x 200 pixels still image to the Ground Control Segment (GCS). The payload will return to the idle mode after the data transmission was completed.

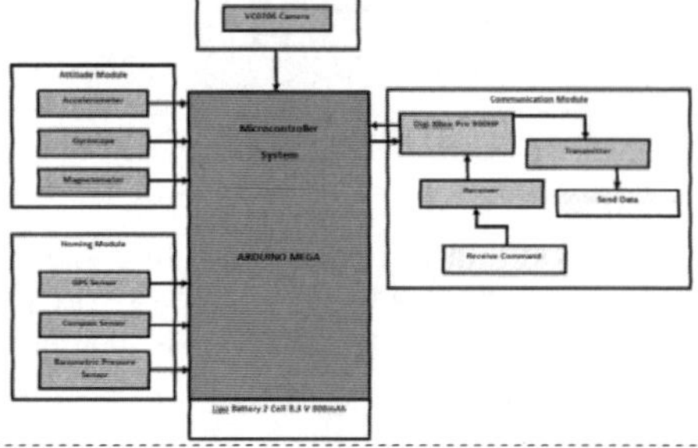

Figure 6. Rocket payload system design

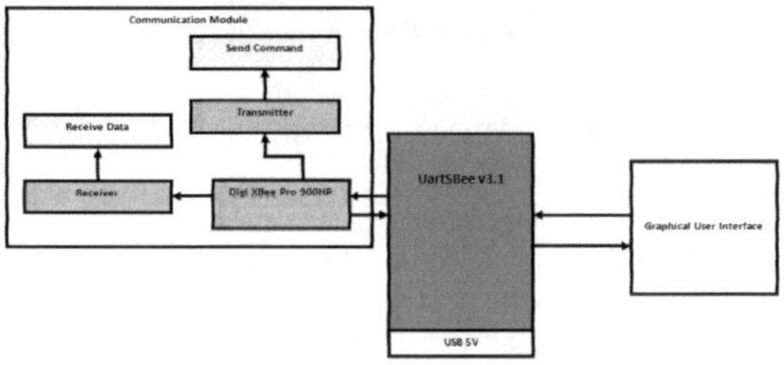

Figure 7. Ground control segment system design

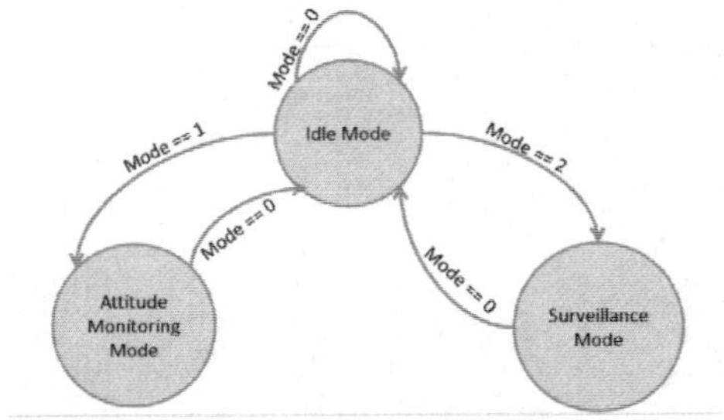

Figure 8. Rocket payload system modes

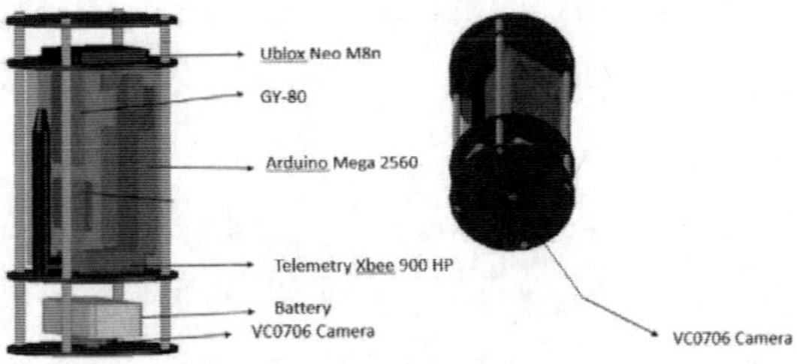

Figure 9. Rocket Payload Design

3.1. Rocket Payload System Design

Rocket payload system consists of Arduino Mega2560, GY-80 module, Ublox Neo M8N GPS module, VC0706 camera module, Lipo Battery 2 cell 8.3 V 800 mAh. It has a mission to process data from GY-80 into a dynamics rocket attitude, and capture the aerial photography also track the rocket through the GPS. These data then sent to the ground control segment by the radio telemetry communication, Digi Xbee Pro 900HP. The following figure shows the rocket payload design

4. Experimental Result and Implementation

Testing system in this research is done experimentally, i.e. by testing functionality, responsiveness and system endurance against some variety of testing methods. Test method is test-based mechanics such as G-Force test and G-Shock test. Function test of telecommand was also performed in this study simultaneously on any mechanical test, i.e. by turning On/Off the telecommand through Ground Control Segment while the hardware on testing. Also, there are some more test, Capture test from camera on the rocket payload and 3-dimensional representation (pitch, roll, yaw).

4.1. G-Force Test

In the G-Force test, the result of the automatic chart plot feature GCS for telemetry data can be seen in the Figure 10. Based on the graph plot, the acceleration of inertia that is measured by the accelerometer sensor is not constant, but varies against time with a maximum value about 2 G. This is an accordance with the actual treatment given to the system during the test, which provides a variation of inertial acceleration through spinning the rocket payload hardware with unstable speed and angular acceleration.

4.2. G-Shock Test

In the G-Shock test, the GCS automatic chart plot for telemetry data accelerometer sensor can be seen in the Figure 11. Based on the graph plot, the accelerometer sensor reads every beat that is applied with a fairly good response as one of the accelerometer sensor functions as a mechanical shock sensor.

IOP Conf. Series: Materials Science and Engineering **336** (2018) 012015 doi:10.1088/1757-899X/336/1/012015

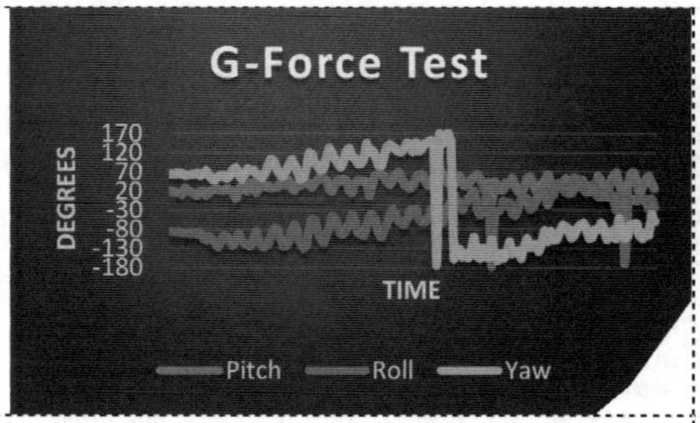

Figure 10. G-Force test chart

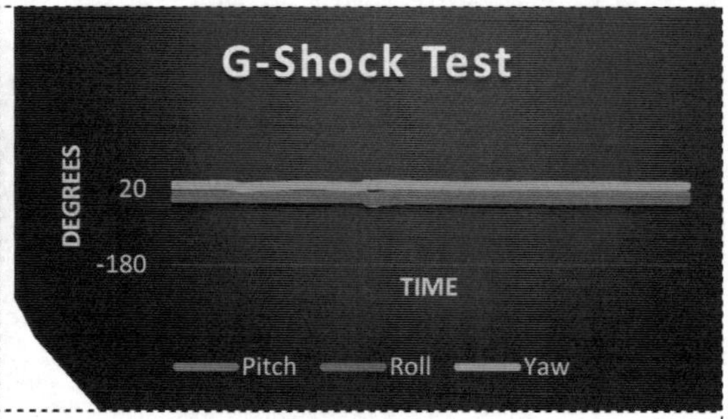

Figure 11. G-Shock test chart

This is evidenced by the emergence of peaks such as spectrum on the graph. This spectrum basically means the magnitude and timing of the pounding. The rise of the spectrum peaks on this result coincides very closely with the time of striking and has a maximum value about 2G. This is also evidenced by the periodicity of the readable signal corresponding to the actual state in which the vibrational frequency applied in the test is constant. In addition, the signal-readability patterns read by the accelerometer sensor in this test can also be interpreted as measurable vibrational characteristics or patterns.

5. Capture Test

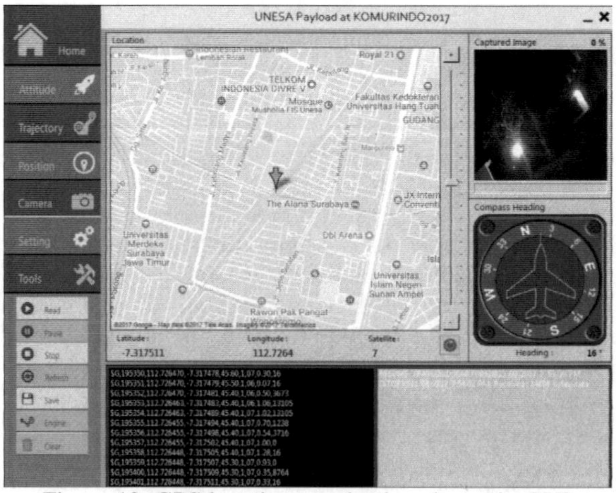

Figure 12. GPS location monitoring shown in GUI

5.1. 3-Dimensional Representation

3-Dimensional representation of rocket payload attitude was displayed on Ground Control Segment and the values of pitch, roll, yaw to determine the movement are also shown in graphic.

5.1.1. Idle Condition

This the idle condition where the rocket payload was not moving yet. All the three-line results just straight line which means, the orientation was 0 degree on all movement.

5.1.2. Pitch Movement

Based on the test results shown in Figure 14, it appears that the gyroscope sensor module can work well to detect angular velocity. This can be seen in the sensor module testing while moving. While pitching, there is a change in blue line. The overall test aims to analyze of the system in the rocket payload in measuring the orientation angle.

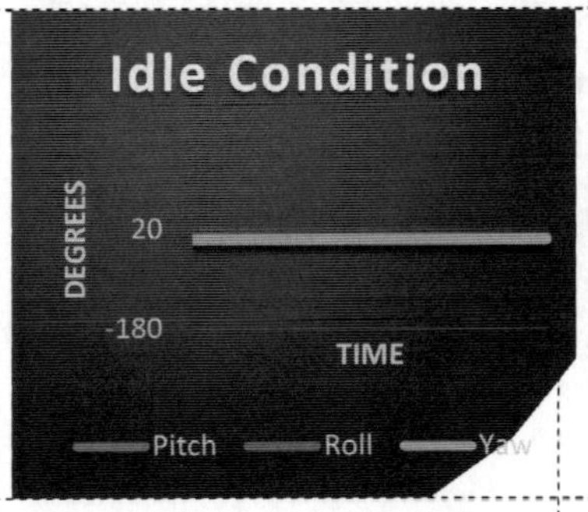

Figure 13. Idle condition chart

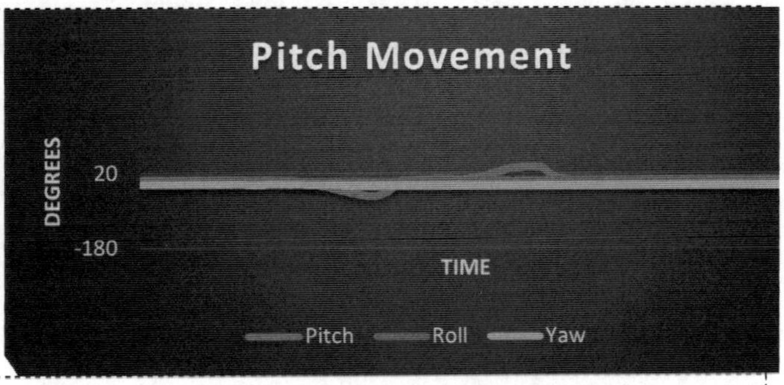

Figure 14. Pitch movement chart

5.1.3. *Roll Movement*

Based on the test results shown in Figure 14, it appears that the gyroscope sensor module can work well to detect angular velocity. This can be seen in the sensor module testing while moving. While rolling, there is a change in red line.

This test is carried out by connecting the circuit according to the block diagram in Figure 6. A protactor is used as a reference to compare the magnitude of the orientation angle of the gyroscope.

5.1.4. *Yaw Movement*

Based on the test results shown in Figure 16 it appears that the gyroscope sensor module can work well to detect angular velocity. This can be seen in the sensor module testing while moving. While pitching, there is a change in green line. The measured angles of orientation are sent to the GCS via a Digi XBee Pro 900HP. Then the data is observed and recorded for analysis for each orientation angle.

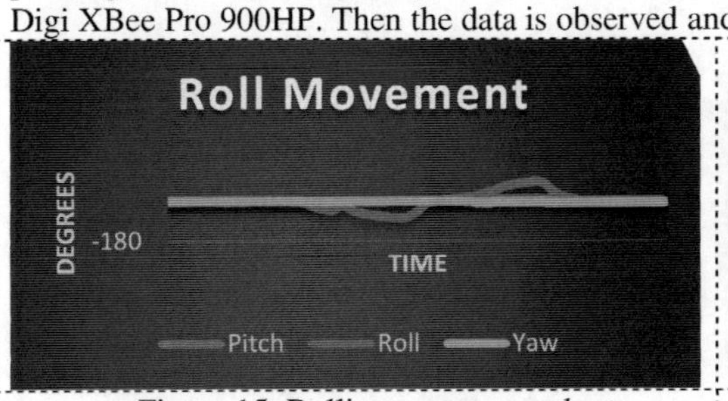

Figure 15. Rolling movement chart

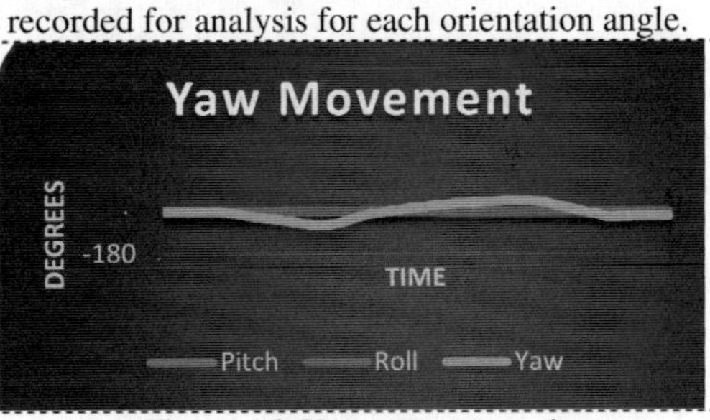

Figure 16. Yaw movement chart

6. Conclusions

Based on the results, it can be concluded that the designed payload system is succeed to monitor the attitude of the rocket and has the ability of telecommand for surveillance. Although the GUI has not meet the 3D movement animation yet, the payload is still functional and can be developed into a better system.

References

[1] Komurindo. 2017. *Buku Panduan Kompetisi Muatan Roket dan Roket Indonesia Kompetisi Muatan Balon Atmosfer*. Jakarta: Lapan.

[2] S Soedianto et al. 2011. *Prototip payload untuk roket uji muatan. Electrical Engineering Journal*, vol.14, pp 66-80.

[3] P Musa et al. 2016. *An Implementation of Direction Cosine Matrix in Rocket Payload Dynamics Attitude Monitoring*. Proc of IEEE International Conference on Informatics and Computing

(ICS).

[4] M Rif'an et al. 2012. *Pemanfaatan 3 axis Gyroscope L3G4200D untuk pengukuran Sudut Muatan Roket*. Jurnal EECCIS Vol. 6, No.2, pp 177-182.

[5] A Irawan et al. 2013. *Attitude Monitoring and Surveillance System for Lapan Payload Test Rocket*. Proc of IEEE 3rd International Conference on Instrumentation Control and Automation (ICA), pp 155-160.

[6] J Prades et al. 2015. *Broadband Sounding Rocket Antenna for Dual-Band Telemetric and Payload Data Transmission*. IEEE Antennas and Wireless Propagation Letters, Vol:15, pp 540-543.

[7] R Reinaldo et al. 2013. *Pengembangan Prototip Sistem Pemantau Sikap Dinamik Roket dengan Visualisasi Grafik dan Animasi Pergerakan 3D secara Real Time*. Proceeding of Seminar Nasional Fisika Universitas Negeri Jakarta.

[8] Neo-M8 u-blox M8 concurrent GNSS modules data sheet. https://www.u-blox.com/en/product/neo-m8-series

[9] Digi XBee Pro 900HP data sheet. https://www.digi.com/products/xbee-rf-solutions/sub-1-ghz-modules/xbee-pro-900hp

IOP Conf. Series: Materials Science and Engineering **336** (2018) 012016 doi:10.1088/1757-899X/336/1/012016

Detection of Lock on Radar System Based on Ultrasonic US 100 Sensor And Arduino Uno R3 With Image Processing GUI

F Baskoro*, B R Reynaldo

Department of Electrical Engineering, Faculty of Engineering Universitas Negeri Surabaya

*faridbaskoro@unesa.ac.id

Abstract - The development of electronics technology especially in the field of microcontroller occurs very rapidly. There have been many applications and useful use of microcontroller in everyday life as well as in laboratory research. In this study used Arduino Uno R3 as microcontroller-based platform ATMega328 as a sensor distance meter to know the distance of an object with high accuracy. The method used is to utilize the function Timer / Counter in Arduino UNO R3. On the Arduino Uno R3 platform, there is ATMEL ATmega328 microcontroller which has a frequency generating speed up to 20 MHz, 16-bit enumeration capability and using C language as its programming. With the Arduino Uno R3 platform, the ATmega328 microcontroller can be programmed with Arduino IDE software that is simpler and easier because it has been supported by libraries and many support programs. The result of this research is distance measurement to know the location of an object using US ultrasonic wave sensor US 100 with Arduino Uno R3 based on ATMega328 microcontroller which then the result will be displayed using Image Processing.

1. Introduction

The use of radar/sonar system is necessary in everyday life. Especially in the field of transportation, where the radar/sonar system is usually used in ships as a pointer or to tell whether there are obstacles from the path of the ship to be bypassed.

Can be imagined if a ship does not know what is in front of it or about the radius of the vessel then it can cause ship bodies hit foreign objects such as rocks or blocks of ice and other bad consequences. In this study used ultrasonic sensors US 100 used to measure distances, Arduino Uno R3 board with microcontroller ATmega328 products from ATMEL, and software used is Arduino IDE that uses C-based programming language with libraries that make it easier in making complicated programs and Processing as the final display (GUI) of the sonar / radar apps to be designed.

The purpose of this research is to make hardware and software to know the location of objects based on object distance with US 100 ultrasonic sensor using Arduino Uno R3 based on ATmega328 microcontroller then the result is shown in Processing software (GUI). benefit of this research is can be used as one of radar learning media and navigation that is by knowing distance of an object. By knowing the distance of an object then the location of the object can be known.

IOP Conf. Series: Materials Science and Engineering **336** (2018) 012016 doi:10.1088/1757-899X/336/1/012016

2. Literature Review

2.1. A. Ultrasonic sensor

Sensors are the "senses" for a system. The sensor obtains data from the environment in the form of mechanical quantities and displays them again in the form of electrical quantities. One type of sensor that is widely used today is the ultrasonic sensor. Ultrasonic sensors work by emitting a wave and then calculating the reflection time of the wave. Ultrasonic sensors work starting frequency 20 KHz up to 20 MHz.

Ultrasonic in its application can be generated using a PZT (piezoelectric) material that acts as a transducer. When the material is supplied with the voltage, it will vibrate and resulting ultrasonic waves (as transmitters). On the other side, the ultrasonic transducer as the receiver will convert ultrasonic waves into electrical quantities. Figure 1 shows the working principle of Ping ultrasonic sensor:

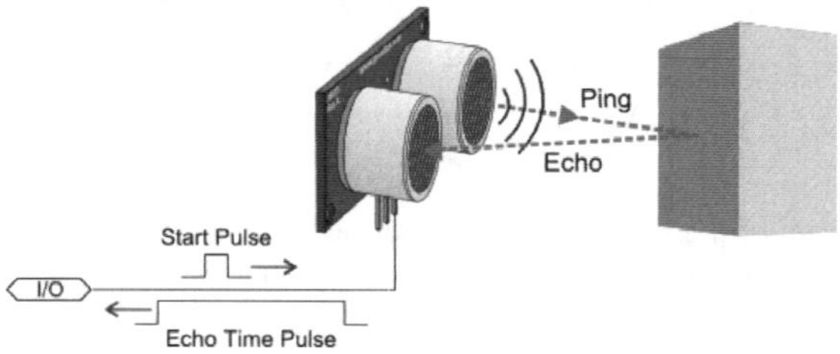

Figure.1 The working principle of ultrasonic distance sensor[8]

In general, based on the way of sending and receiving ultrasonic waves, ultrasonic sensors are divided into 2 kinds,

The results of this sensor measurement is determined by the time and amplitude of the ultrasonic waves. This sensor is usually used to measure the distance of an object, as shown in Figure 1. The results of this sensor measurement is determined by the speed of propagation, deflection or wave biasing, and the magnitude of ultrasonic wave propagation, because this sensor is widely used to measure flow rate, both liquid and gas.

2.2. B. Arduino Uno Board

Arduino is an electronic prototype device based on a flexible and open-source microcontroller. Where hardware and software are easy to use. For beginners using this board will facilitate control with the microcontroller. Arduino can be used to detect the environment by receiving input from various sensors e.g. ultrasonic sensors.

Arduino Uno is an Arduino board that uses ATmega328 microcontroller. Arduino Uno has 14 digital pins (6 pins can be used as PWM output), 6 analogue inputs, a 16 MHz crystal oscillator, a USB connection, a voltage source connector, an ICSP header, and a reset button. Arduino Uno contains everything needed to support a microcontroller.

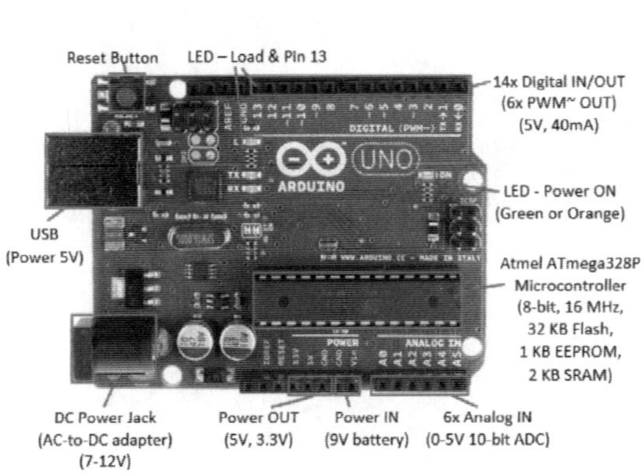

Figure 2 Arduino Uno R3[6]

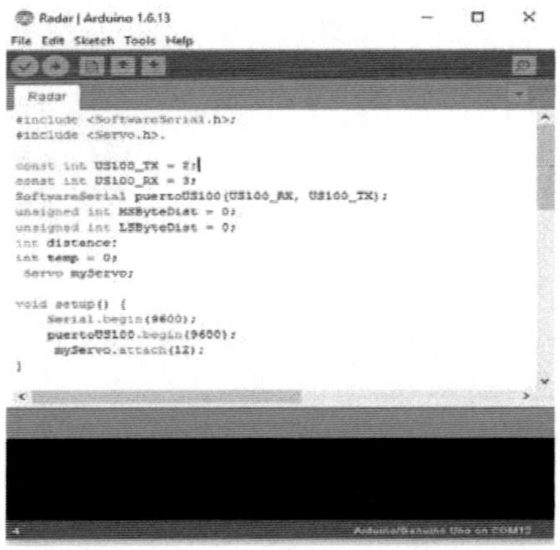

Figure 3. Program example US100

The technical data board Arduino UNO R3 is as follows:

Table 1. Data board Arduino Uno R3

Microcontroller	ATmega328
Operating Voltage	5Volt
Input Voltage (recommended)	7-12 Volt
Digital Pin I / O	14 (6 of them are PWM pins)
Pin Analog input	6
Flash Memory	32 KB with 0.5 KB used for bootloader
Timing Speed	16 Mhz

2.3. *C. Ultrasonic Sensor US 100*

The Ultrasonic sensor module is a convenient way for measuring distances from objects. This module has a lot of applications such as parking sensors, obstacle and terrain monitoring systems, industrial distance measurements, etc. It has a stable performance and high accuracy ranging from 2cm to 450cm. The module sends an ultrasonic signal, eight pulses of 40kHz square wave from the transmitter; the echo is then picked up by the receiver and outputs a waveform with a time period proportional to the distance. The connected microcontroller accepts the signal and performs necessary processing.

Table 2. Specification Ultrasonic sensor US 100

Input voltage	5V
Detection distance	cm-450cm
Dimensions	4.4cm x 2.6cm x 1.4cm
Quiescent current:	less than 2mA

IOP Conf. Series: Materials Science and Engineering **336** (2018) 012016 doi:10.1088/1757-899X/336/1/012016

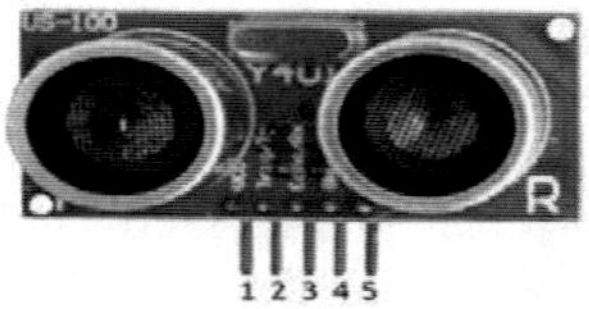

Figure 4 Ultrasonic sensor US 100

Information:
1. VCC: 5 Volt
2. Trig: trigger input
3. Echo: pulse output
4. GND: Ground
5. GND: Ground

2.4. D. Software Processing
In object-detection research on Arduino Uno-based radar and US 100 sensors, the software used to display graphs results from ultrasonic sensor readings, the authors use the 3.2.1 Processing software. Software processing 3.2.1 has advantages that can directly receive input from Arduino Uno R3 board via USB cable. Serial input from ultrasonic sensor readings will be uploaded to the processing software which will be displayed in the form of graphics to be more user-friendly. By using this software, the location of an object along with the distance of the object from the ultrasonic sensor can be detected. The advantages of this software is it is open-source, interactive: it can display output in the form of 2D, 3D and pdf, and can operate on some operating systems such as Windows, Mac, and Linux, and most importantly is the software is supported with libraries that make it easier for us to do programming

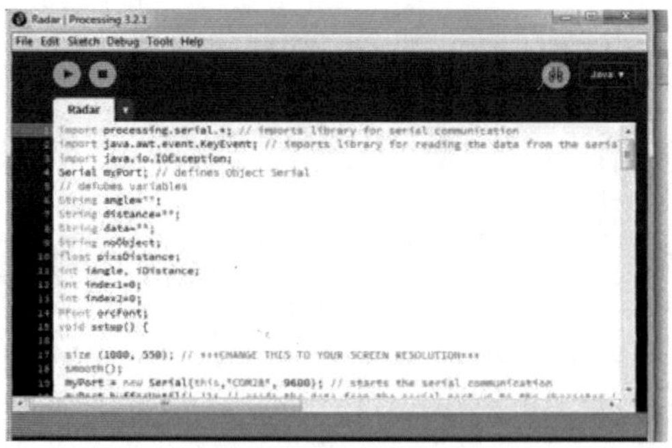

Figure 4. Example program in processing 3.2.1

3. System Model
The overall system block diagram of the design of this tool can be seen in Figure 5

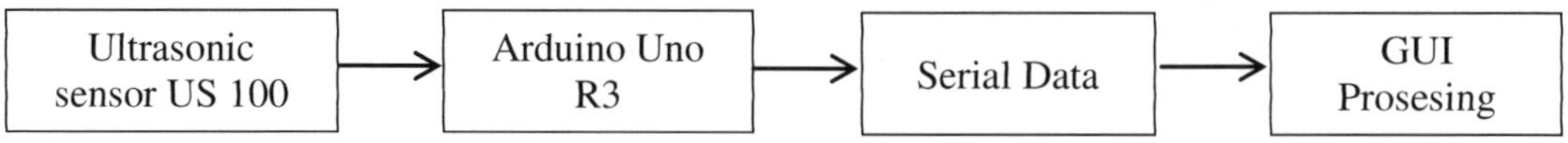

Figure 5. Block diagram system

In this study, US 100 sensors are used to measure the distance through the waves emitted by the sensor US 100. Data from the sensor is processed by Arduino data to get the picture data. Then it will be transferred to GUI processing, the it can be displayed through the GUI

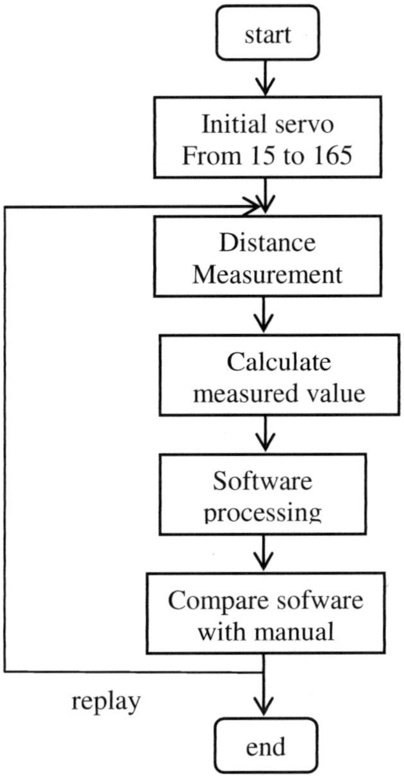

Figure 6. radar flow diagram using US 100 sensors and software processing 3.2.1

Based on the picture 6 it can be seen the initial step is the initialization of servo and programming on the Arduino to determine the distance using the ultrasonic sensors us 100. After the program is completed, then the program is applied. To see the radar display, this study uses US 100 software processing. The results of display the calculation of the distance through software processing then compared with manual calculations and ended by drawing conclusions.

4. Research Result

The testing and analysis chapter of this tool is tested on the system by measuring the distance and the results will be compared through measurement using a regular measuring instrument (meter). In the chapter of testing and analysis is done testing the measurement of the distance of the object from the front of the sensor. After that will be sweeping from 15 - 165 degrees angle which later the data is directly sent to the program display GUI Processing 3.2.1 through serial data communication USB.

In this test, the magnitude of the measured distance on the sensor will be displayed on the LCD screen of the Laptop via Serial Monitor. Then the result of this measurement is compared to the measurement result with the meter measured horizontally toward the front of the sensor. The result of distance measurement from forward direction is shown in table 3.

IOP Conf. Series: Materials Science and Engineering **336** (2018) 012016 doi:10.1088/1757-899X/336/1/012016

Table 3. Measurement results with measuring instruments

The measured distance with ruler **(cm)**	The measured distance in Serial Monitor **(cm)**	**Difference distance (cm)**	**error** (%)
1	2,03	1,03	103
2	2,57	0,57	28,5
3	3,26	0,36	8,66667
4	4,01	0,01	0,25
5	5,37	0,37	7,4
6	6,29	0,29	4,83333
7	7,04	0,04	0,57143
8	8,02	0,02	0,25
9	8,89	0,11	1,222222
10	9,81	0,19	1,9
11	11,05	0,05	0,45455
12	12,10	0,10	0,83333
13	13,01	0,01	0,07692
14	14,26	0,26	1,85714
15	15,37	0,37	2,46667

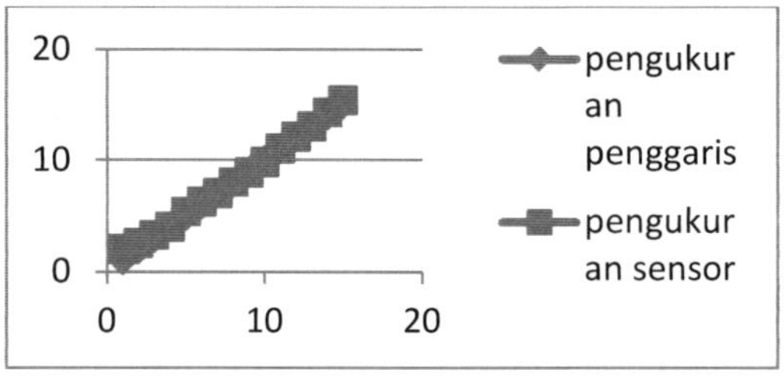

Figure 7 Graph of ultrasonic distance measurement results with actual distance

The measurement results from table 3 can be seen in figure 7. From the calculation can be seen that for very close calculations, under 3 cm measurement using radar has a high error rate. this is due to sensors US 100 has a sensitivity of 2-450 cm. for values above 3 cm the measurement results with the sensor almost closer to the true value. So the US 100 sensor can be recommended as one of the learning media radar system. To display this ultrasonic distance measurement it will sweeping from an angle of 15 - 165 degrees and back again to the beginning as well as repeated continuously and the results will be displayed in graphical visual form. Ultrasonic sensors will detect the presence of objects that exist in front of a radius of 3 - 40 cm during sweeping. The visual appearance of this test is shown in Figure 8 as follows:

IOP Conf. Series: Materials Science and Engineering **336** (2018) 012016 doi:10.1088/1757-899X/336/1/012016

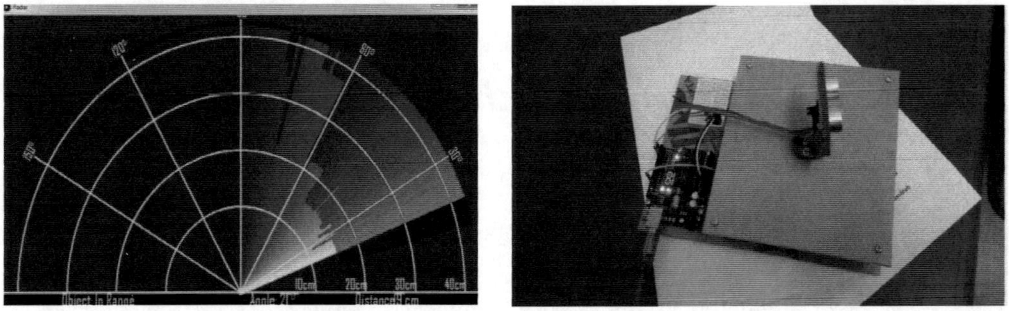

Figure 8. Display of radar applications with US ultrasonic sensors 100

5. Conclusion

On horizontal distance testing the US 100 sensor compared to the measured distance with the meter is relatively close. The percentage error is large if the distance is less than 3 cm. GUI with Image processing able to display the distance with a good illustration and give the appearance in accordance with the Radar. The image illustration gives a good and accurate response when there is an object detected according to the distance and location of the detected object.

6. Suggestion

The recommendation during developing the GUI view with Image Processing is pay attention to computer's screen resolution, BaudRate and COM Port.

Reference

[1] Aldrin S., Sumardi, Setiawan I, Makalah Seminar Tugas Akhir : Pemanfaatan PING)))TM Ultrasonic Distance Sensor dengan Menggunakan Mikrokontroler ATMega8515 Untuk Mengetahui Letak Benda Pada Penderita Cacat Tunanetra, Universitas Diponegoro, Semarang.
[2] Wollard, Barry, Elektronika Praktis, Pradnya Paramitha, Jakarta, 1988.
[3] Kleitz, William, Digital Electronics, a Practical Approach, Fifth Edition, Prentice Hall International, Inc., New Jersey, 1999.
[4] Malvino, A. P., Prinsip-prinsip Elektronik, Edisi kedua, Erlangga, Jakarta, 1999.
[5] Kadir, A., Pemrograman Dasar Turbo C untuk IBM PC, ANDI Offset, Yogyakarta, 1997.
[6] 8-bit AVR Microcontroller instruction set, 2005, http://www.atmel.com,
[7] Instruction Set, http://www.arduino.cc
[8] PING)))^{TM} Ultrasonic Distance Sensor User Manual, http://www.parallax.com

Integration *K-Means Clustering Method and Elbow Method* For Identification of The Best Customer Profile Cluster

[1]M A Syakur, [2]B K Khotimah, [3]E M S Rochman, [4]B D Satoto
[1,2,3,4]Faculty of Engineering, University of Trunojoyo Madura

Abstract. Clustering is a data mining technique used to analyse data that has variations and the number of lots. Clustering was process of grouping data into a cluster, so they contained data that is as similar as possible and different from other cluster objects. SMEs Indonesia has a variety of customers, but SMEs do not have the mapping of these customers so they did not know which customers are loyal or otherwise. Customer mapping is a grouping of customer profiling to facilitate analysis and policy of SMEs in the production of goods, especially batik sales. Researchers will use a combination of K-Means method with elbow to improve efficient and effective k-means performance in processing large amounts of data. K-Means Clustering is a localized optimization method that is sensitive to the selection of the starting position from the midpoint of the cluster. So choosing the starting position from the midpoint of a bad cluster will result in K-Means Clustering algorithm resulting in high errors and poor cluster results. The K-means algorithm has problems in determining the best number of clusters. So Elbow looks for the best number of clusters on the K-means method. Based on the results obtained from the process in determining the best number of clusters with elbow method can produce the same number of clusters K on the amount of different data. The result of determining the best number of clusters with elbow method will be the default for characteristic process based on case study. Measurement of k-means value of k-means has resulted in the best clusters based on SSE values on 500 clusters of batik visitors. The result shows the cluster has a sharp decrease is at K = 3, so K as the cut-off point as the best cluster.

1. Introduction

Clustering is in daily life, because it could not be separated with a number of data that produce information to meet the needs of life. One of the most important tools in relation to data is to classify or classify the data into a set of categories or clusters [1]. One clustering technique is the method of K-means algorithm using the process repeatedly. The K-Means method is the simplest and most common clustering method. K-means has the ability to group large amounts of data with relatively fast and efficient computation time [2]. However, K-Means has a disadvantage depending on the initial cluster center determination. K-Means cluster test results in the form of solutions that are local optimal. From the trial process is expected to have similarities or closeness between data so that can be grouped into several clusters, where among cluster members have a high level of similarity [3]. According to (Celebi et al., 2013) the K-Means algorithm is also versatile, which is easy to modify at every stage of the process, simple in the distance calculation function, and depends on iteration termination criteria. K-Means Clustering is a localized optimization method that is sensitive to the selection of the starting position from the midpoint of the cluster. So choosing the starting position from the midpoint of a bad cluster will result in K-Means Clustering algorithm being trapped in the optimal locale [4]. The K-means method will choose the pattern up to k as the starting point of the centroid randomly or randomly. The number of iterations with the centroid cluster will be affected by the initial centroid cluster at random. So that can be fixed by determining the centroid cluster in the high initial data to get higher performance [5]. Consumer segmentation based on consumer behaviour measured by consumer profiling. Customer profiling is built through information with criteria: age, gender, and residential area information. To process customer profiling using data mining by segmenting consumers can be estimated through the consumer profile, which conducted using clustering algorithm for bank customer segmentation [6]. Kaur et al, 2013 proposed improvements to the classic K-Means algorithm to produce more accurate clusters. The proposed algorithm is based on data separation, to find the initial centroid according to the data distribution. The results of this study have resulted in better clusters in a short calculation time. The main objective of this research is the segmentation of bank customers to find the transactional relationship between customer and company to provide mutual solution [14]. In other words, improving customer relationships and evaluating customer segments by predicting credentials for each group of customers and will provide a more appropriate type of transaction model with the customer [6-7]. Clustering results vary depending on the number of clustering parameter changes. The k-means method is a simple clustering technique and quickly takes care of the problem to determine the exact number of clusters in the data set. Customer segmentation research is proposed for the use of various regulations for various customers with high risk to the bank [8]. Elbow proposes several ways to determine k as the number of dynamically formed clusters, one of which is the elbow method [9].

This research will combine K-Means with elbow method to determine the actual number of clusters Customer profiling segmentation in SME. The value on k will continue to increase in each process and decrease with great value. The graph shows the elbow n of all the k values obtained. This research will find the best value of k by using elbow method. The elbow method is easy to implement by looking at the ideal k value graph with the position on the elbow along with the SSE (Sum of Square Error) which is less than 1. The best cluster k result will be the basis for clustering. The smaller the value of SSE and the elbow graph decreases the better the cluster results.

2. Literature Review

2.1. Customer Segmentation

One of the ways used to know customer characteristics information by doing customer segmentation. One way that can be used to manage the relationship between customers and companies is to provide different treatment according to customer characteristics of each segment. Customer segmentation is done with data mining to know the customer characteristics information hidden inside. The way to find out the customer segments of a company is clustering analysis. Clustering is the process of forming segments of a set of data by measuring similarities between data with other data [10]. The purpose of segmentation is to customize the products, services, and marketing messages for each segment. The segmentation process puts customers in line with the characteristics of similar customer groups. Customer segmentation is a preparatory step to classify each customer according to a defined customer group. Customer segmentation based on market research and demography requires understanding the characteristics of all customers to be more effective. Customer segmentation also identifies segmentation in customer behaviour. In addition, customer payment transaction data is used to gain insight into customer behaviour. Customer segmentation to form groups based on their income and expenses. It can be used to identify high-value customers and prioritize service [11].

2.2 K-Means Clustering

The K-means algorithm is one of the algorithms with partition, since K-Means is based on determining the initial number of groups by defining the initial centroid value [12]. The K-Means algorithm requires precise numbers in determining the number of clusters k, since the initial cluster centre may change so that this event may result in unstable grouping of data [13]. The output of K-Means depends on the selected centre values on clustering. This algorithm the initial value of the cluster's centre point becomes the basis for the cluster determination. The initial cluster centroid cluster randomly assigns an impact to the performance of the cluster (14-16). K-Means Clustering algorithm is one of the clustering methods by partitioning from set data into cluster K. It is a distance-based clustering algorithm that divides data into a number of clusters in numerical attributes.

1. Determine the number of clusters K and the number of maximum iterations.
2. Perform the initialization process K midpoint cluster, then the equation of centroid count feature:
$$C_i = \frac{1}{M}\sum_{j=1}^{M} x_j$$
Equation 1 is done as much as p dimensions from i = 1 to i = p
3. Connect any observation data to the nearest cluster. Euclidean distance spacing measurements can be found using equation 2.
$$d = \sqrt{(x_1 - x_2)^2 + (y_1 - y_2)^2}$$
4. Reallocation of data to each group based on comparison of distance between data with each group's centroid [9].
$$a_{ij}\begin{cases}1 & d = \min\{D(x_i, c_i)\\ 0 & otherwise\end{cases}$$
5. Recalculate the cluster midpoint position.
a_{ij} is the value of the membership of point x_i to the centres of the group c_1, d is the shortest distance from the data x_i to the group K after being compared, and c_1 is the centre of the group to 1. The objective function used by this method is based on the distance and the value of the data membership in the group. The objective function according to MacQueen (1967) can be determined using equation.
$$J = \sum_{i=1}^{n}\sum_{l=1}^{k} a_{ic}D(x_i, c_1)^2$$
n is the amount of data, *k* is the number of groups, a_{i1} is the membership value of the data point x_i to the c_1 group followed *a* has a value of 0 or 1. If the data is an anngota of a group, the value a_{i1} = 1. If not, the value a_{i1} = 0.
6. If there is a change in the cluster midpoint position or number of iterations <the maximum number of iterations, return to step 3. If not, then return the clustering result.

2.3 Elbow Criterion

Illustration of K value on Elbow combination with K-Means was graph of cluster relationship with error decreasing, increasing value of K then graph will decrease slowly until result of value of K is stable. For example, the value of the cluster K = 2 to K = 3, then from K = 3 to K = 4, it shows a drastic decrease to form the elbow at point K = 3 then the ideal cluster k is K = 3 [11]. The combined Elbow and K-Means Methods can determine the value of K at the best cluster.

1. Find k as the number of clusters formed. This study will use the elbow criterion method to select the number of k clusters to be used for grouping data on the K-Means algorithm. The elbow method is expressed by Sum of Squared Error (Irwanto, et al, 2012):

$$SSE = \sum_{K=1}^{K} \sum_{x_i \in S_K} \|X_i - C_k\|_2^2$$

With k = many clusters formed Ci = the i-th cluster, x = the data present in each cluster.

2. Determine the cluster's center point at the beginning at random. Early centroid determination is done randomly from the available objects as much as cluster k, then to calculate the next i-cluster centroid, by the following formula:
   $v = \frac{\sum_{i=1}^{n} x_i}{n}$; $i = 1,2,3, \ldots\ldots, n$
3. Calculate the distance of each object to each centroid using the Euclidian Distance.
   $d(x, y) = \|x - y\| = \sqrt{\sum_{i=1}^{n}(x_i - y_i)^2}$; $i = 1,2,3, \ldots\ldots, n$
   with x_i : Variable on Object x to-i and y_i : Variable output y
   n : The number of objects
4. Allocate each object into the nearest centroid.
5. Allocation of objects into each cluster at iteration with k-means. Where each cluster member object has been measured the proximity distance to the cluster's center point.
6. Perform iteration, then process determine the position of new centroid by using equation (2.2).
7. Repeat step 3 if the new centroid position with the old centroid is not the same.

3. Results and Discussion

K-means clustering research is done by taking data during the last 1 month for profiling customer with parameters as attribute in the form of customer criteria. Where each criterion has a predetermined category range, for gender criteria and activity has sub criteria, following sub criteria for gender and activity. Customers have been filled out the questionnaire using 6 criteria, according to table 2.

Table 1. Sampel data profiling of customer

No	User Name	Age	Profession	Income (million)	Education	Quality	Gender
1	Moawi	25	Employee	1.500.000	high school	3	Female
2	Windaryanti	47	Teacher	7.000.000	S1	1	Male
3	Sulistiara	60	Pension	3.000.000	S1	5	Female
4	Mailinda	70	Entrepreneur	12.000.000	Junior high school	6	Female
5	Fania Dara	56	Farmer	1.000.000	primary school	4	Male
6	Fleni Dwi	63	Government employees	6.000.000	S1	2	Female
7	Imroatun	48	Trader	11.000.000	primary school	6	Female
8	Marjatul	20	Student	500.000	high school	1	Male
9	Yayuk	29	Employee	6.000.000	D3	3	Female
10	Siti Zainab	42	Teacher	5.000.000	S1	1	Female

Table 2. Process of data transformation with age group

No.	Age	Frequency/Initial
1	15-25	14
2	25-35	13
3	35-45	2
4	45-55	31
5	55-65	25
6	x≥65	16

IOP Conf. Series: Materials Science and Engineering **336** (2018) 012017 doi:10.1088/1757-899X/336/1/012017

The data contained in table 1 would not be directly processed, because there is a large number of numbers between the variables. The difference in distance or the magnitude of this number can be quite difficult in the process of grouping. One of the solutions used to minimize the number of variables between the variables with the equation 5 [11].

$$\text{Normalization}\frac{(x_i - x_{min})}{(x_{max} - x_{min})}$$

The value of the variables is normalized to the range 0 - 1. Normalization of numbers on each variable before the calculation process is done so that the centroid value on the parameter does not exist that predominates in the calculation of the distance between data [11]. The data used in the form of data on the number of visits batik sales with criteria : Profession, Income, Quality of Batik, and Gender Education. The result combined system test of K-Means method with Elbow according to Table 3.

Table 3. Result of anaisa cluster of K-Means data method with Elbow

No	Age	Profession	Income (million)	Education	Quality	Gender	Cluster	Number
1	25≤x≤40	Employee, Teacher	x≤2	High School	3	Female	3	132
2	41≤x≤55	Teacher, Entrepreneur	5≤x≤10	S1	1	Male	2	223
3	15≤x≤24	Student	2≤x≤5	S1	5	Female	3	112
4	x≤65	Pension, Entrepreneur	x≥10	Junior High School	6	Female	1	25

Furthermore, this research performs the best cluster optimization on K-means with Elbow method. The determination of k value with Elbow Method by choosing the number of clusters or k values by seeing the decrease significantly.

Table 4. Sum of Square Error results from each cluster for 100 and 300 Customer Profiling

Cluster	Result of Sum Square Error for 100 data	The difference between SSE	Result of Sum Square Error for 300 data	The difference between SSE
C2	845,432	845,432	786,563	786,563
C3	532,133	313,299	665,434	121,129
C4	431,429	100,704	301,341	364,093
C5	367,232	64,197	234,453	66,888
C6	334,451	32,781	278,842	-44,389
C7	243,341	91,110	289,678	-10,836
C8	231,211	12,130	234,563	55,115

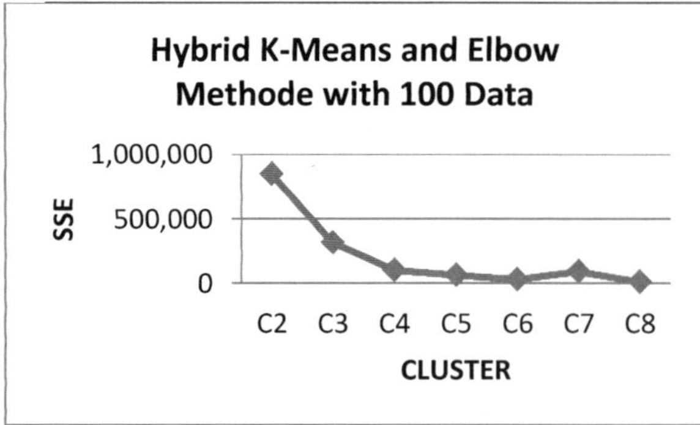

Figure 1. Graph of Sum of Square Error 100 Data

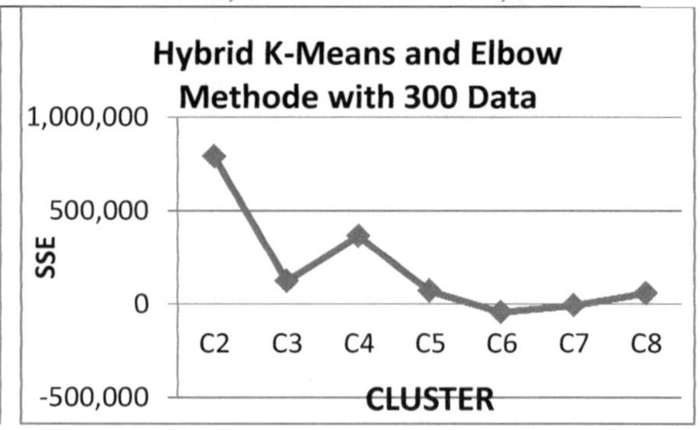

Figure 2. Graph of Sum of Square Error 300 Data

Test Data used k-means clustering and uses the Elbow method at 100 and 300 Customer Profiling. The K-Means Clustering process uses the Elbow method to determine the value of k. The result of the cluster formed will be labelled or named to facilitate the company in considering the characteristics of its customers. Performance tests have used 100 and 300 customer data purchases of goods. The results of Sum of Square Error calculations of each cluster have experienced the greatest decrease in K = 3 can be seen in Table 4, Figure 1 and Figure 2. In this test will find the performance of each number of clusters that are adjusted to the range of values on the Elbow method. The graph contained the SSE value in the experimental number of clusters between 2 and 8. The number of segments of 3 SSE values is 313,29, the value is not the lowest SSE value but the lower value.

4. Conclusion

The results obtained from the process in determining the best number of clusters with elbow and K-Means methods that the determination of the best number of clusters can produce the same number of clusters K on the amount of different data. The result of determining the best number of clusters with elbow method will be the default for characteristic process based on case study.

References

[1] Bataineh K M, Naji M and Saqer M 2011 A Comparison Study Between Various Fuzzy Clustering Algorithms. Jordan Journal of Mechanical and Industrial Engineering (JJMIE) **5** p335

[2] Bain K K, Firli I, And Tri S 2016 Genetic Algorithm For Optimized Initial Centers K-Means Clustering In SMEs, Journal of Theoretical and Applied Information Technology (JATIT) **90** p 23

[3] Bain K K 2015 Customer Segmentation of SMEs Using K-Means Clustering Method and modeling LRFM, International Conference on Vocational Education and Electrical Engineering, Universitas Negeri Surabaya

[4] Celebi M E, Kingravi H A, and Vela P A 2013 A comparative study of efficient initialization methods for the k-means clustering algorithm, Expert Systems with Applications **40** p 200

[5] Neda S and Farzad M S 2016 Customer Segmentation of Bank Based on Discovering of Their Transactional Relation by Using Data Mining Algorithms, Modern Applied Science **10** p 283

[6] Gursharan S, Harpreet K and Gursharan S 2014 A Novel Approach Towards K-Mean Clustering Algorithm With PSO, International Journal of Computer Science and Information Technologies (IJCSIT) **5** p 5978

[7] Antreas D A 2000 Customer Satisfaction Cues To Support Market Segmentation and Explain Switching Behavior, Journal of Business Research **47** p 191

[8] Madhulatha T S 2012 An Overview On Clustering Methods, IOSR Journal of Engineering **4** p 719

[9] Joshi K D and Nalwade P S 2013 Modified K-Means for Better Initial Cluster Centres. International Journal of Computer Science and Mobile Computing II **7** p 2

[10] Singh H and Kaur K 2013 New Method for Finding Initial Cluster Centroids in K-means Algorithm. International Journal of Computer Applications, LXXIV **6** p 27

[11] Sujatha S and Sona A S 2013 New Fast K-Means Clustering Algorithm using Modified Centroid Selection Method. International Journal of Engineering Research & Technology II **2** p1

[12] Eltibi M F and Ashour W M 2011 Initializing K-Means Clustering Algorithm using Statistical Information. K-means clustering algorithm is one of the best known, XXIX **7** p 51

[13] Aristidis L, Nikos V, Jacob J V 2011 The global k-means Clustering algorithm IAS technical report series, IAS-UVA-01-02

[14] Cosmin M P, Marian C M, Mihai M An Optimized Version of the K-Means Clustering Algorithm, Proceedings of the 2014 Federated Conference on Computer Science and Information Systems (ACSIS) **2** p 695

IOP Conf. Series: Materials Science and Engineering **336** (2018) 012018 doi:10.1088/1757-899X/336/1/012018

Automated Temperature Control with Adjusting Outlet Valve of Fuel in the Process of Cooking Palm Sugar

H. Aripin[1*], Nurul Hiron[1], Edvin Priatna[1], Nundang Busaeri[1], Asep Andang[1], Suhartono[2], Svilen Sabchevski[3]

[1]Department of Electrical Engineering, Faculty of Engineering, Siliwangi University, Tasikmalaya 46115, Indonesia.
[2]Department of Agrotechnology, Faculty of Agriculture, Siliwangi University, Tasikmalaya 46115, Indonesia.
[3]Lab. Plasma Physics and Engineering, Institute of Electronics of the Bulgarian Academy of Sciences, Sofia 1784, Bulgaria.

*aripin@unsil.ac.id

Abstract. In this paper, a real-time temperature control system for coconut sugar cooking is presented. It is based on a thermocouple temperature sensor. The temperature in the closed evaporator is used as a control variable of the DC servo control system for opening and closing of a valve embedded in a gas burner. The output power level, which is necessary in order to reach the target temperature is controlled by the microcontroller ATMega328P. A circuit module for control of the valve and temperature sensors as well as software for data acquisition have been implemented. The test results show that the system properly stabilizes the temperature in the closed evaporator for coconut sugar cooking in the range from room temperature to 110°C. A set point can be reached and held with an accuracy of ±0.75°C at a temperature of 110°C for 60 minutes.

1. Introduction

Coconut sugar is produced from the sap of the coconut palm. Nowadays, most of its production is based on the following traditional method. The sap is poured in a large wok and heated on a wood-fired stove [1]. At the beginning of cooking, the fire must be intensive in order to accelerate the evaporation of water from the sap. The sap is then boiled at a temperature of over 100°C for a few hours until its concentration increases and sugar syrup is obtained. Afterwards, this substance is stirred manually using large wooden spoon until it becomes viscous. The producer then determines the final product quality assessing the intensity of its brown colour, thickness and viscosity of the liquid during the on-going process. The traditional method is laborious and requires a long time to evaporate water until the final product of coconut sugar is produced. In order to reduce the thermal degradation during the evaporation, it is necessary to minimize the heating temperature and the processing time. Efforts have been made to overcome the problem by using the method of vacuum evaporator [2-3]. Palm sap of 15 litres was concentrated by using a vacuum evaporator at 70 and 80°C until its total soluble solids reached 70°Brix to obtain palm sugar syrup. The TSS of finished palm sugar syrup should be not less than 65°Brix according to the Thai industrial standards institute of the ministry of industry. However, these studies have demonstrated that the cooking time estimation cannot be controlled properly. It should be underlined that the sufficiency of heating during the process is a very important factor because it affects such quality factors of the produced sugar as the moisture content, total sugar content, reducing sugar, colour and texture. During the heating process, various chemical and physical changes occur. During a long heating process, a sugar with hard texture and dark brown colour is produced while a short heating provides a soft and easily melted product. Therefore, it is necessary to control the heating temperature and the time during the cooking process.

Ho *et al*., (2008) [4] have developed a method to monitor the properties of the palm sugar syrup during the heating process by open skillet with a controlled temperature. In this process, the cooking temperature of coconut sugar is measured using an infrared thermometer. The measurement results show that after 30 minutes, the cooking temperature is 78.5°C and rises to 147.6°C at the cooking end. The water content was reduced from 84.66% to 12.11%. The advantage of the cooking temperature control using an infrared thermometer is that the measurement is contactless and the exposure is less specific to the temperature of the measured object. Furthermore, Hasan *et al.,* (2015) [5] have studied a new method for palm sugar cooking and compared it with the traditional method. The palm sap is placed in the cooking pot and heated by using an electrical heater. The temperature is maintained by using a controller in order to produce a more uniform process. In our study, the coconut sugar syrup is produced by evaporating coconut sap in the closed evaporator and heating is performed using a gas burner. The gas output from the LPG gas cylinder is regulated by an output valve installed between the gas burner and the gas cylinder. Prior to this, only manual operated valve was used to open and close the gas flow to the gas burner, such that it is difficult to adjust the desired speed. The speed of opening and closing of the valve affects the size of the fire, especially when the sap is concentrated. At such conditions, a stable temperature is necessary for producing the best quality of sugar in colour and texture. Therefore, an automatic control valve with high accuracy opening valve area is needed to get an appropriate speed and to maintain the gas flow before its entering the gas burner. In the present work, we have developed an automatic control valve by using a servo motor at the inlet of the gas burner that has an adjustable opening area in order to get the appropriate flow rate while raising and lowering the heating of the sap in a closed evaporator as necessary. The aim of this paper is to describe the main features of the developed temperature control system based on a thermocouple temperature sensor as well as to provide the experimental verification validating the performance of this cost-effective solution.

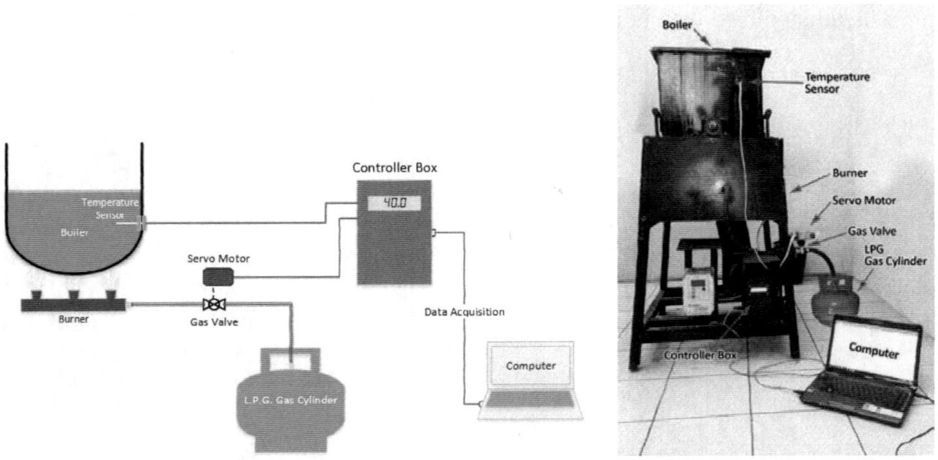

(a) (b)

Figure 1. (a) Schematic of the coconut sugar cooking system and (b) photograph of the test setup.

IOP Conf. Series: Materials Science and Engineering **336** (2018) 012018 doi:10.1088/1757-899X/336/1/012018

2. Experimental Procedures

2.1. The coconut sugar cooking system

Figure 1 shows schematic of the coconut sugar cooking system. The main components of the system are: (i) closed evaporator (boiler); (ii) thermocouple as a sensor to read the temperature; (iii) gas burner; (iv) LPG gas cylinder; (v) servo motor; (vi) controller box; and (vii) a computer. The burner module is a self-made gas stove. It consists of three stainless steel pipes with diameter of 2 in and mounted lengthwise under the evaporator. The diameter of the hole for the gas output on the pipe is 3 mm and the number of holes in each pipe is 20. The gas from the cylinder flows through the hose and generates a flame on the gas burner. The gas that flows into the burner is controlled using a gas valve. In fact, the valve regulates the size of the gas flow, which ultimately regulates the size of the flame. The size of the gas flow is controlled by a valve rotation driven automatically by the servo motor. The similar methods have also been used in the previous studies for producing charcoal from coconut shell waste [6].

2.2. Temperature Control System

2.2.1 Temperature Sensor. Figure 2 shows the circuit of the temperature control system in the coconut sugar cooking system set on the controller box on Fig. 1. The temperature sensor uses a thermocouple of the K type module MAX6675 as a temperature gauge. The output of MAX 6675 has a resolution of 0.25°C and has a temperature reading range between 0°C and 1024°C. It has the ability to reduce the noise error in the thermocouple conductor. A noise can arise from the power supply source of MAX 6675. In order to minimize it, a ceramic capacitors is installed between these two components. The thermocouple detects the temperature in the closed evaporator and sends a voltage signal to the IC MAX 6675. The received voltage signal is converted into a digital signal form by MAX 6675 and then it is sent to the microcontroller ATMega328P based Arduino UNO.

2.2.2 Microcontroller Circuit. Arduino UNO is a microcontroller board system based on the microcontroller ATMega328P. It can read data from MAX 6675 by sending a low bit-rate signal to CS pin of MAX 6675, a clock signal to SCK and retrieve data from pin SO MAX 6675. The temperature is displayed by Arduino in real-time using 16x2 LCD. The latter uses I2C communication to save pin I / O Arduino. The temperature data is processed by the Arduino and is compared with the user defined set point (SP).

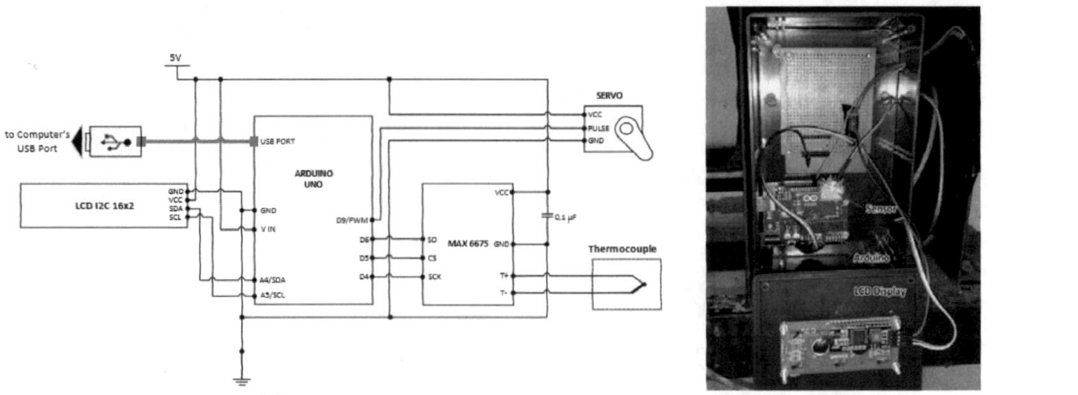

(a) (b)

Figure 2. (a) Temperature control system circuit of the coconut sugar cooking system and (b) photograph of control system circuit.

IOP Conf. Series: Materials Science and Engineering **336** (2018) 012018 doi:10.1088/1757-899X/336/1/012018

2.2.3 Servo motor circuit. This circuit is an automatic switch that regulates the opening and the closing of the gas valve, in accordance with the commands provided by ATMega328P microcontroller through the program installed in it. In this system, the servo is connected to the gas valve to regulate the heat generated by the flame in the gas burner. Rotations from 90° to 0° and from 0° to 90° are used to open and close the gas valve, respectively [7]. The work procedure of the servo motor circuit is as follows. As the temperature rises above the set point, the program sends an appropriate control signal to the Arduino and the servo rotates from 90° to 0° to close the gas valve. On the contrary, when the temperature falls below this point, the program sends another control signal through Arduino making the servo to rotate in the opposite direction from 0° to 90°, i.e. to open the gas valve.

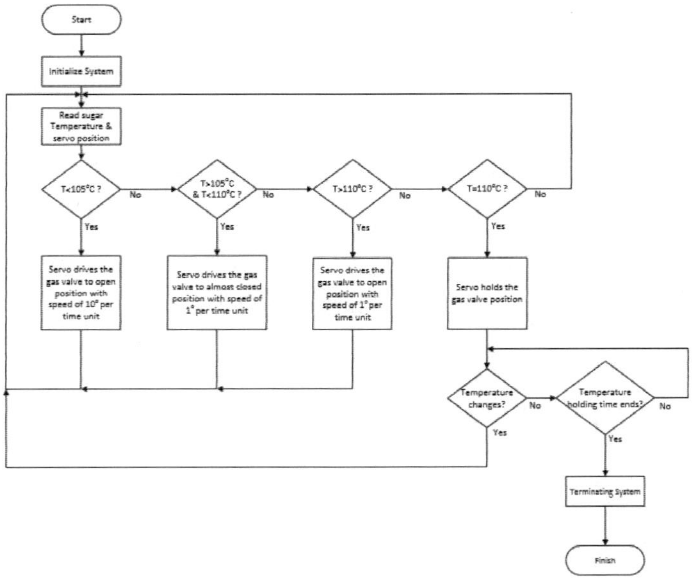

Figure 3. Flowchart of the processes involved in the temperature control.

Table 1. Precision of the thermocouple K type sensor module IC MAX6675 for measuring the temperature of cooking of coconut sugar in the temperature range from 30 to 110°C.

No	Time (min)	Detected temperature (°C)	Actual temperature (°C)
1	1	34.50	35.50
2	5	41.00	42.00
3	10	52.75	54.50
4	15	64.75	66.50
5	20	77.00	78.00
6	25	87.25	88.50
7	30	96.75	98.00
8	35	104.00	105.00
9	90	104.75	105.50
10	250	108.25	109.00
11	255	109.50	110.00
12	256	110.00	111.00

IOP Conf. Series: Materials Science and Engineering **336** (2018) 012018 doi:10.1088/1757-899X/336/1/012018

3. Results and Discussion

Table 1 shows a comparison between the temperature measured by the thermocouple K type sensor module IC MAX6675 and the actual temperature registered by the thermocouple Mastech MS6514. The measured temperature is the temperature of the coconut sap from first minute to 90th minute and the temperature of sugar syrup from 250th minute to 256th minute. The temperature was measured at the closed evaporator centre. The maximum error deviation of the temperature is less than 2°C. If the test results meet the corresponding standard requirements, it means that the overall performance of the system meets the imposed requirements.

Figure 4a and 4b show open and close cycles of the gas valve controlled by servo rotation and the changes in the temperature around the set-point temperature of 110°C and 115°C as a function of the cooking time, respectively. The working principles for the test results shown in Figure 4a and 4b follow the flowchart presented in Fig. 3. The 0° angle is for the position of completely open valve. At this condition, the cooking evaporator is heated to a set-point temperature of 110°C. As the temperature rises above 110°C, the servo rotates from 0° to 57° at a speed of 60° per milliseconds to partially close the gas valve so that there is still a flame on the gas burner. As the temperature falls below 110°C, the servo rotates from 57° to 0° to open the gas valve, so that the flame is enlarged. This cycle continues until the sugar becomes concentrated. For the case of the set-point temperature of 115°C shown in Figure 4b, it is the same as for 110°C, but the speed of the rotation is 10° per milliseconds. Furthermore, we compare the temperature response around 110°C and 115°C in Fig. 4 and 5, respectively. It can be seen that the temperature fluctuations are ±0.75°C and ±3.0°C for the set-point temperatures of 110°C and 115°C, respectively. The fluctuation around 110°C is much lower than that at 115°C. This behaviour can be related to the speeds of the servo rotation to open and to close the gas valve.

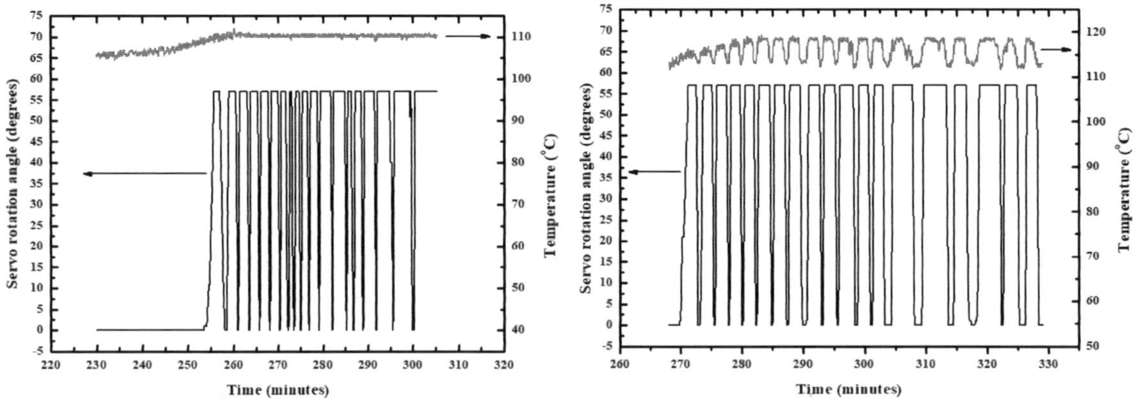

(a) (b)

Figure 4. Open and close cycles of gas valve by servo rotation and changes in temperature around the set-point temperature of (a) 110°C and (b) 115°C as a function of the cooking time.

4. Conclusion

In this work, a temperature control system for cooking of coconut sugar was developed using a DC servo control. The experimental tests have confirmed its efficiency and operational performance. The results have demonstrated that the temperature sensor works with an accuracy better than 2°C to stabilize the temperature in the range from room temperature to 110°C. A significant effect of the servo rotation on the detected temperature in the closed evaporator is clearly observed at higher speeds of the servo rotation. At this condition, the temperature can be held at 110°C with an accuracy of ±0.75°C for 60 minutes. Such accurate temperature control in a closed evaporator for cooking of coconut sugar has been realized for the first time.

Acknowledgments
This research was supported by a fund of the Directorate of Research and Community Service, Minister of Research, Technology and Higher Education. Republic of Indonesia through the Project of Penelitian Produk Terapan in 2017 (Contract number: 28/E/KPT/2017). The authors would like to thank Mr. Muhammad Ilyas and Mr. Apang Kurnia for kindly helping to prepare the test of the evaporator.

Reference

[1] Iskandar, A., Yuliasih, I., Haryanto, B., 2014. Effect of Temperature and Time on Dry Granulation Process of Arenga Palm Sugar, International Journal of Research in Engineering & Technology (IMPACT: IJRET), 2, 33-44.

[2] Naknean, P., Meenune, M., and Roudaut, G., 2009. Changes in physical and chemical properties during the production of palm sugar syrup by open pan and vacuum evaporator, Asian Journal of Food and Agro-Industry, 2, 448-456.

[3] Naknean, P., Meenune, M. and Roudaut, G., 2013. Changes in properties of palm sugar syrup produced by an open pan and a vacuum evaporator during storage, International Food Research Journal, 20, 2323-2334.

[4] Ho, C.W., Wan Aida, W.M., Maskat, M.Y., and Osman, H., 2008. Effect of Thermal Processing of Palm Sap on the Physico-Chemical Composition of Traditional Palm Sugar. Pakistan Journal of Biological Sciences, 11, 989-995.

[5] Hasan, R., Shamsudin, S.A., Muhammad, N., Tamaldi, N., 2015. Development of innovation in Malaysian traditional palm sugar cooking machine, Proceedings of Mechanical Engineering Research Day, 143-144.

[6] Langue, M., Sawidin, S., and Mappadang, J., 2016. Monitoring the Charcoal Furnace Control System of Coconut Shell Waste Treatment Becoming Charcoal with ATMega 328P Microcontroller, International Journal of Computer Applications, 156, 22 – 28.

[7] Blum, J., 2013. Exploring Arduino Tools and Techniques For Engineering Wizardy, Published by John Wiley & Sons Inc, 80-86.

Measurement of Non-Invasive Blood Glucose Level Based Sensor Color TCS3200 and Arduino

Humaidillah Kurniadi Wardana[1], Elly Indahwati[1], Lina Arifah Fitriyah[3]

[1]Department of Electrical Engineering, Faculty of Engineering, Universitas Hasyim Asy'Ari Tebuireng Jombang
[3] Department of Natural Science, Faculty of Science Education Universitas Hasyim Asy'Ari Tebuireng Jombang

Abstract. Design and measurement of Arduino-based urinary (non-invasive) urine glucose using RGB tcs3200 sensor. This research was conducted by making use of the urine in diabetes patients detected by sensor colours then measured levels of colour based on the RGB colour of the urine of diabetics. The detection is done on 4 urine samples with each consisting of 3 diabetics and 1 non-diabetics. Equipment used in this research, among others, Arduino Uno, colour sensor tcs3200, LCD 16x4. The results showed that the detection of RGB values in diabetics 230 with blue and not diabetics 200 with red.

1. Introduction

Diabetes Mellitus (DM) is one of the largest number of sufferers of the disease around the world. According to the WHO, approximately 415 million people are affected by the disease of DM [1]. Whereas in Indonesia by the year 2015 was ranked seventh with number of sufferers ±10 million [2]. The disease is closely related to the rise in blood glucose levels or Hyperglycaemia [2]. Resulting in abnormalities in the metabolism of the body because the body cannot use the insulin it perfectly or does not produce any insulin at all. In the process of digestion, the food eaten to make blood glucose rise and stimulate the pancreas to producing insulin. Insulin moves sugar into the cells make to converted into energy. If too much glucose in the blood then the pancreas difficult making insulin work well [3]. Normal range blood sugar levels around ≤ 120 mg/dl [4].

Checking blood sugar levels in the diagnosis distinguished into two categories, namely minimally invasive and non-invasive techniques. Minimally invasive technique is to measure levels of glucose by way of hurting the patient's blood inside the body are taken using a needle. The weakness of this method is the use of a needle that results in pain in people with DM. Non-invasive use of biofluids techniques namely saliva, sweat, tears, and urine, is used to measure glucose levels so that this technique was felt more comfortable by diabetics [5], [6]. Research the use of urine as a basis for measuring glucose levels include: using spectroscopic method using infrared light (NIR) and photodiode using the Arduino and microcontroller, which are brought on the finger of a patient DM [7], [8], [9], [10], [11].

Each colour can be composed from the basic colours. On the basis of the colour of light has its compilation is the colour red, green and blue, or better known as the RGB (Red, Green, Blue). The colour spectrum is contained within a perfect light (white). The RGB colour model is a model based on the concept of adding strong colour radiance primer i.e. Red, Green and Blue. In a space that is absolutely no light, then the room is dark. There is no signal light waves are absorbed by the eye or RGB (0, 0, 0). When adding the red light in the room, then the room will change colour to red for instance RGB (255, 0, 0), all objects in the room only can be seen in red [12]. TCS3200 is a programmable IC which is the way it works to convert light into colour frequency with output signal box-shaped. The main constituent of this colour is the sensor and photodiode current to frequency converter. Sensor colour TCS3200 colour sensor light is equipped with a filter and a light base colour RGB [13].

This research created tool to measure sugar levels without having to hurt patients (non-invasive) and can be a good alternative for monitoring blood sugar levels without any pain with the use of TCS3200 sensors to detect the color of urine at sufferers of DM and not a sufferers of DM connected to the Arduino and microcontroller, as displayed on the LCD.

2. Theoretical Review

2.1. Urine

Urine or urine is liquid waste excreted by the kidneys that would then be removed from the body through the process of the Urineasi. The excretion of urine is needed to dispose of the rest of the molecule is a molecule in the blood is filtered by kidneys and to maintain the homeostasis of bodily fluids. The urine is filtered by the kidney, brought through the ureter toward the bladder, finally being dumped out of body through the urethra (s. Nojoatmojo, 2000). In macroscopic tests include urine colour and clarity (a normal urine is yellow and crystal clear clarity or slightly turbid), smell (the smell of urine is normally a characteristic caused by volatile organic acid), the volume of urine (people normal adult urine production of approximately 1500 ml/24 hours, is useful for determining the presence of impaired kidney abnormalities Faal balance of body fluids), gravity or BJ (gives an impression about the sensitivity of urine, urine concentrations with BJ >1,030 indicates the possibility of glucose in the urine. The limit of the normal range from 1,003 – 1,030 BJ).

2.2. Colour Sensor of TCS3200

TCS3200 colour sensor is the way it works to change the colour into the frequency. Made from a mixture of silicone and current to frequency converter on one circuit integration. This gave rise to a wave of censorship output box with 50% duty cycle with a proportional frequency. Pin configuration TCS3200 can be seen in Figure 1

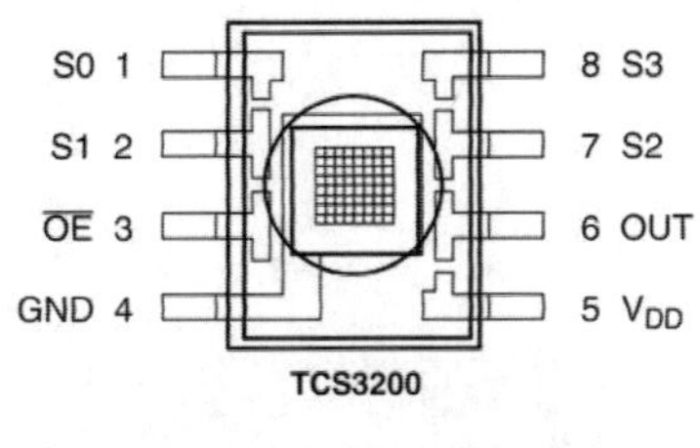

Figure 1. Pin TCS3200

A full-scale frequency can be set with one of the three initial values through two control input pin. On TCS3200, the sensor light to frequency converter read 8 x 8 array of photodiodes, 16 blue filter photodiodes, 16 green photodiode filters, and the last is red filter photodiode. Example of the colour sensor can be seen in Figure 2

Figure 2. Sensor TCS3200

IOP Conf. Series: Materials Science and Engineering **336** (2018) 012019 doi:10.1088/1757-899X/336/1/012019

PIN S2 and S3 is used to select the photodiode that want enabled. The combination of inputs for each colour can be seen in table 1.

Table 1. Sensor TCS3200

S2	S3	Color
L	L	Red
L	H	Blue
H	L	Clear
H	H	Green

Terminal function TCS3200 can be seen the table 2

Table 2. Sensor TCS3200

Name	No	*I/O*	Description
GND	4		*Ground*
OE	3	I	Active Low
OUT	6	O	Output Frequency
S0, S1	1,2	I	Output Frequency at scale
S2, S3	7,8	I	Type photodiode
VDD	5		*Power Supply*

2.3. Arduino UNO

Arduino UNO is a microcontroller board based on ATmega328. Arduino has a 14 pin input/output, 6 pin which can be used as PWM outputs, 6 analogue input, a 16 MHz crystal oscillator, a USB connection, power jack, head of the ICSP, and the reset button. Arduino UNO is able to download support microcontroller; can be connected to the computer using the USB cable and can be supplied with an AC adapter to the DC or battery to get started. Figure 2.3 the Board Arduino UNO.

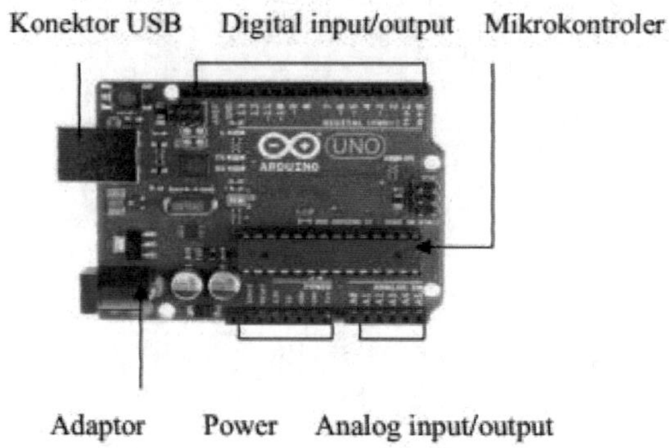

Figure 3. Board Arduino Uno

Board Arduino UNO has the following features:

a. Pin out 1.0: SDA and SCL pins plus closer to AREF pin and two new pin the other is placed close to pin RESET, IOREF which allows shield-shield to adjust the voltage supplied from the board. For the future, the shield will be made compatible/match the board which uses an AVR that operates with a voltage of 5V Arduino Due and with operating voltages 3 .3V. The second is a pin is not connected, which provided for the purpose in the future.

b. Reset Circuit.
c. Atmega16U2 replaces Atmega8U2.

As for the Arduino UNO specification summary is as follows:

a. Microcontroller: ATMEGA328b.
b. Operating voltage: 5V
c. Input voltage (recommended): 7 - 12 V
d. Input voltage (limit): 6-20 V
e. Digital pin I/O: 14 (six of them is PWM pin)
f. Analog input pin: 6
g. DC current I/O pin: 40 mA
h. DC current for pin 3.3 V: 150 mA
i. Flash Memory: 32 KB with 0.5 KB to use for the bootloader
j. EEPROM: 1 KB
k. SRAM: 2 KB
l. The timing speed: 16 Mhz

3. Research Methodology

On the design of measuring glucose levels in urine, there are 2 main stages, namely designing hardware and designing software. The stages of design as follows:

3.1. Design of Hardware

Tool test levels of glucose in the urine as described in Figure 4 blocks

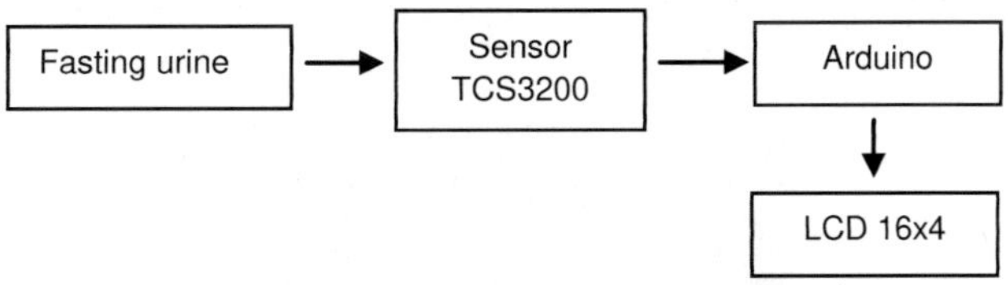

Figure 4. Block Diagram Tools Urine Glucose Levels

a. Fasting Urine
Testing tool to measure levels of glucose in urine is done twice experiment using 4 samples of urine that is fasting urine 4. Fasting urine used consists of 3 DM urine and 1 is not a DM urine. Fasting Urine taken after first wake up in the morning before eating or drinking anything.
b. Sensor TCS3200
The function to change the colour into an electric current, electric current is then converted into the form of a frequency signal box shape. Direct sensor bias output data is read by the series of Arduino UNO.
c. Arduino UNO
A microcontroller is used the Arduino Uno. That serves as a data processing unit later in the show to the output.
d. LCD 16x4
To display the output in the form of letters or numbers used LCD display 16 x 4, where the LCD is composed in 16 columns and 4 rows.

3.2. Design of Software

Software test tools glucose levels in urine include reading frequency of the colour sensor and then displays the RGB value of the colour sensor output TCS3200 by setting a filter on the sensor mode through pin S2 and S3.

4. Result And Discussion

Realization of measuring glucose levels in urine as a whole looks in Figure 5

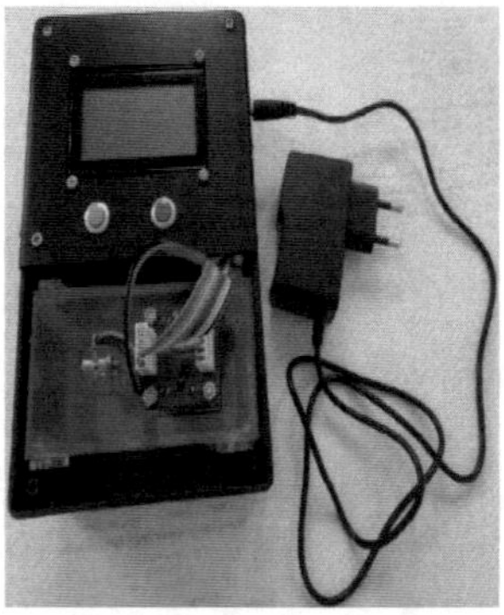

Figure 5. Fasting Urine Levels of measurement

Measuring blood sugar levels in urine is composed of several sections, among others; the power supply as a voltage source to switch on the measuring instrument, the calibration button is green is used as the initial calibration by using the normal urine (not a DM) before making measurement, the urine is used as shelter samples of urine, TCS3200 sensor serves to detect urine color based on RGB values, and 16 x 4 LCD is used to display data for measuring results.

Initial display on the gauge levels of glucose in urine is visible in Figure 6

Figure 6. Fasting urine Measurement Result

Overall tool testing done by pouring a pot of urine into the box urine and close the box made of urine in a meeting. TCS3200 sensors to detect the colour based on RGB visible in Figure 7

Figure 7. Fasting urine Measurement Result

The following data result measurement levels of glucose in the urine compared to lab test result are shown in Table 3

Table 3. Fasting Urine Measurement Result

Kode	Color Value						Result Value RGB	Lab Results	Result Value Lab
	R		G		B				
	I	II	I	II	I	II			
A	214	214	224	224	236	236	Blue	++	
B	211	209	217	214	230	230	Blue	+++	
I	218	215	221	221	234	232	Blue	++++	
J	205	203	193	192	188	188	Red	-	

From Table 3 Result of reading levels of glucose in urine is obtained as a result of reading the sensor colour TCS3200 colour of urine that was diagnosed with DM is blue while not diagnosed DM are red. Specific value of each sample is as follows:

a. RGB values for sample code A on the first test and both have the same value i.e. R = 214, G = 224, B = 236 with the results a foregone conclusion is blue while the lab results showed the two positive (++) which means it is among 51 mg/dl up to 100 mg/dl and diagnosed DM.
b. Sample code B on the first test value of R = 211, G = 217, B = 230, in the second test of the value of R = 209, G = 230, B = 214 with the conclusion is blue with the results of the lab shows the positive three (+++) which means the value of DM medically among 101 mg/dl to with 300 mg/dl. The value of R and G on RGB to sample B has different values between the first and second experiments with difference of 3 numbers.
c. For the code samples I in the first test of the value of R = 218, G = 221, B = 234, while in the second test of the acquired value of R = 215, G = 221, B = 232 with the result of the reading is blue and shows a positive lab four hail (++++) with the range of value of DM 301 mg/dl to with 1000 mg/dl. There is a difference between the value of R and B in RGB at first and second experiments with difference value for R_I and R_{II}3 numbers as well as B_I and B_{II} 2 numbers.
d. On the samples I RGB values in each experiment the first value R = 205, G = 193, B = 188. For second experiment the value R = 203, G = 192, B = 188 with a conclusion that is red

which means not diagnosed DM. Lab Results also showed a negative value (-). The experiment result between the first and second experiments had a slight difference in the value of R_I and R_{II} 2 numbers as well as G_I and G_{II} 1 numbers.

5. Conclusion

Based on the results of the research conducted, it can be drawn the conclusion that the gauge levels of glucose in the urine has been successfully created and analyzed. The result of two times the measurement reading RGB colour obtained results that are diagnosed DM in each urine sample produced a reading of the blue colour of the urine sample while in the not diagnosed DM are red.

Acknowledgements

Researchers say many thanks to some parties who have helped in finishing this research, such as:

1. LPPM Unhasy who have given permission and instructions in doing research lecturer beginners.
2. DRPM Ristekdikti who gave the funds in this research so that research is completed.
3. RSUD Dr. H. Moh. Sumenep Regency Anwar as partners in making research data.

References

[1] WHO. 2016. *Diabetes Fakta dan Angka.* Artikel:http://www.searo.who.int/indonesia/topics/8-whd2016-diabetes-fact-and numbers indonesian.pdf?=1.

[2] Kemenkes RI. 2014. *Tahun 2030 Prevalensi Diabetes Melitus di Indonesia Mencapai 21,3 Juta Orang.* Artikel. www.depkes.go.id.

[3] Akhbar, Zheni. 2014. *Rancangan Prototype Instrumen Ukur Kadar Gula Darah dengan Metode Pengenalan Warna dan Tingkat Kekeruhan Urine dalam Uji Benedict Berbasis Visual Basic.Net.* Skripsi Jurusan Teknik Elektro Fakultas Teknik Universitas Jember.

[4] Aulia Haque, Aditta Putri dkk. 2013. *Analisis Profil Potensial Listrik Pada Titik Akupuntur Untuk Diagnosis Diabetes Mellitus.* Jurnal Fisika dan Terapannya Fisika Unair. Vol. 1. No. 4. Hal. 1. Desember 2013.

[5] C.P. John et al. 2005. *In vivo Glucose Monitoring: The Clinical Reality and The Promise, Biosens.* Bioelectron. 20 (2005) 1897-1902.

[6] H.D. Park et al. 2005. *Design of portable Urine Glucose Monitoring System for Healthcare, Comput.* Biol. Med. 35 (2005) 275-286.

[7] Anas et al. 2012. *Non-Invasive Blood Glucose Measurement.* IEEE Conference on Sustainable Utilization and Development in Engineering and Technology (Student) Universiti Tunku Abdul Rahman, Kuala Lumpur Malaysia. 978-1-4673-1705-4/12/$31.00.

[8] Buda, R.A. and Mohd.Addi, M. 2014. *A Portable Non-Invasive Blood Glucose Monitoring Device.* IEEE Conference on Biomedical Engineering an Sciences Sarawak Malaysia. 978-1-4799-4084-4/14.

[9] Lawand, Komal et al. 2015. *Design And Development of Infrared LED Based Non Invasive Blood Glucometer.* IEEE Indicon. 1570186243.

[10] Tamridho, Reza. 2010. *Rancang Bangun Alat Pengukur Kadar Gula Darah.* Skripsi Fakultas Teknik Program Studi Teknik Elektro UI Depok.

[11] Vivekanandan, S. dan Devanand, S. 2015. *Remote Monitoring for Diabetes Disorder: Pilot Study Using InDiaTel Prototype.* European Research in Telemedicine 2015 4, 63 - 69. http://dx.doi.org/10.1016/j.eurtel.2015.04.002. 2212-764X/c 2015 Elsevier Masson SAS.

[12] Sutisna, Dedi., Ihsanto, Eko. 2014. *Perancangan dan Pembuatan Alat Pendeteksi Warna Menggunakan Sensor TCS3200 Pada Proses Produksi Kaleng Berbasis Arduino.* Skripsi Teknik Elektro Universitas Mercu Buana.

[13] Sutisna, Dede., and Ihasanto, Eko. 2015. *Perancangan dan Pembuatan Alat Pendeteksi Warna Menggunakan Sensor TCS3200 Pada Proses Produksi Kaleng Berbasis Arduino.* Tugas Akhir: Program Studi Teknik Elektro Universitas Mercu Buana Jakarta.

[14] Notoatmojo, S. 2000. *Metodelogi Penelitian Kesehatan.* Jakarta: Rineka Cipta.

IOP Conf. Series: Materials Science and Engineering **336** (2018) 012020 doi:10.1088/1757-899X/336/1/012020

The Design of Artificial Intelligence Robot Based on Fuzzy Logic Controller Algorithm

M S Zuhrie [1], Munoto [1], E Hariadi [2], S Muslim [1]

[1]Department of Electrical Engineering, Universitas Negeri Surabaya, Indonesia

[2] Department of Computer Engineering, Universitas Negeri Surabaya, Indonesia

*zuhrie@unesa.ac.id

Abstract. Artificial Intelligence Robot is a wheeled robot driven by a DC motor that moves along the wall using an ultrasonic sensor as a detector of obstacles. This study uses ultrasonic sensors HC-SR04 to measure the distance between the robot with the wall based ultrasonic wave. This robot uses Fuzzy Logic Controller to adjust the speed of DC motor. When the ultrasonic sensor detects a certain distance, sensor data is processed on ATmega8 then the data goes to ATmega16. From ATmega16, sensor data is calculated based on Fuzzy rules to drive DC motor speed. The program used to adjust the speed of a DC motor is CVAVR program (Code Vision AVR). The readable distance of ultrasonic sensor is 3 cm to 250 cm with response time 0.5 s. Testing of robots on walls with a setpoint value of 9 cm to 10 cm produce an average error value of -12% on the wall of L, -8% on T walls, -8% on U wall, and -1% in square wall.

1. Introduction

In recent years technological developments in the field of robotics have become a serious concern. This is primarily in the role of robots that can replace human jobs, especially in hazardous environments. Wheeled robots are mechanical devices capable of moving in environments with a certain degree of autonomy. Autonomous navigation is associated with the availability of external sensors that get information from the environment through visual images or distances or measurement approaches. The most commonly used sensors are proximity sensors (ultrasonic, laser, infrared, etc.) that can detect obstacles and measure the distance to the wall closest to the robot. When autonomous robot navigation moves forward in a room environment (industrial or civil building), the robot must have the ability to move through the corridor, to follow the wall, to change the angle and enter the open space area

This study uses ultrasonic sensors HC-SR04 to measure the distance between the robot with ultrasonic wave-based wall. The advantage of ultrasonic sensors is to calculate the distance to the object by calculating the time interval between the trigger signal and the echo signal. HC-SR04 has the ability to transmit and receive ultrasonic signals with an angle of reach of 15 degrees to the right and left of the direction with maximum distance to detect objects of four meters but has the possibility of error or unreadability. Maximum of sensor sensing rate is 20Hz. Fuzzy logic is an appropriate way to map an input space in an output space and has a continuous value. Excess fuzzy logic exists on the ability of language reasoning. Thus, in designing it does not require complex mathematical equations of objects to be controlled. The reason to use fuzzy logic as a controller is the fuzzy logic concept is easy to understand. The mathematical concept underlying fuzzy logic reasoning is very simple and easy to understand. In addition, fuzzy logic has a tolerance of improper data.

2. Theoritical Framework

2.1. Artificial Intelligence Robot
Artificial Intelligence Robot is an automatic robot which its motion follows the boundary walls of the track. One of the advantages is it does not need any guideline or a specific mark as the director for the robots. It works by reading and detecting a barrier or wall against the robot with the use of proximity sensor. When a change occurs, then the robot will move to then adjust the distance again. This process will be done repeatedly.

2.2. Ultrasonic Sensor HC-SR04
Ultrasonic sensor is a sensor that serves to change the physical quantity (sound) into electrical quantities and vice versa. The way the sensor works is based on the principle of the reflection of a sound wave so that it can be used to interpret the existence (distance) of an object with a certain frequency. It is referred to as an ultrasonic sensor because it uses ultrasonic waves (ultrasonic sound). HC-SR04 has 2 main components as the compiler is ultrasonic transmitter and ultrasonic receiver. The function of the ultrasonic transmitter is to emit ultrasonic waves with a frequency of 40 KHz then ultrasonic receiver captures the ultrasonic wave reflection results of an object

2.3. AVR ATmega 16 Microcontroller
AVR Microcontroller ATmega 16 has an 8-bit architecture, where all instructions are packed in 16-bit code and most instructions are executed in one clock cycle.

2.4. 12 V DC Motor
DC motor is an electric motor that requires a direct current voltage supply on the field coil to be converted into mechanical energy motion. The field coil on the dc motor is called the stator (the non-rotating part) and the anchor coil is called the rotor (the rotating part). Direct current motors, as the name implies, use indirect current or direct-unidirectional currents

2.5. DC L298 DC Motor Driver
IC H-Bridge DC L298 motor driver has two H-Bridge circuit in it, so it can be used to drive two DC motors. H-Bridge DC motor driver L298 can each deliver current up to 2A. However, in its use, H-Bridge DC L298 motor driver can be used in parallel, so the ability to deliver from H-Bridge DC motor driver L298 current to 4A

2.6. 16X2 LCD Display
LCD Display a liquid crystal or electronic device that can be used to display numbers or text. There are two main types of LCD screens that can display numerically (watches and calculators) and display alphanumeric text (coffee and mobile phones). In this numerical display the crystals are formed into bars, and in alphanumeric displays, the crystal is only set into a dot pattern. When the crystal is off, the crystal light looks the same as the background material, so the crystal cannot be seen. But when an electric current passes through a crystal, it changes shape and absorbs more light. This makes the crystals look darker than the sight of the human eye so that the point or bar shape can be seen from different backgrounds

2.7. Code Vision AVR
Code Vision AVR The use of microcontroller can work if it has filled a program, this program filling can be done using the compiler which is then downloaded into the microcontroller using downloader. Code Vision AVR has an advantage over other compilers, the existence of codewizard, this facility allows us to initialize the microcontroller we will use, codevision has provided configurations that can be set on each microcontroller chip that we will use, so we do not need to see the datasheet to simply configure the microcontroller.

2.8. Fuzzy Logic Controller
Fuzzy logic Controller calculates the output based on the number of weighted membership values in each set. The information flowing requires a fuzzy setting system. Before it becomes system output, the system requires three transformations for system input

a. Fuzzification

Fuzzification is the process of decomposing an input and / or output system into one or more fuzzy sets

b. Collection of Fuzzy Rules

After input and output are decomposed into fuzzy sets, we need a rule base that governs the behaviour of the system for each input combination. Each rule consists of one condition and one action. Conditions are interpreted from the input of fuzzy sets and actions are determined by the output of the fuzzy set

c. Defuzzification

Defuzzification is the change of a fuzzy quantity to a numerical value. The fuzzy process output can be either a logical unit of two or more fuzzy membership functions and defined in the output universe setDefuzzification is the change of a fuzzy quantity to a numerical value. The fuzzy process output can be either a logical unit of two or more fuzzy membership functions and defined in the output universe set.

2.9. ATmega 8 Microcontroller

ATMEGA 8 is a low-power 8-bit CMOS microcontroller based on enhanced RISC architecture. Most instructions are performed on one clock cycle, ATMEGA 8 has a throughput of close to 1 MPS per MHz making the design of the system to optimize power consumption versus process speed. The array of pins of the ATMEGA 8 microcontroller IC is shown in the figure below. This IC is composed of 28 pins that have certain functions

3. Research Method

3.1. Hardware Design

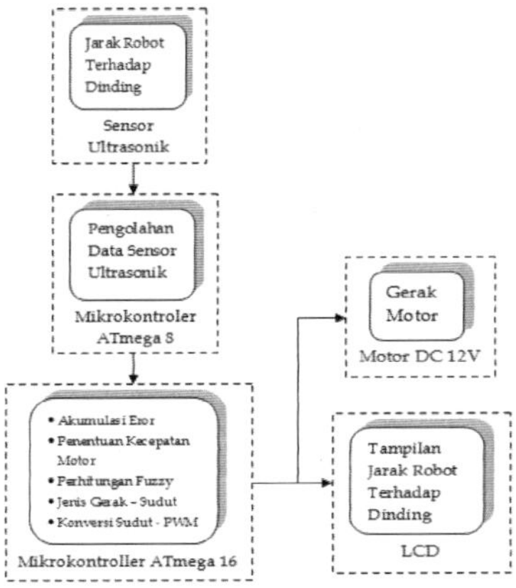

Figure 1. Artificial Intelligence Block diagram of Robotmotor DC which displayed on the LCD

Figure 1. is a block diagram of an Artificial Intelligence Robot system in which an ultrasonic sensor detects the robot's distance to the wall. Ultrasonic sensor data processing on ATmega microcontroller 8. Next microcontroller ATmega 16 duty to calculate fuzzy and determine speed of DC motor based on rule-rule on fuzzy. Sensor data and DC motor speed are displayed on the LCD

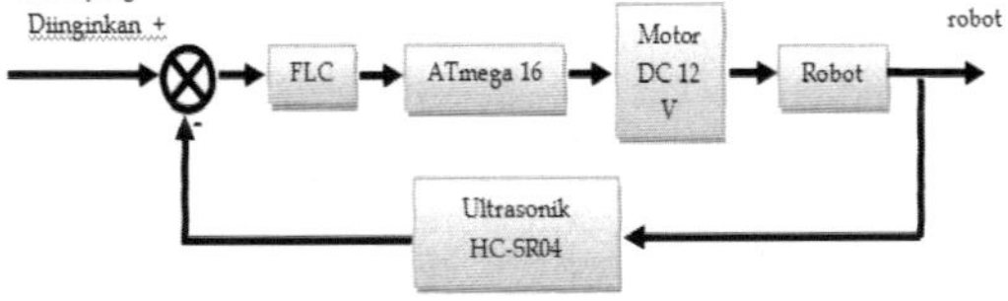

Figure 2. Artificial Intelligence Block diagram Robotmotor DC which displayed on the LCD

IOP Conf. Series: Materials Science and Engineering **336** (2018) 012020 doi:10.1088/1757-899X/336/1/012020

The 12 V DC motor is the actuator. Figure 2 shows the block diagram of a closed loop control system on Artificial Intelligence Robot. With the desired distance value is 9 cm to 10 cm. Sensor data received ATmega 8 then go into ATmega 16 which acts as a controller. Fuzzy logic is used to calculate the motor speed obtained from the calculation of the distance of the robot against the wall. Each motor has a different speed. If the robot moves less than the set point, meaning that the robot is in the near state of the wall then the right motor moves faster than the left motor. If the robot moves more than the set point, meaning that the robot is far from the wall then the right motor moves more slowly than the left motor. 12 V DC motor is the actuator and its plant is a robot with output value of DC motor speed.

Robot with output value of DC motor speed. robot with output value of DC motor speed: 1)If the distance of the sensor against the wall> 3 and ≤5 Then right motor rotation 50 PWM, left motor rotation 5 PWM. 2)If the distance of the sensor against the wall> 5 and ≤6 Then the motor rotation right 50 PWM, motor rotation left 10 PWM. 3) If the distance of the sensor against the wall> 6 and ≤7 Then the motor rotation right 50 PWM, motor rotation left 15 PWM. 4)If the distance of the sensor against the wall> 7 and ≤8 Then right motor rotation 50 PWM, left motor rotation 30 PWM. 5)If the distance of the sensor against the wall> 8 and ≤9 Then the motor rotation right 50 PWM, motor rotation left 45 PWM. 6)If the distance of the sensor against the wall> 9 and ≤10 Then the motor rotation right 50 PWM, left motor rotation 47 PWM. 7)If the distance of the sensor against the wall> 10 and ≤11 Then right motor rotation 50 PWM, motor rotation left 48 PWM. 8) If the distance of the sensor against the wall> 11 and ≤12 Then the motor rotation right 50 PWM, motor rotation left 50 PWM. 9)If the sensor distance against wall> 12 and ≤13 Then right motor rotation 15 PWM, motor rotation left 50 PWM. 10)if the sensor distance against wall> 13 and ≤14 Then right motor rotation 10 PWM, motor rotation left 50 PWM.11)if the sensor distance against wall> 14 and ≤250 Then right motor spin 15 PWM, motor rotation left 70 PWM.

4. Result and Discussion

4.1. Analysis of Artificial Intelligence Robot Movement Against the Wall

Analysis of the Artificial Intelligence Robot movement against the wall is done on the wall of the letter L, the letter T, the letter U, and the square with the set point 9 cm and 10 cm against the wall. Error values are calculated from robot motions that do not match the set point. The percentage of error values obtained with the following formula:

$$\text{Error value} = \frac{\textit{hasil yang didapat} - \textit{hasil yang dinginkan}}{\textit{hasil yang dinginkan}} \text{ X } 100$$

Table 1. The Testing Result of Artificial Intelligence Robot on an L-letter wall

Robot's Position	Distance	Conformity to the setpoint	Error Value
L1	9-10 cm	Appropriate	0%
L2	9-10 cm	Appropriate	0%
L3	9-10 cm	Appropriate	0%
L4	9-10 cm	Appropriate	0%
L5	9-10 cm	Appropriate	0%
L6	9-10 cm	Appropriate	0%
B1	9-10 cm	Appropriate	0%
B2	8 cm	-1 cm	-11,11%
B3	5 cm	-4 cm	-44,44%
L7	7 cm	-2 cm	-22,22%
L8	8 cm	-1 cm	-11,11%
L9	9-10 cm	Appropriate	0%
L10	9-10 cm	Appropriate	0%
B4	4 cm	-5 cm	-55,55%
L11	6 cm	-3 cm	-33,33%
L12	7 cm	-2 cm	-22,22%
L13	8 cm	-1 cm	-11,11%
L14	9-10 cm	Appropriate	0%

B5	9-10 cm	Appropriate	0%
B6	6 cm	-3 cm	-33,33%
B7	7 cm	-2 cm	-22,22%
L15	7 cm	-2 cm	-22,22%
L16	7 cm	-2 cm	-22,22%
L17	8 cm	-1 cm	-11,11%
L18	9-10 cm	Appropriate	0%
L19	9-10 cm	Appropriate	0%
L20	9-10 cm	Appropriate	0%
L21	9-10 cm	Appropriate	0%
B8	7 cm	-2 cm	-22,22%
B9	8cm	-1 cm	-11,11%
The average value of error			-12%

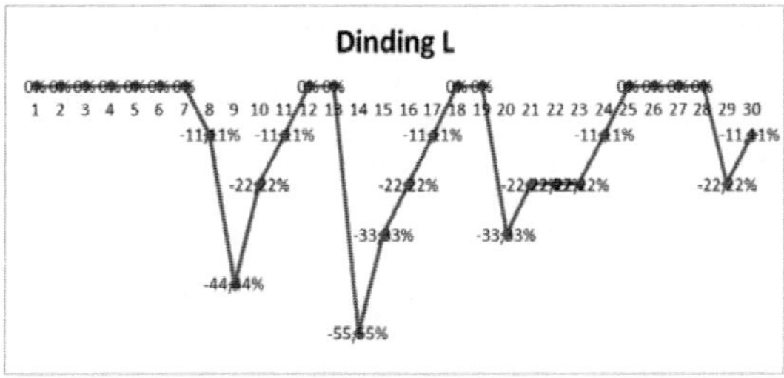

Figure 3. Graph of Error Robot Value on Wall Based Letter L

Table 2. Test Result Artificial Intelligence Robot movement on Wall T

Robot Position	Distance of Robot	Compatibility with set point	Error Value
L1	9-10 cm	Appropriate	0%
L2	9-10 cm	Appropriate	0%
L3	9-10 cm	Appropriate	0%
L4	9-10 cm	Appropriate	0%
L5	9-10 cm	Appropriate	0%
L6	11 cm	+2 cm	22,22%
B1	9-10 cm	Appropriate	0%
B2	7 cm	Appropriate	-22,22%
B3	7 cm	-2 cm	-22,22%
L7	9-10 cm	Appropriate	0%
L8	9-10 cm	Appropriate	0%
L9	9-10 cm	Appropriate	0%
L10	9-10 cm	Appropriate	0%
B4	5 cm	-4 cm	-44,44%
L11	6 cm	-3 cm	-33,33%
L12	7 cm	-2 cm	-22,22%
L13	8 cm	-1 cm	-11,11%
L14	9-10 cm	Appropriate	0%
B5	9-10 cm	Appropriate	0%
B6	6 cm	-3 cm	-33,33%
B7	9-10 cm	Appropriate	0%
L15	9-10 cm	Appropriate	0%
L16	9-10 cm	Appropriate	0%
L17	9-10 cm	Appropriate	0%

L18	9-10 cm	Appropriate	0%
B8	9-10 cm	Appropriate	0%
B9	5 cm	-4 cm	-44,44%
L19	6 cm	-3 cm	-33,33%
L20	8 cm	-1 cm	-11,11%
L21	9-10 cm	Appropriate	0%
L22	9-10 cm	Appropriate	0%
B10	9-10 cm	Appropriate	0%
B11	7 cm	-2 cm	-22,22%
B12	7 cm	-2 cm	-22,22%
L23	8 cm	-1 cm	-11,11%
L24	9-10 cm	Appropriate	0%
L25	9-10 cm	Appropriate	0%
L26	9-10 cm	Appropriate	0%
L27	9-10 cm	Appropriate	0%
L28	8 cm	-1 cm	-11,11%
L29	9-10 cm	Appropriate	0%
L30	9-10 cm	Appropriate	0%
The average value of error			-8%

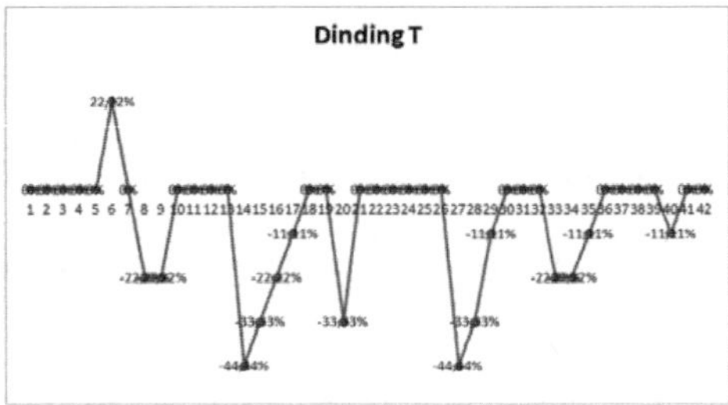

Figure 4. Graph of Error Robot Value on T-Shaped Wall T

Table 3. Test Result Artificial Intelligence Robot movement on Wall U

Robot Position	Distance of Robot	Compatibility with set point	Error Value
L1	9-10 cm	Appropriate	0%
L2	9-10 cm	Appropriate	0%
L3	9-10 cm	Appropriate	0%
L4	9-10 cm	Appropriate	0%
L5	9-10 cm	Appropriate	0%
L6	9-10 cm	Appropriate	0%
B1	9-10 cm	fit	0%
B2	7 cm	-2 cm	-22,22%
B3	8 cm	-1 cm	-11,11%
L7	8 cm	-1 cm	-11,11%
L8	8 cm	-1 cm	-11,11%
L9	9-10 cm	Appropriate	0%
L10	9-10 cm	Appropriate	0%
B4	5 cm	-4 cm	-44,44%
L11	5 cm	-4 cm	-44,44%
L12	7 cm	-2 cm	-22,22%

L13	9-10 cm	Appropriate	0%
B5	7 cm	Appropriate	-22,22%
L14	8 cm	-1 cm	-11,11%
B6	7 cm	-2 cm	-22,22%
B7	8 cm	-1 cm	-11,11%
B8	7 cm	-2 cm	-22,22%
L15	8 cm	-1 cm	-11,11%
L16	9-10 cm	Appropriate	0%
L17	9-10 cm	Appropriate	0%
L18	9-10 cm	Appropriate	0%
L19	9-10 cm	Appropriate	0%
B9	8 cm	-1 cm	-11,11%
L20	9-10 cm	Appropriate	0%
L21	9-10 cm	Appropriate	0%
L22	9-10 cm	Appropriate	0%
L23	9-10 cm	Appropriate	0%
L24	9-10 cm	Appropriate	0%
L25	9-10 cm	Appropriate	0%
L26	9-10 cm	Appropriate	0%
B10	9-10 cm	Appropriate	0%
B11	9-10 cm	Appropriate	0%
The average value of error			-8%

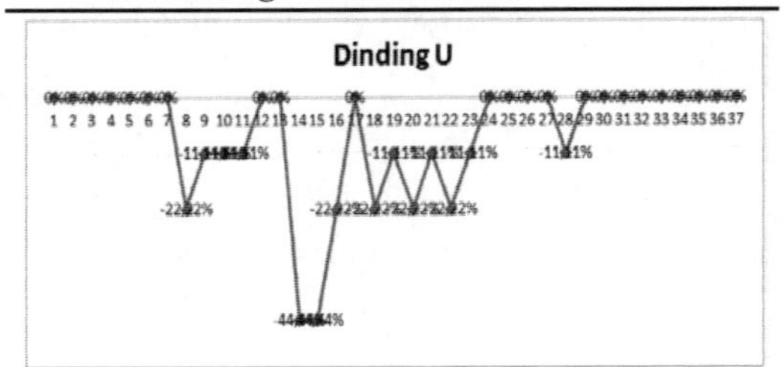

Figure 5. Graph of Error Robot Value on U-Shaped Wall

Table 4. Test results of Artificial Intelligence Robot movement on Square Wall

Robot Position	Distance of Robot	Compatibility with set point	Error Value
L1	9-10 cm	Sesuai	0%
L2	9-10 cm	Sesuai	0%
L3	11 cm	+2 cm	22,22%
L4	11 cm	+2 cm	22,22%
L5	9-10 cm	Sesuai	0%
L6	9-10 cm	Sesuai	0%
B1	9-10 cm	Sesuai	0%
B2	11 cm	+2 cm	22,22%
L7	9-10 cm	Sesuai	0%
L8	9-10 cm	Sesuai	0%
L9	9-10 cm	Sesuai	0%
L10	9-10 cm	Sesuai	0%
L11	6 cm	-3 cm	-33,33%

B3	9-10 cm	Appropriate	0%
B4	9-10 cm	Appropriate	0%
L12	9-10 cm	Appropriate	0%
L13	8 cm	-1 cm	-11,11%
L14	9-10 cm	Appropriate	0%
L15	9-10 cm	Appropriate	0%
L16	9-10 cm	Appropriate	0%
B5	9-10 cm	Appropriate	0%
L17	8 cm	-1 cm	-11,11%
L18	8 cm	-1 cm	-11,11%
L19	9-10 cm	Appropriate	0%
L20	9-10 cm	Appropriate	0%
L21	9-10 cm	Appropriate	0%
L22	9-10 cm	Appropriate	0%
L23	9-10 cm	Appropriate	0%
B6	9-10 cm	Appropriate	0%
B7	7 cm	-2 cm	-22,22%
The average value of error			-1%

Figure 6. Graph of Error Robot Values on Square-shaped Walls

Robot Reaction When Given Disisturbance

Figure 9 shown a robot image when given interference. The result of the robot can run stable following the wall.

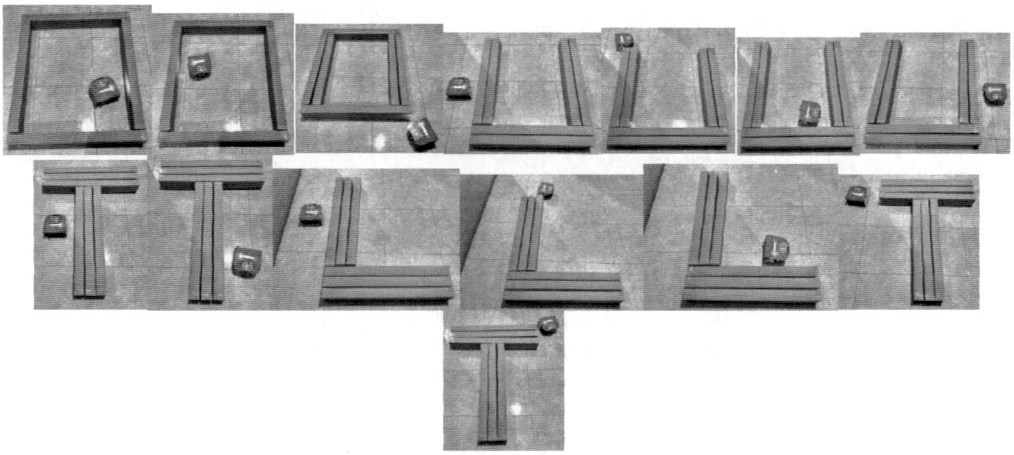

Figure 7. a robot image when given interference

IOP Conf. Series: Materials Science and Engineering **336** (2018) 012020 doi:10.1088/1757-899X/336/1/012020

5. Conclusion

Based on the results of research that has been done can be concluded that the Artificial Intelligence Robot can detect T-shaped wall produces an average error value of -8%. The U-shaped wall produces an average error value of -8%. On rectangular walls yield an average error value of -1%. So a better wall as a barrier wall is a square-shaped wall.

The robotic reaction when given a disorder of various patterns that is, the pattern of the square, the pattern of the letter U, the pattern of the letter T, and the pattern of the letter L can run balanced and not hit the wall.

References

[1] Budiharto Widodo. 2008. *Elektronika Digital + mikroprosesor.* Yogyakarta: Andi Offset.

[2] Hanafi Dirman dkk. 2012. *Artificial Intelligence Autonomous Robot Development ApplyingFuzzy Incremental Controller.* Universiti Tun Hussein Onn Malaysia. http://dx.doi.org/10.4236/ica.2013.41003. Pandjaitan Lanny. 2007. *Dasar-dasar Komputasi Cerdas.* Yogyakarta: Andi Offset.

[3] Saelan Athia. 2009. *Fuzzy logic.* Institut teknologi Bandung. Santoso Hari. *Cara Kerja Sensor Ultrasonic, Rangkaian, dan Aplikasinya.* Solichah Anna. 2016. *Rancang Bangun Simulator Sinyal Elektrokardiogram (EKG).* Surabaya: UNESA.

[4] Suharto Budi. 2013. *Sistem Navigasi Wall Following Robot Krpai Divisi Berkaki Menggunakan Kontroler Pid.* Universitas Brawijaya.

IOP Conf. Series: Materials Science and Engineering **336** (2018) 012021 doi:10.1088/1757-899X/336/1/012021

Early Leakage Protection System of LPG (Liquefied Petroleum Gas) Based on ATMega 16 Microcontroller

Sriwati[1*], Nur Ikhsan Ilahi[2], Musrawati[3],Syarifuddin Baco[4], Ansar Suyuti'Andani Achmad,Ejah Umrianah[5]

[1,2]Electrical Engineering , [3]Industri Engineering, [4] Informatic Engineering of Makassar Islamic University, [5]Electrical Engineering of Hasanuddin University. Perintis Kemerdekaan Street Km.9 No.29 Makassar

*sriwati.dty@uim-makassar.ac.id

Abstract. LPG (Liquefied Petroleum Gas). LPG is a hydrocarbon gas production from refineries and gas refinery with the major components of propane gas (C3H8) and butane (C4H10). Limit flame (Flammable Range) or also called gas with air. Value Lower Explosive Limit (LEL) is the minimum limit of the concentration of fuel vapor in the air which if there is no source of fire, the gas will be burned. While the value of the Upper Explosive Limit (UEL), which limits the maximum concentration of fuel vapor in the air, which if no source of fire, the gas will be burned. Protection system is a defend mechanism of human, equipment, and buildings around the protected area. Goals to be achieved in this research are to design a protection system against the consequences caused by the leakage of LPG gas based on ATmega16 microcontroller. The method used in this research is to reduce the levels of leaked LPG and turned off the power source when the leakage of LPG is on the verge of explosive limit. The design of this protection system works accurately between 200 ppm up to 10000 ppm, which is still below the threshold of explosive. Thus protecting the early result of that will result in the leakage of LPG gas.

1. Introduction

LPG (Liquefied Petroleum Gas). LPG is a hydrocarbon gas production from refineries and gas refinery with the major components of propane gas (C3H8) and butane (C4H10). At atmospheric pressure, a gaseous LPG, but for ease of distribution, LPG converted to a liquid phase by applying pressure. In liquid form, LPG is distributed in tubes or tanks.

LPG gas is in the tube in a liquid state and partly intangible steam. The ratio between the volumes of gas when evaporating gas in the liquid state varies depending on the composition, pressure and temperature, but is typically around 250: 1. The ability of the gas could be transformed into liquid is an excess of material that is gas volume can be reduced. Excess is applied mainly to store and deliver gas in the tank, which in this way is economically very profitable.

Limit flame (Flammable Range) or also called burst limit (Explosive Range) is the ratio of the mixture (in percentage) between the gas with air, where in that range may occur fire or explosion. To be able to fire or occur explosion, the magnitude of the ratio between gas vapor and air does not have a single value (number), but the values that have a lower limit and upper limit.

So if there is a mixture of gas and air in the range bottom and top value, there will be a fire or explosion. Lower flame limit value is also called the Lower Explosive Limit (LEL), is the minimum limit of the concentration of fuel vapor in the air where if there is no source of fire, the gas will be burned. While the upper limit value or Upper Explosive Limit (UEL), which limits the maximum

IOP Conf. Series: Materials Science and Engineering **336** (2018) 012021 doi:10.1088/1757-899X/336/1/012021

concentration of fuel vapor in the air where if there is no source of fire, the gas will be burned. Limit flame (Flammable Range) for propane is between 2.4% to 9.6%, while the butane between 1.9% to 8.6%. This means that for example, when a mixture of propane 2.4% with 97.6% air, then the mixture will be lit, but the amount of propane gas is a minimal amount. If the amount is less than 2.4% propane, the flame will not occur. And vice versa, if the amount is more than 9.6% propane also will not turn on, even though there are sources of ignition.

2. Fundamental theories

LPG (Liquefied Petroleum Gas) is made with gas, but for ease of distribution, LPG converted to a liquid phase by applying pressure. In liquid form, LPG is distributed in a tube or tank (Pertamina, 2012). The blast, which occurred on LPG gas cylinder caused by several factors such as leaks in tubing, valves, regulators, and accessories. In addition to not meet the standards, tools and supplies LPG gas tube was also damaged by the mixing activities from certain elements.

2.1. Gas Propane

Propane gas properties colorless, odorless, and flammable. This substance has a specific gravity of 1.56 and is heavier than air. An alkane series of hydrocarbons, with the formula C3H8. Made from crude oil, natural gas, and as a byproduct of the refinery cracking gases during petroleum refining. At higher temperatures, propane caught fire in the air, and produce carbon dioxide and water as end products. The atmosphere during combustion, occurs smoke production normally.

2.2. Butane Gas

This gas is colorless and has properties flammable. Butane has a specific gravity of 2.01 and is heavier than air. Butane is one of two hydrocarbon saturated with the chemical formula C4H10. In the n-butane (normal), the chain is continuous and unbranched whereas in i-butane (iso), one of the carbon atoms branched off to the side. Differences in the structure is small, but differ in terms of nature. N-butane melts at -138.3 ° C (-216.9 ° F) and boils at -0.5 ° C (31.1 ° F), and i-butane melts at -145 ° C (-229 ° F) and boils at -10.2 ° C (13.6 ° F). Both butane is formed from natural gas, petroleum, and gas refineries. They showed little chemical reactivity at ordinary temperatures, but it is easy to burn when ignited in air or oxygen. In the production of atmospheric combustion fumes.

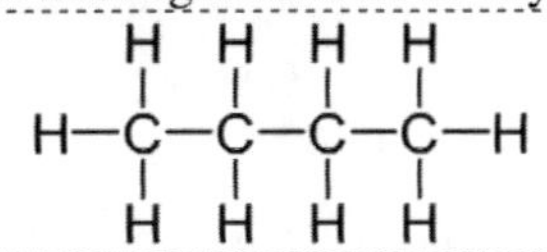

Figure 1. Propane Chemical Structure

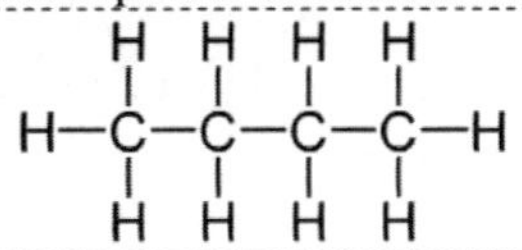

Figure 2. Butane Chemical Structure

2.3. LPG Gas Low Explosive Level

LPG gas leakage can occur through a **hose connection** that is not waterproof or the hose itself is porous pore so that it can be penetrated by a gas due to inadequate quality hose. This occurs through the valve of the tube itself that not fit on the stand or it could be like the writer ever experienced when cook gravy food that gone overflow, so as to put the fire in the stove. This causes the gas gush continue unchecked so that the gas fills kitchen.

In case of LPG gas leak in the kitchen, will be very dangerous because of the disadvantages that LPG is heavier than air. If the air is 1 per unit weight of the LPG , that means gas is 2 per unit weight.

The volume of gas may unexpectedly have accumulated and are in the explosive mixtures. LPG gas mixture to air up to 1.8% despite ignited or burned with a cigarette lighter flame that will not be an explosion or ignite. But at the **gas content between 1.8% - 10%, it will explode very severe if there is no source of fire or electric static.** In the **LPG content of 10%, then it will only light up.** That is because the nature of LPG gas mixture itself (SoehatmanRamli, 2005).

LPG explosion in the content of 1.8% - 10% are categorized perfectly, so it is very terrible destruction of the power mains, and also the blast power is depends on the amount of the mixture exploded. At the time of bursting, all of the oxygen in the area will be used up and becomes a vacuum, so that if there are people in the surrounding area, in addition got burns also be difficulty breathing. The surrounding buildings will bore the brunt of the air back and forth. **LPG exploded on**

the content of 1.8% to 10%, essentially was not followed by a fire. If followed by a fire means that the gas content is already 10% on (flammable) instead of an explosion.

2.4. Fire

There is some definitions of fire, as follows:

1. Fire is uncontrolled flame beyond human ability and desire (Ramli, 2010).
2. According to the Labor Department, fire is an exothermic oxidation reaction that progresses rapidly from a fuel accompanied by the emergence of fire or ignition (Wisaksono Department of Labour, 2011).
3. Fire is a chemical reaction involving rapid oxidation (combustion) of fuel (Fire Risk UK).

There are several theories that explain the fires, as follows:

a. Fire Triangle

According to this theory, fire occurred because of 3 factors into the fire element, namely:

- Fuel
- Heat source
- Oxygen

Fires can occur when all three elements of the fire is mutually react with each other (Ramli, 2010).

2.5. Tetrahedron Theory

The fire triangle theory then developed into the theory tetrahedron where a fire could occur if there is a fourth element called a chain reaction. Without a combustion reaction, the fire will not be able to live constantly (Ramli, 2010).

Source: fire-extinguisher-indonesia.com, December 2015

Figure 3. Fire Triangle

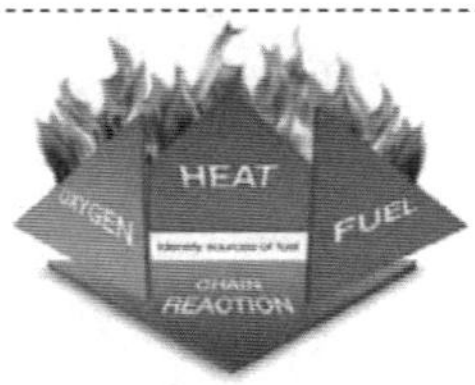

Source: CVCC, 2010

Figure 4. Tetrahedron Theory

2.6. ATMega16 Microcontroller

Microprocessor technology has been progressing. The same thing happened on microcontroller technology. If the previous microprocessor technology *complex instruction-set computing (CISC)* processors such as the Intel 386/486, then the microcontroller ATMEL uses types of MCS (AT89C51, AT89S51 and AT89S52).

After growing, technology microprocessors and microcontrollers to increase that occurred in the range of 1996 to 1998. ATMEL issuing microcontroller technology the most manifold AVR *(Alf and Vegard's Risc processor)* technology that uses a RISC *(Reduce instruction Set Computer)* with the advantages of more than its predecessor.

Microcontroller ATmega16 is a low-power CMOS 8-bit microcontroller based on the AVR enhanced RISC architecture. With control of the execution of the instructions in a single clock cycle, ATmega16 approaches 1 MIPS per MHz allowing the system designed to optimize processing speed than what it used to consume.

ATmega16 uses Harvard architecture with separate *bus* between the memories to maximize the ability and speed. Instructions in the program memory are executed with a *single level pipelining*.

When one instruction is executed, the next instruction is taken from the program memory. This concept led to the instructions executed in each *clock cycle*. CPU consists of 32 x 8 *general purpose registers* that can be accessed quickly in a single clock cycle, resulting in operation *Arithmetic Logic*

IOP Conf. Series: Materials Science and Engineering **336** (2018) 012021 doi:10.1088/1757-899X/336/1/012021

Unit (ALU) can be performed in a *single cycle*. In ALU operation, two operands come from the register, then the operation is executed and the results are stored back in the register in one clock cycle. Arithmetic and logic operation in ALU will change the bits that contained in *Status Register* (SREG). The process of taking instructions and executing instructions are running in parallel, and can be seen in the Figure below:

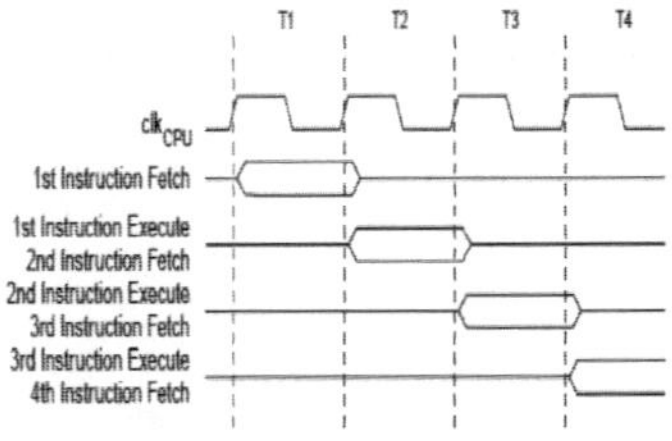

Figure 5.The Process of Taking Instructions and Executing Instructions in Parallel

2.7. LPG MQ-6Sensor

To keep scanning the gas from LPG, ATmega16 microcontroller requires external devices such as an input device or a sensor input module LPG MQ-6. This sensor will read the content of LPG gas in the air.

MQ-6 has a gas sensor used to detect LPG, iso-butane, and propane with high sensitivity. MQ-6 gas sensor has a small sensitivity to a drug alcohol and cigarette smoke. MQ-6 gas sensor is a sensor that has a rapid response to the LPG. The sensor is also stable, durable, and can be used in a simple drive circuit.

Figure 6.LPG MQ-6 Gas Sensor

Source:http://www.etlogics.com/LPG_Gas_Sensor_MQ-6.html,2015

2.8. The ISIS Proteus 7 Professional Application

Proteus ISIS application is an electronic schematic creation program that can simulate circuit. Proteus ISIS program can also be used to design printed circuit boards (PCB). However, the writer only use Proteus ISIS as a maker of application circuit schematic and simulate it.

To use Proteus ISIS, it must be installed by following the installation prompts. How to use Proteus ISIS as follows:

a. Run the application on the computer by selecting the application's shortcut or menu in Windows.
b. Create a file by clicking "new design".
c. Find the required key componentsin the button P.
d. Arrange the components and create lines by connecting feet of components schema.

2.9. The PCB Wizard Application

The PCB Wizard Application is an application for designing printed circuit boards (PCB).

IOP Conf. Series: Materials Science and Engineering **336** (2018) 012021 doi:10.1088/1757-899X/336/1/012021

3. Research method

3.1. Hardware Design

Flowchart Research

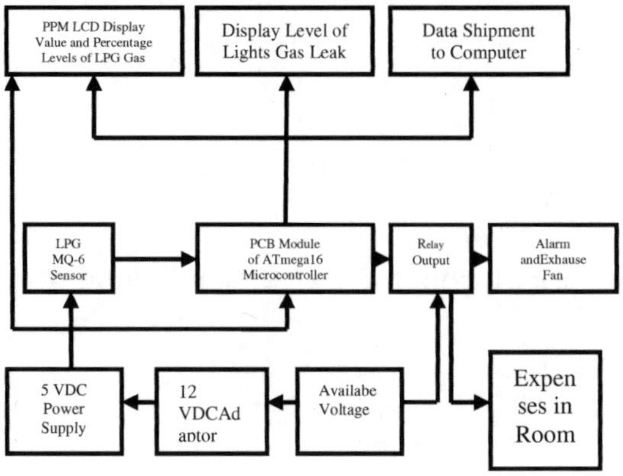

Figure 7. Flowchart Research

3.2. Software Design

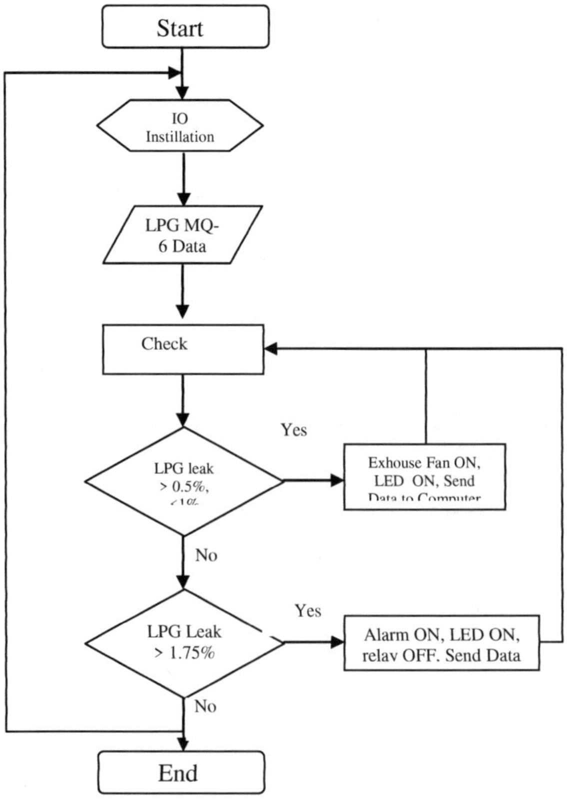

Figure 8. Software Design

IOP Conf. Series: Materials Science and Engineering **336** (2018) 012021 doi:10.1088/1757-899X/336/1/012021

4. Result and analysis

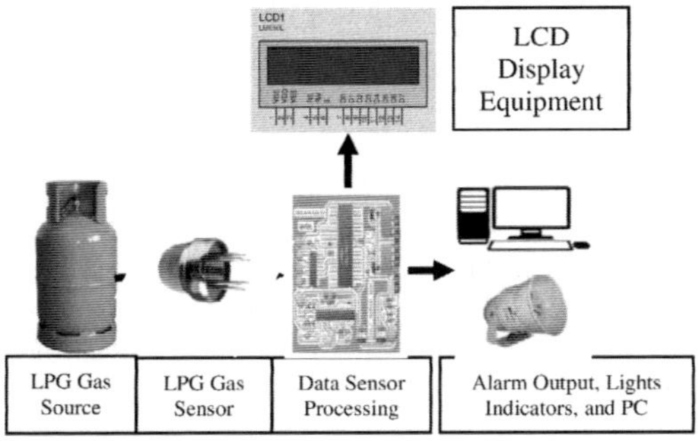

Figure 9. LPG Gas Detection Groove

LPG Gas scannings taken through the LPG (MQ-6) sensor data in the form of analog voltage is then converted into digital data 10 bit value from 0 to 1023. From this data, it will be the initial data to perform the calibration process that embedded within the microcontroller program ATmega16.

1. PCB Designing

PCB Designing uses PCB Wiz Application.

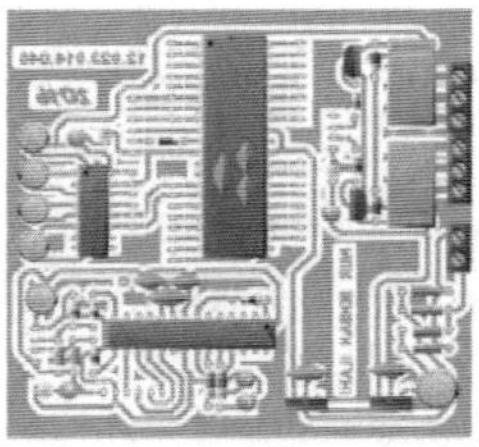

Figure 10. Component Llayouts and PCB Track

2. Circuit Power Supply Testing

Measurement aims to determine the power supply output voltage generated by the power supply, when it has no load and when connected to a load. It is also useful as a tool for sensor calibration calculations of MQ-6. Power supply is used to provide voltage to the sensor MQ-6, the minimum system microcontroller, and LED & LCD. From the results of 5 times tests, the result can be seen in the table.

The results of each output measured voltage when using a power supply load and no-load, that the value is found eligible for the supply voltage to the existing circuit. And measures can be seen in the Figure below.

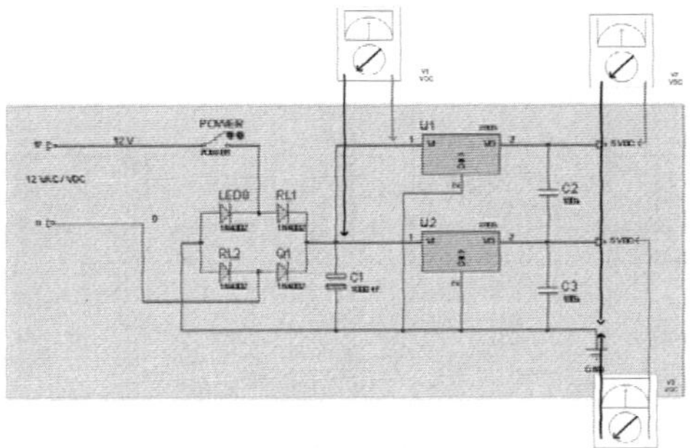

Figure 11. Power Supply Measurement

The Figure above is the power supply voltage measurement points that are entered into the circuit system. Measurement aims to determine whether good or bad the power supply to use.

Table 1. Power Supply Measurement without Load

Measurement	**Voltage 1 (V1)** *Output DC*	**Voltage 2 (V2)** *Output DC*	**Voltage 3 (V3)** *Output DC*
1	12,71	5,00	4,98
2	12,36	5,01	4,99
3	12,16	4,98	5.00
4	12,03	5,00	5,00
5	11,97	4.89	5,01

Source: Primary Data, 2016

The table above is a measurement table results without using a power supply load. The expenses in the form include sensors, LCD, microcontroller ATmega16, indicator lights, and relay output.

Table 2.Power Supply Measurement with Load

Measurement to	**Voltage 1 (V1)** *Output AC*	**Voltage 2 (V2)** *Output DC*	**Voltage 3 (V3)** *Output DC*
1	12,61	4,90	4,82
2	12,27	4,85	4,94
3	12,34	4,89	4,95
4	12,23	4,93	4,89
5	12,03	4,87	4,90

Source: Primary Data, 2016

The table above measurement table results using a power supply with a load, which aims to determine the voltage drop of the power supply, and to determine whether the power supply available to supply the used load.

Table 3. Data Graph,Data Sheet

No	**Data from Data Sheet Graphic**		**Percent(%) LPG**
	PPM	**Rs / Ro**	
1	200	2	0,02
2	300	1,8	0,03
3	400	1,6	0,04
4	500	1,5	0,05
5	600	1,4	0,06
6	700	1,3	0,07
7	800	1,2	0,08
8	900	1,1	0,09
9	1000	1	0,1
10	2000	0,75	0,2
11	3000	0,65	0,3
12	4000	0,56	0,4
13	5000	0,50	0,5
14	6000	0,48	0,6
15	7000	0,45	0,7
16	8000	0,43	0,8
17	9000	0,40	0,9
18	10000	0,30	1.0

Source: Primary Data,2016

The data tables above represent primary data taken from the data sheet of LPG sensor MQ-6 which became the initial data to perform the calibration process sensors in the microcontroller program.

Searching for Ro value in clean air

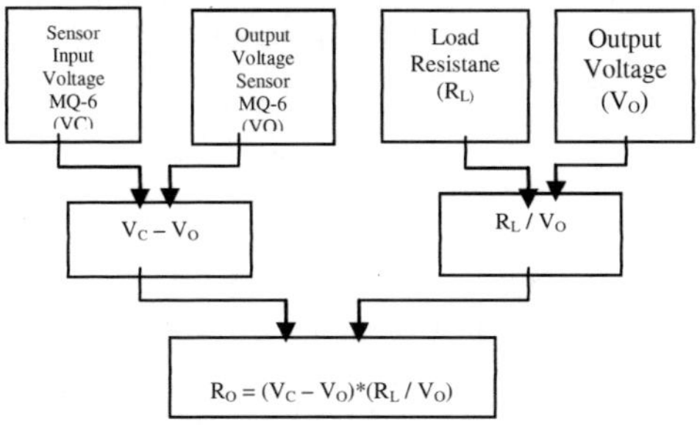

Figure 12. Searching for R_O Value in Clean Air

Ro = (Vc - Vo) * (RL / Vo)
Ro = (4,84 – 0,35) * (20000 / 0,35)
Ro = (4,49) * (57142,85)
Ro = 256571,40 Ω

Ro is a sensor resistance value when the condition is worth Rs 1000 ppm. The explanation of condition 1000 ppm is currently normal air conditions that is not mixed by propane gas and butane gas, because the state of 1000 ppm is the initial state or a minimum of Rs before detecting the levels of the gas, so that if the initial conditions, the RL produced was 0.35 volts. Rated voltage of 0.35 volts taken from a minimum value issued by the sensor currently worth Rs 1000 ppm. Vc formula is used to activate the circuit. Formula RL is resistor used this tool to measure the RL at 20K.

Ro value has been obtained, that is 256,571.40 Ω, resulting from the Ro value can be used to find the value of Rs/Ro. With the value of Rs, the condition of the sensor can detect a wide variety of gas levels in air. Therefore, The Bascom Program states as follows:

Ro = 256571.40
Voltage = Gas_ref/1023
*Voltage = Voltage *4.84*
X = 4, 84 – Voltage
Y = X / Voltage
*Rs = Y * 20000*
D = Rs / Ro

D is value of Rs/Ro where the D value to be converted into the PPM via a conversion formula data from the data sheet that provides charts for LPG concentration in ppm against Rs/Ro, to the condition of 20 C and 65% RH. If it matches the curve to a given value, it was found that:

Rs/Ro = 18,446 * (ppm LPG) ^ -0,421

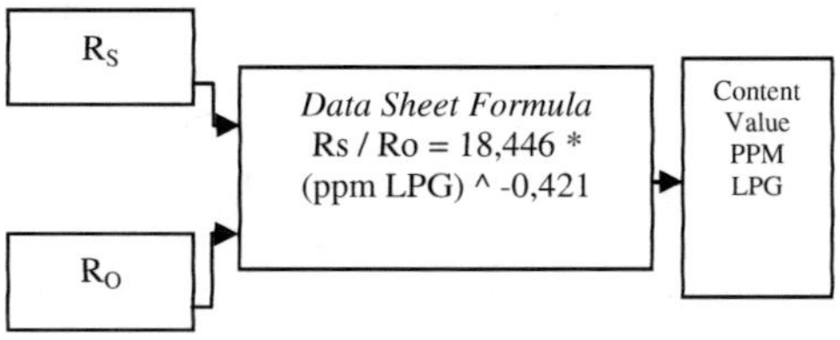

Figure 13. Sensor Calibaration

From the above conversion process, it was discovered the formula to change the value of the voltage that has been converted to Rs and Ro into PPM. The formula is as follows:

PpmLPG=[(Rs/Ro)/18,446]^(1/-0,0421)

In the Bascom program, AVR Rs/Ro is defined as variable D. And the result of the division between 24.83 to Rs/Ro defined as variables E. The following is how Bascom explains the formula.

$D = Rs / Ro$
$E = D / 18{,}446$
$Ppm = E \wedge (-2.375)$

PPM levels have been obtained using a formula that is displayed on the LCD screen that to be calibrated again by adjusting the data sheet chart, whether or not formula display value of Rs/Ro compared with PPM gas levels as follows.

From the table above, it is found the increment ppm conduction chart data with the results of the calculation formula. Minimal difference between the value chart with the formula is 0.41% and the maximum difference of 5.87%. This was due to a lack of accuracy in the data collection charts. Means that the formula can be used in the calibration program but with the accuracy ± 6%.

3. Program Making

The program making will be embedded in the microcontroller using BASCOM-AVR and Khazama AVR application.

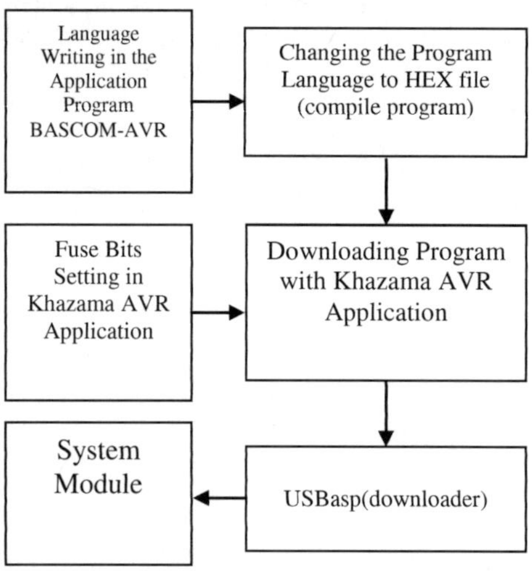

Figure 14. Charging Program of Microcontroller ATmega16

Table 4. Graphic Data and Formula Comparison

No	Data Graph Data Sheet		Data With Formula		Difference	Percent
	Rs / Ro	PPM	Rs / Ro	PPM		
1	2,00	200	2,00	195,691	4,30	2,15
2	1,70	300	1,70	287,873	12,13	4,04
3	1,50	400	1,50	387,526	12,47	3,11
4	1,35	500	1,35	497,708	2,29	0,46
5	1,25	600	1,25	597,526	2.47	0,41
6	1,17	700	1,17	699,159	0,84	0,12
7	1,10	800	1,10	809,487	9,49	1,19
8	1,05	900	1,05	904,052	4,05	0,45
9	1,00	1000	1,00	1015,121	15,12	1,50
10	0,75	2000	0,75	2010,237	10,23	0,51
11	0,65	3000	0,65	2823,895	176,11	5,87
12	0,56	4000	0,56	4023,196	23,20	0,57
13	0,50	5000	0,50	5265,795	265,80	5.32
14	0,48	6000	0,48	5801,894	198,11	3,30
15	0,45	7000	0,45	6762,979	237,02	3.39
16	0,42	8000	0,42	7967,107	32,89	0,41
17	0,40	9000	0,40	8945,925	54,07	0,60
18	0,38	10000	0,38	10104,890	104,89	1,05

4. Charging Program

Steps charging module using the application program to KHAZAMA AVR.

a. Connect the USB cable to the module and computer.
b. Open KHAZAMA AVR application.
c. Select ATmega16 microcontroller.

IOP Conf. Series: Materials Science and Engineering **336** (2018) 012021 doi:10.1088/1757-899X/336/1/012021

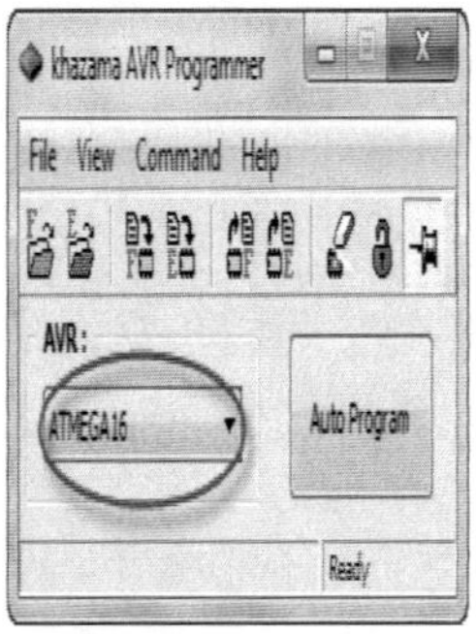

Figure 15. Choosing the Type of Micorcontroller

Operation of Equipment/System

The system works by detecting levels of LPG gas which is in the room source/engine using LPG gas. If there is a LPG gas is detected, the tool will first be issued the indicator of LPG PPM value on LCD, and if it increases, the leakage indicator light will turn on the appropriate level. Furthermore, leakage of data will be sent to the computer via Hyper Terminal and if the leak continues to rise, then try to flutter exhaust fan to reduce the levels of LPG in the room. However, if the LPG gas levels continue to rise and reach the threshold of explosive/fire, then the system will shut off the power source and exhaust fan, and also turn on the alarm.

5. Conclusion

Based on the result of design and research, it can be concluded as follows:

1. System tools have been successfully applied to the methods of reducing levels of gas in the room.
2. If the sensor detects the MQ-6 LPG gas at a certain level, the system displays on the computer, turn on the alarm, and turn off the power source.
3. Calibration sensor MQ-6 inserts into programming microcontroller ATmega16 for an accurate scanning.
4. A maximum load of alarm and exhousefan 1 A for 120V specified relay system output.

Suggestion

Further suggestion:

1. Installation of equipment is carried out outdoors given the equipment is not immune to the explosion, as well as the equipment did not follow to die with the death of the power source in the room.
2. Installation of LPG gas sensor MQ-6 placed as close as possible to the source or sources of LPG gas is most vulnerable to leakage.
3. Use exhouse fan larger for a big room.
4. Replace relay output for bigger output power use.

References

[1] https://www.sparkfun.com/datasheets/Sensors/Biometric/MQ-6.pdf, 2015), diakses Januari 2017.

[2] Praveen Kumar , R. Kaushalendra Rao , N. Hemalatha Reddy, Sustained uptake of LPG as cleaner cooking fuel in rural India: Role of affordability, accessibility, and awareness, World Development Perspectives 4 (2016) 33–37

[3] https://www.wikiwand.com/id/Propana, diakses Desember 2016.

[4] https://www.savvymicrocontrollersolutions.com/index.php?sensor=mq-gas-sensors, diakses Desember 2016

[5] Bambang Eko Soemarsono.1, Evi Listiasri1, Gilang Candra. (2015) *Early Detection*

IOP Conf. Series: Materials Science and Engineering **336** (2018) 012021 doi:10.1088/1757-899X/336/1/012021

Equipment against LPG JURNAL TELE Volume 13 Nomor 1 Edisi Maret 2015

[6] Dian Sartika K. 2012. Consequential Analysis Of Gas Distribution, Fire, and Explosion due to 12 Kg LPG Tubes In the Manggarai Southware Village In 2012 By Using Breeze Incident Analyst Software . Final Project Report Published 2012. Universitas Indonesia

[7] Sidik Nurcahyo. (2012). *Aplikasi dan Teknik Pemprograman Mikrokontroler AVR Atmel.*Penerbit Andi

[8] AfrieSetiawan.(2011) *20 Aplikasi mikrokontroler Atmega 8535 & Atmega 16 Menggunakan Bascom-AVR.* Penerbit Andi

[9] HeriAndrianto. (2008). *Pemprograman Mikrokontroler AVR Atmega16 Menggunakan Bahasa C (CodeVision AVR).* Penerbit Informatika

[10] Z. Ning, T.L. Chan, On-road remote sensing of liquefied petroleum gas (LPG) vehicle emission measurement and emission factors estimation, Atmospheric Environment 41 (2007) 9099–9110

[11] Suhata, ST. 2005. *Aplikasi mikrokontroler sebagai Pengendali Peralatan Elektronik Via Line Telepon.* PT Elex Media Komputindo. Jakarta

[12] L.C. Shirvill, Efficacy of water spray protection against propane and butane jet fires impinging on LPG storage tanks, Journal of Loss Prevention in the Process Industries 17 (2004) 111–118

IOP Conf. Series: Materials Science and Engineering **336** (2018) 012022 doi:10.1088/1757-899X/336/1/012022

Classification of Children Intelligence with Fuzzy Logic Method

Syahminan*, Permata ika Hidayati

University Kanjuruhan Malang , Indonesia

*syahminan@unikama.ac.id

Abstract. Intelligence of children s An Important Thing To Know The Parents Early on. Typing Can be done With a Child's intelligence Grouping Dominant Characteristics Of each Type of Intelligence. To Make it easier for Parents in Determining The type of Children's intelligence And How to Overcome them, for It Created A Classification System Intelligence Grouping Children By Using Fuzzy logic method For determination Of a Child's degree of intelligence type. From the analysis We concluded that The presence of Intelligence Classification systems Pendulum Children With Fuzzy Logic Method Of determining The type of The Child's intelligence Can be Done in a way That is easier And The results More accurate Conclusions Than Manual tests.

1. Introduction

Children are the heart of every parent of all the attention and affection devoted to the delight of the baby, sometimes all the child's desire is always filled to please his baby, the role of parents greatly affect the development of children's abilities in the hope that the child grows with the ability in Want, look smart and smart (smart visibility) This is the old faithful hope.

In this sense intelligence is the inferred entity, term or construct we use to explain the difference in behavior to predict differences in future behavior. Has repeatedly emphasized the fact that all tests, IQ tests will produce different intelligences, when taken from a sample test that [1]

Therefore, the development of children is very important to note while the level of ability of children have diverse. No exception in children. Today at an early age have been prepared to interact globally with the changes. Therefore, the child's intelligence fertilization often

This means that intelligence or skill is only understood as the ability to solve problems mathematically according to existing theory. Of course, the assumption about this mathematical ability raises the logical consequence of teachers in assessing their students using IQ parameters only. So the impact on the provision of the highest awards on students who have logical intelligence (*"smart logic")*. In contrast, children who are unskilled or unskilled in mathematics are often regarded as lacking in intelligence, so the most fundamental question is whether if a student gets a score of 0 in a mathematics course, can it be concluded that the student is stupid (IQ squatting)? (Agustin Leony: 2)

1.1. Purpose

The purpose of this study with the topic "Grouping of Children's intelligence, with Fuzzy Logic approach" in early childhood between the ages of 5 to 12 this year are:

1. The realization can know the level of intelligence of a child at an early age.

IOP Conf. Series: Materials Science and Engineering **336** (2018) 012022 doi:10.1088/1757-899X/336/1/012022

2. As input for decision making with regard to child intelligence type and how to optimize its potential.

2. Fuzzy Logic Method

Fuzzy Logic is an appropriate way to map an input space into an output space. One example of such mapping is in graphical form as shown on the picture below explains that between the input and output there is a link that must be mapped *input* to *output* accordingly.

Representation of the Triangle Curve is basically a combination of two linear lines as shown at figure 1.

$$\mu[x] = \begin{cases} 0; & x \leq a \\ (x - a) / (b - a); & a < x \leq b \\ (c - x) / (c - b); & b < x < c \\ 1; & x \geq c \end{cases}$$

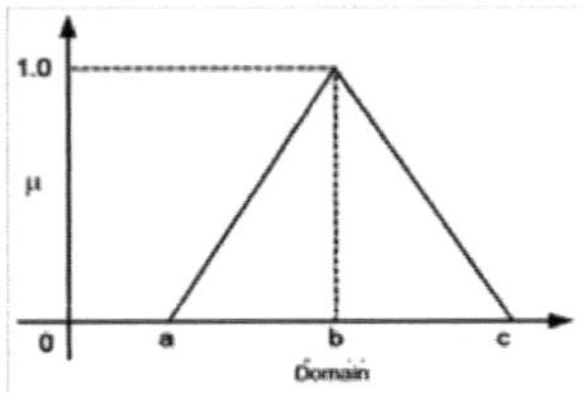

Figure 1. Representation of the Triangle Curve

2.1. Membership Function on Fuzzy Logic

Membership function *(MF)* is a curve that shows the mapping of points *of input* data into membership values (often called the degree of membership) which has the interval between 0 and 1 (Sri Kusumadewi, 2004: 160). One way that can be used to get a membership value is through a function approach. Some commonly used functions include linear representation, triangular curve representation, trapezoid curve representation, shoulder curve representation, curve S representation, and bell curve representation. The trapezoid curve is basically a triangular shape, but there are some points that have a membership value of 1.

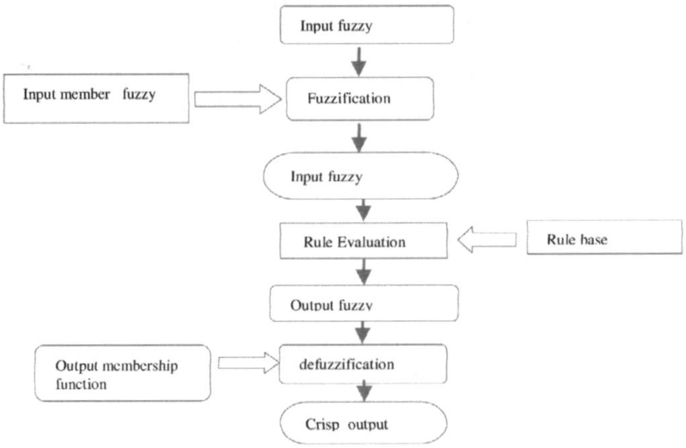

Figure 2. Flowchart System

IOP Conf. Series: Materials Science and Engineering **336** (2018) 012022 doi:10.1088/1757-899X/336/1/012022

2.2. Fuzzyfication

Fuzzyfication is the process of mapping the input values (crisp input) derived from a controlled system (non fuzzy quantity) into the fuzzy set according to its membership function

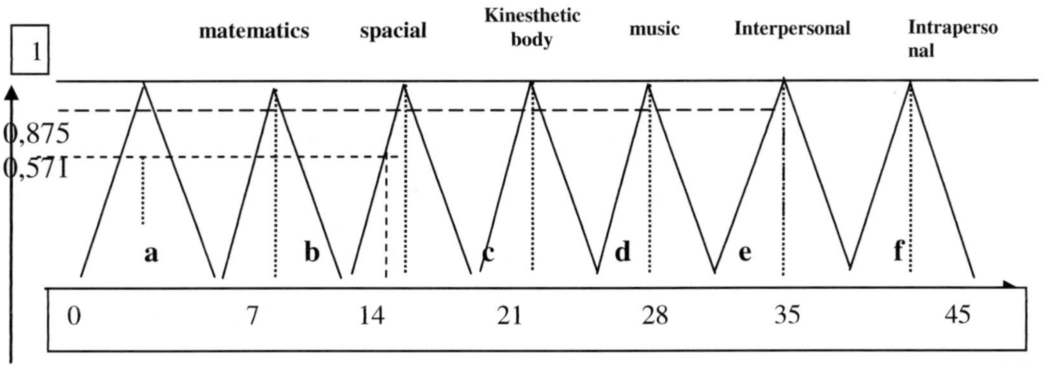

Figure 3. the trapezoid curve

Based on the trapezoid curve in Figure 3 above, the fuzzy logic membership function can be written as follows

$x \leq a$ atau $x \geq d$

$(x-a) / (x-b)$;

$a \leq x \leq b$

1. $b \leq x \leq c$

$(d-x) / (d-c)$; $x \geq d$

2. $c \leq x \leq c$

$(c-x) / (d-c)$; $x \geq d$

3. $c \leq x \leq e$

In the book called Master Faster, Collin Rose [4] noted that in 1983 Howard Gardner *Theory of Multiple Intelligence* displays that strengthen the perspective of human cognition. Here's a description of seven intelligences according to Gardner:

1. Linguistic Intelligence *(Linguistic Intelligence)*
   Linguistic intelligence is the ability to think in terms of words and use language to express and appreciate complex meanings.
2. Logical-Mathematical intelligence (mathematical *Logical- Intelligence)*
   Logic-mathematical intelligence is the ability to calculate, measure and consider propositions and hypotheses, as well as complete mathematical operations
3. Spatial intelligence *(Spatial Intelligence)*
   Spatial intelligence awakens the capacity to think in three dimensional ways. This intelligence enables one to feel the external and internal shadows, re-paint the shadows, and objects through the room, which can produce information graphically.
4. Kinaesthetic Intelligence Body *(Bodily-kinaesthetic Intelligence)*
   Kinaesthetic intelligence allows one to move objects and subtle physical skills. And accustomed to express themselves through the movement of the body.
5. Musical intelligence *(Musical Intelligence)*
   Musical intelligence is clearly visible in a person who has a sensitivity to the pattern of pitch, melody, rhythm, and tone. Not only with composers or musicians, but also sensitive listeners.
6. Interpersonal intelligence *(Interpersonal Intelligence)*

Interpersonal intelligence is the ability to understand and interact with others effectively. As western culture begins to introduce the relationship between mind and body, it is necessary to realize again the importance of the value of expertise in interpersonal behaviour.

7. Intrapersonal intelligence *(intrapersonal Intelligence)*
   Intrapersonal intelligence is the ability to make accurate perceptions about oneself and use such knowledge in planning and directing one's life. Some individuals who have this kind of intelligence are theologians, psychologists, and philosophers.

2.3. Children's Intelligence

A person's intelligence can be detected at the age of the children. Because at an early age that the ability to think a child is growing. Benefits of detecting children's intelligence from an early age include:

1. Efficient in choosing readings, toy tools and courses to suit his talents.
2. Focus and more details on the child's primary needs.
3. Confident and more confident of his superior talent as well as not feeling down for the weak talent.
4. More definitely looking for an environment that suits his superior talent. Planned early career

The place chosen to make this observation is on the learning guidance agency Primagama Quantum Kids Sulfate, Insan Permata Kindergarten Malang, and Early Childhood Education (PAUD) Permata Iman Malang.

With the number of students who have done the research as many as 10 students was already provided

Table 1. Description Characteristics and Weighting the value of intelligence of children

No.	Characteristic features	Check
1	Hear / respond to every sound, colour, word expression	
2	Learn through listening, reading, discussion	V
3	Understand, interpret what has been read well	V
4	Speak effectively to various listeners	
5	Great interest in journalism, poetry, story telling	V
6	Easy to remember melodies of songs that are listened to, following the rhythm of music	V
7	Achieving in the field of music or can play a musical instrument	
8	More can or like to learn with in music accompaniment	
9	Can create original compositions using certain musical instruments	
10	Feel the keen sensitivity to visual detail, balance, colour, line, shape and space	
11	Uses visual representation as a tool to remember information	
12	Creating a new form of visual-spatial media or artwork	
13	Loved the habit of seeing movies, slides and pursue fields	

3. Research result

Here is the result of fuzzy search using mamdani

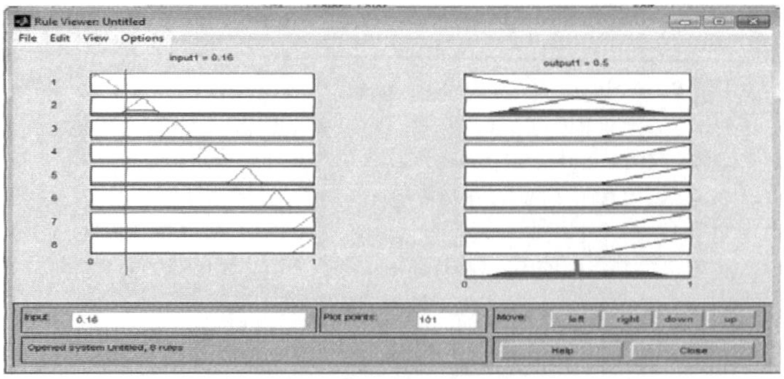

Figure 4. fuzzy grouping

Table 2. Testing Results Child Intelligence Grouping System

No	Name	Age	Intelligence Fuzzy logic method with matlab approach
1	Salsa	5	Linguistics
2	Octa	5	Interpersonal
3	Dhana	6	Linguistics
4	Dhani	6	Music
5	Aura	6	Music
6	Farah	5	Spatial
7	Khansa	6	KinestetikTubuh
8	Soraya	6	Math logic
9	Hildan	5	Math logic
10	Above	6	KinestetikTubuh
11	Fahri	5	Linguistics

4. Conclusion.

After analyzing the program and testing the object, it can be concluded as follows:

Child intelligence grouping system by using fuzzy logic method can do the reasoning with fuzzy logic method to determine child intelligence type with percentage of truth test result 80%. With the data accuracy error of 20% with the percentage calculation as follows:

From the number of data inputs of 100 true children of 80 then obtained 80/100 = 0.8x100 then obtained the results as mentioned above

References

[1] Verauli R, Measuring child intelligence, Vol.2 juli 2009
[2] Sujono YN, The basic concept of early childhood education, 5th Jakarta Indonesia: Gramedia 2009.
[3] Agustin Leoni, Smart Psychotest and TPA, August, Bandung, Gramedia, 2008
[4] Srikusumadewi, *Artificial Intelligence,* Yogyakarta, Central Java: Gramedia 2007
[5] Ari Firmanto, "intelligence, creativity, *task commitment* as preditor student achievement," Journal of the Science and Practice of Psychology, Vol.1, No. 2303-2936, pp. 26-36, 2013.
[6] Vijay K Mago, *"Supporting meningitis diagnotis amongst infants and Children Through the use of fuzzy," BMC Medical Informatics,* Val ix, No. 12, p.12, September 2012.
[7] *Collin Rose,* "Master Faster" Jakarta Pustaka 2010
[8] M. Ali Ramdani, "Jakarta's decision support system": Andi ofset 2002

.

IOP Conf. Series: Materials Science and Engineering **336** (2018) 012023 doi:10.1088/1757-899X/336/1/012023

Model Multi Criteria Decision Making with Fuzzy ANP Method for Performance Measurement Small Medium Enterprise (SME)

E Rahmanita[1*],V T Widyaningrum[2], Y Kustiyahningsih[3], J Purnama[4]

[1]Informatic Engineering, Faculty of Engineering, Trunojoyo University.
[2]Mechatronic Engineering, Faculty of Engineering, Trunojoyo University.
[3]Information System, Faculty of Engineering, Trunojoyo University.
[4]Industrial Engineering, Department of Engineering, UNTAG, Surabaya.

*ezarahmanita@gmail.com

Abstract. SMEs have a very important role in the development of the economy in Indonesia. SMEs assist the government in terms of creating new jobs and can support household income. The number of SMEs in Madura and the number of measurement indicators in the SME mapping so that it requires a method.This research uses Fuzzy Analytic Network Process (FANP) method for performance measurement SME. The FANP method can handle data that contains uncertainty. There is consistency index in determining decisions. Performance measurement in this study is based on a perspective of the Balanced Scorecard. This research approach integrated internal business perspective, learning, and growth perspective and fuzzy Analytic Network Process (FANP). The results of this research areframework a priority weighting of assessment indicators SME.

1. Introduction

Currently, the number of SMEs in the city of Bangkalan Madura reached more than 125 thousand units. SMEs provide opportunities for economic growth, reduced unemployment, and poverty. It absorbs approximately 79.04 million workers or 99.4% of the total workforce (Bank Indonesia, 2006). SME productivity is also an important factor affecting the progress of SMEs, increasing the income of the community, and play a role in realizing national stability.There are many indicators or criteria in determining performance measurement and business strategic plan, so it takes a decision-making method many criteria. The method used in this research is Fuzzy Multi Criteria Decision Making (FMCDM). Fuzzy Method Multi-criteria is a method used to determine the criteria weighting of several alternatives based on one's assessment.

Fuzzy Multi Criteria Decision Making (FMCDM)method used in this research is Fuzzy Analytical Network Process (FANP). Analytic Network Process Method (ANP) is a method that can represent the importance of various parties by considering the interconnection between existing criteria and sub-criteria [1][2]. ANP is a development of AHP, has a more complex system of analysis and the consistency of index in the assessment of questioner [3]. Previous research has been done performance measurement according to indicator contained in balance Scorecard [3]. In this study more detail about the measurement of risk to SMEs. There are several performance assessment criteria: employee training, owner education, owner training, net income, sales transactions, ownership (shop), variations of batik motive, employee maturity, buyer satisfaction, preferred motives, raw material price increases, weather conditions, production. There are difficulties during qualitative performance measurements such as preferred motives, weather conditions, and others. In measuring qualitative factors using fuzzy numbers make decisions easier and get more realistic results [4].

This research, FANP method is used to determine the best criteria and decision-making based on the existing criteria, both qualitative and quantitative. The use of Fuzzy in this study to accommodate the vague nature of decision-making by giving consideration that can overcome the uncertainty in qualitative criteria [4] [5].

2. Research Method

2.1. Fuzzy Theory

Fuzzy theory is used to overcome uncertainty due to imprecision and vagueness. Fuzzy contributed his ability to represent vague data[6]. The set is marked by a membership function (characteristic), which assigns each object a membership value ranging between zero and one [7]. A triangular fuzzy number(TFN), x is shown in Fig. 2.1. A TFN is denoted simply as (l/m, m/u) or (l, m, u). The parameters l, m and u, respectively, denote the smallest possible value, the most promising value, and the largest possible value that describe a fuzzy event. Membership Function (MF) or the degree of membership is a curve showing the mapping point of data input into the value of membership.

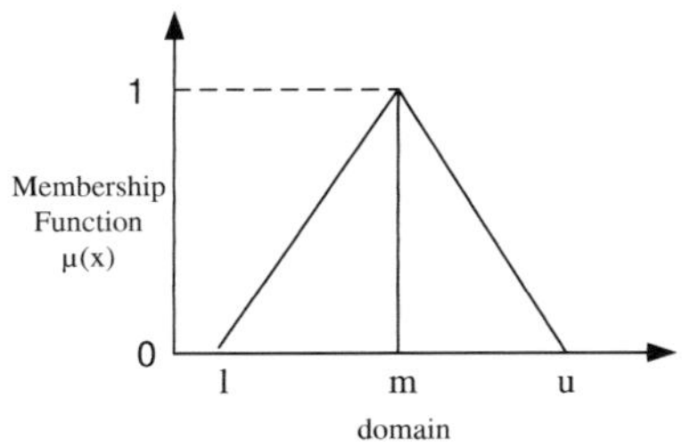

Figure 1. Triangular Fuzzy Number

Fuzzy Membership Function [4]

$$\mu(x) = \begin{cases} 0; & Otherwise \\ (x-l)/(\mathrm{m}-l); & l \le x \le m \\ (\mathrm{u}-x)/(\mathrm{u}-m); & m \le x \le u \end{cases} \quad (1)$$

2.2. Fuzzy Analytical Network Process (FANP)

Fuzzy ANP method is applied for an extension of the AHP and ANP by combining the fuzzy set theory. In the ANP Fuzzy, Fuzzy ratio scale used to indicate the relative strength of the factors on which the relevant criteria. The fuzzy decision so that a matrix can be formed. Kahit of alternatives are also presented in the Figures Fuzzy [8]. Based Chang each object of each criterion and sub-criteria to be considered and extend the analysis to obtain a goal executed. This means it is possible to obtain the analysis which can extend the value indicated by the notation as follows [9].

$$\mathrm{M}^1_{gi}, \mathrm{M}^2_{gi}, \mathrm{M}^3_{gi}, \mathrm{M}^4_{gi}, \mathrm{M}^5_{gi} \ldots \mathrm{M}^m_{gi} \quad (2)$$

Set asa goal (1,2,3,…,m), and M^j_{gi} (j= 1,2,3,...,m) areTriangular Fuzzy Number, after identifying initial assumptions, extend the analysis of Chang can be described with the following stages:

1. Decision Modeling with fuzzy ANP by specifying criteria and sub-criteria.

2. Development B pairwise comparison matrix between all the elements / criteria, sub-criteria of the fund each dimension criteria in a hierarchical system based on an assessment of linguistic variables.

$$B=\begin{bmatrix} & C_1 & C_2 & C_3 & \dots & C_n \\ C_1 & 1 & b_{12} & b_{13} & \dots & b_{1n} \\ C_2 & b_{21} & 1 & b_{23} & \dots & b_{2n} \\ C_3 & b_{31} & b_{32} & 1 & \dots & b_{3n} \\ M & M & M & M & M & M \\ C_n & b_{n1} & b_{n2} & b_{n3} & \dots & 1 \end{bmatrix}$$

Figure 2. Pairwise Comparison Matrix

Where
n = number of criteria to be evaluated
Ci = i. Criteria
bij = interests of i. criteria based j

3. Normalization Matrix, determine the consistency value of the index, if Consistency Ratio (CR) <= 0.1 then proceed to step number 4, otherwise it will be re-questioner.
4. Change the linguistic variables in the form of fuzzy numbers. Questionnaire data in the form of linguistic variables fuzzy numbers are converted to forms. TFN Chang fuzzy numbers to be seen (the scale of the fundamental interests of Relative ANP) with a different level of importance.

Table 1. Triangular fuzzy number (TFN) scale and linguistic variables scale conversion

Linguistic Scale	Values interest	TFN Scale	TFN inverse scale
Equally important	1	(1,1,1)	(1,1,1)
A little more important	3	(1,3,5)	(1/5,1/3,1/1)
More important	5	(3,5,7)	(1/7,1/5,1/3)
Very important	7	(5,7,9)	(1/9,1/7,1/5)
The most important	9	(7,9,11)	(1/11,1/9,1/7)

5. Enter the search criteria and the weighting formula contained in steps as follows [9,10] :
   a. Determining the value of synthetic extend (the) associated with the object to i then represented as follows

$$si=\sum_j^m =1\, M_{gi}^j \quad \otimes \quad \left[\sum_{j=1}^n \sum_{j=1}^m M_{gi}^j\right]^{-1} \tag{3}$$

To get this $M_{gi}^j \sum_{j-1}^m$ done adding fuzzy operation of m with particular matrix

$$\sum_{j=1}^m M_{gi}^j=\left[\sum_{j=1}^m L_j ,\sum_{j=1}^m m_j ,\sum_{j=1}^m u_j \right] \tag{4}$$

To get $\left[\sum_{j=1}^n \sum_{j=1}^m M_{gi}^j\right]^{-1}$ surgery fuzzy value from (j = 1,2,3....m)

$$\sum_{i=1}^n \sum_{j=1}^m M_{gi}^j = \left(\sum_{j=1}^n l_j, \sum_{i=1}^n m_j ,\sum_{i=1}^n u_j\right) \tag{5}$$

$$\left[\sum_{i=1}^{n}\sum_{j=1}^{m} M_{gi}^{j}\right]^{-1} = \left(\frac{1}{\sum_{i=1}^{n} u_i}, \frac{1}{\sum_{i=1}^{n} m_i}, \frac{1}{\sum_{i=1}^{n} l_i}\right)$$

At the end of the first step of the determinant of the inverse vector (6)

b. Determining the degree of likelihood (degree of possibility) and fuzzy set m2 = (l2, m2, u2) ≥ M1 = (L1, M1, U1) is defined as

$$V\ (m2 \geq M1) = \text{supy} \geq x\ [\min\ (\mu m1\ (x), \mu m2\ (y))]$$

x and y are the value on the axis of each membership function. Applied to the theory and applications of fuzzy TFN with 3-type of low, medium and upper (l, m, u) and membership functions have been formed with the following equation :

$V\ (m_2 \geq m_1) = hgt\ (m_1 \cap m_2) = \lambda\, m_2(d)$

$$V\ (m_2 > m_1) = \begin{cases} 1, & if\ m_2 \geq m_1 \\ 0, & if\ l_1 \geq u_2 \\ \dfrac{l_1 - u_2}{(m_2 - u_2) - (m_1 - l_1)} & otherwise \end{cases} \quad (7)$$

c. Determining the degree of likelihood for Confex fuzzy number is greater than k at Confex fuzzy number for Mi = (i = 1,2,...,k) can be defined as:
V = (M ≥ M1, M2,..., Mk)
= V [(M ≥ M1)] and [M ≥ M2] and (M ≥ Mk)]
V = min (M ≥ Mi)
It is assumed that d` = min V (Si ≥ Sk)
For k = 1,2, ..., n k ≠ i then the weight vector used
W` = (d` (Ai), d` (A2), d` (A3), ..., d` (An)) T
Where Ai (i = 1,2,3, ..., n) is an element n

b. Through normalization, weighting vector normalization
W` = (d (A1), d (A2), ..., d (An)) T
Where we are nonfuzzy numbers.
c. Determination of global sub-criteria weights matrix by multiplying matrix interdependence with WT2 (weighted sub-criteria)
d. Measurement of sub-criteria using linguistic variables by multiplying the weight of global (global weight) with a value scale (scale value) each sub-criteria.
e. Consistency test is done by looking at the value of l, m and u. Value $l \leq m \leq u$ shows fuzzy consistent ratings. Fuzzy comparison matrix that consists of two dimensions, consistency index are always consistent.
f. Hitung Bobot akhir Kriteria
g. Jika bobot tidak sesuai maka dapat dilakukan looping lagi dengan menentukan skala dan kuisioner ulang.

Framework model for measurement of the SME is provided in Fig. 2.3.

3. Results and Analysis

This stage consists of modeling, simulation, and analysis of results. The modeling stage is theidentification of MCDM problems by determining the number of variables to be used in the study (criteria, alternatives, and respondents) as shown in Figure 3.1. Standard indicator this stage will be made the network structure of the criteria of measurement with the concept of multi-criteria and a decision maker. Some of the measurement indicators of SME in this study are determined based on

IOP Conf. Series: Materials Science and Engineering **336** (2018) 012023 doi:10.1088/1757-899X/336/1/012023

integrated internal business perspective, learning and growth perspective [11]. The next stage is Simulation and Analysis. Based on the framework model, the simulation and analysis of the model have been made based on existing indicators in SME. This is done to determine the optimal solution in decision-making and the smallest threshold value to determine the recommendation of Strategy map SME, SWOT, and clustering SME.

The stages of the simulation of this research program is

1. Doing questioners to some expert people to determine the comparative matrix assessment.
2. Calculating the consistency matrix of pairwise matrices, if the consistency ratio (threshold) is less than 0.1 then the matrix is considered consistent.
3. Determining the linguistic scale by using TFN.
4. Conversion scale TFN
5. Synthetic Extend Calculate
6. Normalize Criteria Weight
7. Perform an iteration to determine the criteria weights with different scales
8. Determine the final weight.

Based on results from table 1. Pairwise Comparison and table 2. Criteria Weight, the authors conclude that the most influential weights on the progress of SMEs are the number of production (0.902) and the weather (0.534).

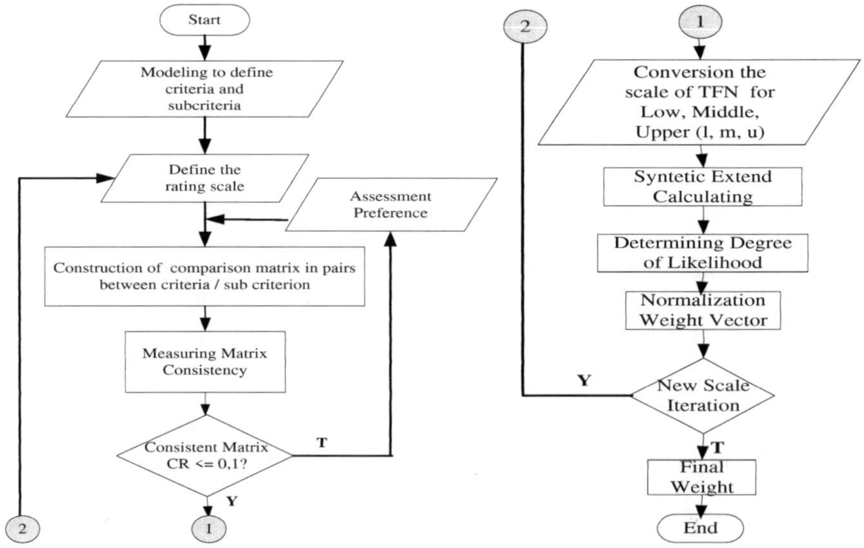

Figure 3. Framework model for measurement of the SME

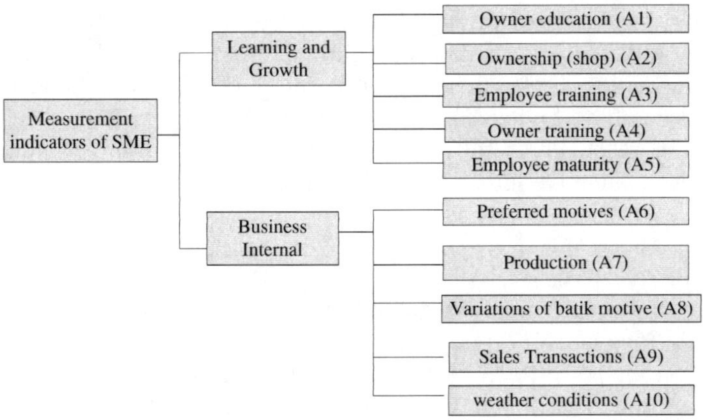

Figure 4. SME Criteria

Table 2. Pairwise Comparison

Criteria	A1	A2	A3	A4	A5	A6	A7	A8	A9	A10
A1	1	5	3	1	0,33	1	1	3	3	1
A2	0,20	1	0,33	0,20	0,20	0,33	0,14	3	3	0,20
A3	0,33	3	1	0,14	0,20	3	0,14	1	3	0,14
A4	1	5	7	1	1	7	1	5	5	5
A5	3	5	5	1	1	3	1	3	3	3
A6	1	3	0,33	0,14	0,33	1	0,2	3	5	0,2
A7	1	7	7	1	1	5	1	3	3	0,2
A8	0,33	0,33	1	0,20	0,33	0,33	0,33	1	1	0,33
A9	0,33	0,33	0,33	0,20	0,33	0,20	0,33	1	1	1
A10	1	5	7	0,20	0,33	5	5	3	1	1

Table 3. Criteria Weight

Criteria	Weight
A1	0,35706
A2	0,03269
A3	0,1986
A4	0,12444
A5	0,40038
A6	0,08506
A7	0,90165
A8	0,04688
A9	0,28328
A10	0,53465

4. Conclusions

Based on this research can be concluded that the recommendation SME performance can be determined based on framework model measurement SME with integrated internal business perspective, learning and growth perspective and fuzzy Analytic Network Process (FANP). This method determines the weighting of criteria based on the level of importance of the SME. This framework is adaptive and dynamic. The advantages of this Framework is to provide benefits for decision makers in determining the fuzzy scale and iterating dynamically to get the optimal decision. The resulting decisions can have high accuracy if there are preliminary data. Decisions are taken also consider the opinions of respondents by checking the consistency ratio. Future research on SME measurements can be used Adaptive Interval Fuzzy AHP and Adaptive Interval Fuzzy ANP.

References

[1] Skondras, E., Sgora, A. & M, A., 2014. An ANP and Trapezoidal Interval Value Fuzzy Technique for order preference by similarity to ideal solution Network access selection method. *Int. J. Commun. Syst,* pp.

[2] Meade, L.M., Sarkis, J., 1999. Analyzing organizational project alternatives for agile manufacturing processes: an analytical networkapproach. International Journal of Production Research 37, 241–261.

[3] Kustiyahningsih, Y., Rahmanita, E. & Purnama, J., 2016. Integration Balanced Scorecard and Fuzzy Analytic Network Process (FANP) for Measuring Performance of Small Medium Enterprise (SME). *Journal of Theoretical and Applied Information Technology (JATIT),* 94(2), pp. 343-352.

[4] DagdevirenMetin, YükselIhsan, A fuzzyanalytic network process (ANP) model for measurement of the sectoralcompetition level(SCL), Expert Systems with Applications 37,2010, pp. 1005-1014.

[5] Saaty T.L,Vargas L.G . Decision Making withthe Analytic Network Process (Economic,Political, Social and TechnologicalApplications with Benefits,Opportunities, Costsand Risks).United States of America Springer.

[6] Zadeh, L.A., 1965. Fuzzy sets. Information and Control 8, 338–353.

[7] Kahraman, C., Ruan, D., Dog ˘an, I., 2003. Fuzzy group decision-making for facility location selection. Information Sciences 157, 135–153.

[8] Chen, J. K., & Chen, I. S., Select innovativeindex of higher educational institutions byFAHP. Journal of American Academy ofBusiness, Cambridge, 13(1), 2008, pp. 151–157.

[9] Chang YD , "Applications of the extent analysismethod on fuzzy AHP", European Journal ofOperational Research, vol. 95, 1996, pp. 649-655.

[10] Gupta Manish, NarainRakesh , a Fuzzy ANPbased approach in the selection of the best e-business strategy and to assess the impact of e-procurement on organizational performance,information technology, and management,volume 16, issue 4, December 2015, pp 339-349.

[11] Kaplan, R. S. and D.P. Norton, Alignment:Using the Balanced Scorecard to CreateCorporate Synergies, Boston: HBSPress,2006a

IOP Conf. Series: Materials Science and Engineering **336** (2018) 012024 doi:10.1088/1757-899X/336/1/012024

Arduino Based Weather Monitoring Telemetry System Using NRF24L01+

Rafi Sidqi, Bagus Rio Rynaldo, Satya Hadi Suroso, Rifqi Firmansyah*

Department of Electrical Engineering, Universitas Negeri Surabaya

* rifqifirmansyah99@yahoo.co.id

Abstract-Weather is an important part of the natural environment, thus knowing weather information is needed before doing activity. The main purpose of this research was to develop a weather monitoring system which capable to transmit weather data via radio frequency by using nRF24L01+ 2,4GHz radio module. This research implement Arduino UNO as the main controller of the system which send data wirelessly using the radio module and received by a receiver system. Received data then logged and displayed using a Graphical User Interface on a personal computer. Test and experiment result show that the system was able to transmit weather data via radio wave with maximum transmitting range of 32 meters.

1. Introduction

Weather is an important part of the natural environment. It directly or indirectly affects many of our life activities. Thus, knowing weather information is needed before performing an activity, especially activities that directly relate to nature such as agricultural activities and so on. [1]

Furthermore, Weather or climate plays an important role in human life. The thermal comfort of human being is known to be influenced mostly by six parameters, i.e., air temperature, radiation, air flow, humidity, activity level and clothing thermal resistance. The advancement in technology has made these small and reliable electronic sensors capable of monitoring environmental parameters more favourably. [2]

In order to describe the weather accordance to the atmospheric conditions, certain key weather elements must be known. Some of those are temperature, sun intensity, relative humidity, and atmospheric pressure. So in this case, weather monitoring system is necessary to meet the needs of describing weather condition. [3]

However, the lack of wireless technology in conventional weather monitoring system is an efficiency and effectiveness issue. Therefore to overcome this issue a system which capable to transmit weather information wirelessly is needed? [4]

This paper aims to build a weather monitoring telemetry system which capable to acquire and record data wirelessly. Those ability may be beneficial in providing a portable and practical weather monitoring system. The proposed system has three sensors that measure temperature, relative humidity, atmospheric pressure, and light intensity respectively. Sensory data then

IOP Conf. Series: Materials Science and Engineering **336** (2018) 012024 doi:10.1088/1757-899X/336/1/012024

transmitted via an NRF24l01 2.4 GHz transmitter. Transmitted data then received and displayed in a graphical user interface and logged to a tab delimited text file.

2. Literature Review

This weather monitoring system is implemented using Arduino Uno Board. The board is used to interface the DHT11 sensor, LDR sensor, BMP180 Barometric Pressure and the NRF2401 module. The Arduino accepts the data from the sensors and displays the output on the PC using serial communication. The NRF2401 module is responsible for transmitting the data obtained from the sensors DHT11, LDR and BMP180.

2.1. Arduino UNO R3

Arduino is an open source device, a prototyping board consisting of ATmega328P microcontroller providing a 5V and 3.3V output volatage options. It takes input voltage from either connecting USB to your computer or either using a coaxial cable using a portable power supply. The arduino board is also capable of reading Twitter messages and respond in order to that. On the arduino you can upload sketches using Arduino IDE. Arduino comes in various flavours and according to needs like Uno, Mega, Yun etc.

In this instrument I've used Arduino unoboard. It is cheap and flexsible. Also it is good to start as a beginner. It has 14 digital input/output pins, 6 analog inputs and a reset button. It takes input voltage in between 7-12V.

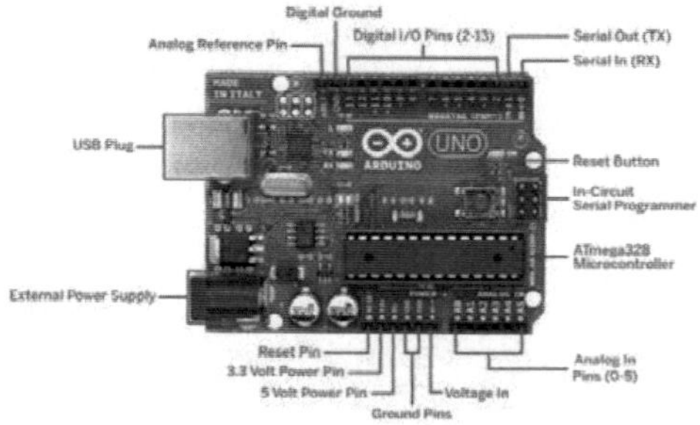

Figure 1. Arduino UNO R3

Related technical specification shown in Table 1.

Table 1. Arduino Uno spesifications.

Microcontroller	Atmega328P
Operating Voltage	5V
Input Voltage (recommended)	7 - 12 V
Input Voltage (limit)	6 – 20 V
Digital I/O Pins	14
PWM Digital I/O Pins	6
Analog Input Pins	6
DC Current per I/O Pin	20 mA
DC Current for 3,3 V Pin	50 mA
Flash Memory	32 KB
SRAM	2 KB
EEPROM	1 KB
Clock Speed	16 MHz
Length	68,6 mm
width	53,4 mm
Weight	25 g

2.2. Digital Humidity Temperature Sensor

DHT11 is a digital temperature and humidity sensor which provides a calibrated digital signal output of the temperature and humidity. Due to its exclusive digital modules and

IOP Conf. Series: Materials Science and Engineering **336** (2018) 012024 doi:10.1088/1757-899X/336/1/012024

the temperature and humidity sensing technology, it makes sure that the output obtained is of high reliability and stability. The sensor consists of a resistive sense of wet components and an NTC temperature measurement devices, and connected with a high-performance 8-bit microcontroller. DHT11 sensor ensures fast response, low consumption of power, low cost. In addition to this, it also has the ability to avoid interference and can transmit data to a long distance. The communication and synchronization between microcontroller unit and the DHT11 sensor takes place using a single wire two way connection and requires single bus data format for the same. The information packet transmitted by the sensor is of 40 bits. The sensor operates in 2 different modes-low power consumption mode and running mode. The start signal signifies the transformation from low power mode to running mode. As the start signal is completed, the sensor sends a response signal containing the information regarding the relative temperature and humidity. [5]

For weather station, the sensor used is a 3 pin module consisting of Vcc, Data and Ground (GND).The Data pin is connected to the A0 pin of the Arduino board whereas the Vcc is connected to the 5V pin of Arduino. The sensor transmits the information of temperature and humidity to the Arduino board which is later displayed on the serial monitor using serial communication.

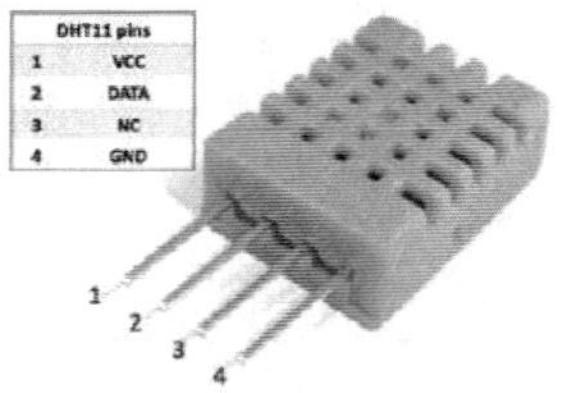

Figure 2. DHT 11

Figure 3. BMP 180

2.3. BMP180

BMP180 sensor It is a barometric pressure sensor which senses with an I2C interface As the pressure varies with altitude it can used to measure the altitude too. It has 4 pins SDA, SCL, GND and Vin. [6]

Table 2. BMP180 Spesifications

Pressure range	300 – 1100 hPa
Supply Voltage	1,8 – 3,6 V
Operating temperature	-40 to +85 ℃
Absolute accuracy pressure (0 to +65 ℃)	-4,0 to +2,0 hPa
Supply current	3 – 32 μA
Absolute accuracy temperature (0 to +65 ℃)	-2,0 to + 2,0 ℃
Dimensions	0.4 x 0.4 x 0.4 inches

2.4. Light Dependent Resistor (LDR)

An LDR is a component that has a (variable) resistance that changes with the light intensity that falls upon it. This allows them to be used in light sensing circuits. A light-dependent resistor (LDR) is a light-controlled variable resistor. The resistance of this decreases with increasing incident light intensity; in other words, it exhibits photo-conductivity. An LDR can be applied in light-sensitive detector circuits, and light- and dark-activated switching circuits. An LDR is made of a high resistance semiconductor. In the dark, an LDR can have a resistance as high as a few mega ohms (MΩ), while in the light, an LDR can have a resistance as low as a few hundred ohms. If incident light on an LDR exceeds a certain frequency, photons absorbed by the semiconductor give bound electrons enough energy to jump into the conduction band. The resulting free electrons (and their whole partners) conduct electricity, thereby lowering resistance. The resistance range and sensitivity of an LDR can substantially differ among dissimilar devices. [7]

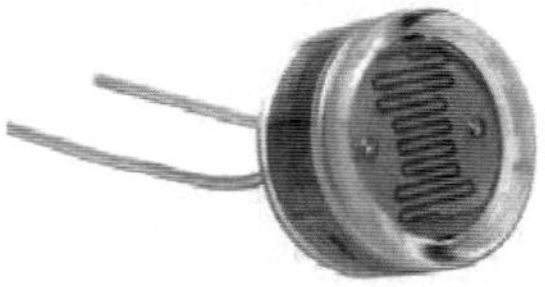

Figure 4. LDR

Figure 5. NRF24L01+

2.5. NRF24L01+

The nRF24L01 is a single chip radio transceiver for the global, license-free 2.4 GHz ISM band. The low cost nRF24L01 is designed to merge very high speed communications (up to 2Mbit/s) with extremely low power (the RX current is just 12.5mA) [4]. The transceiver consists of a fully integrated frequency synthesizer, a power amplifier, crystal oscillator, demodulator, modulator and Enhanced ShockBurst protocol engine. In addition, the nRF24L01 also offers an innovative on-chip hardware solution called 'MultiCeiver' that can support up to six simultaneously communicating wireless devices. This makes it ideal for building wireless Personal Area Networks in a wide range of applications. The PCB of this WSN node is circular, having two inches diameter. [8]

Table 3. nRF24L10+ Spesification

Transceiver model	nRF24L01+
Frequency	2,4 GHz
Input Voltage	1,9 - 3,6 V
Modulation	GFSK
Data rate	250 kbps, 1 Mbps, and 2 Mbps
Output current (transmitter)	11,3 mA at 0 dBm
Output current (Receiver)	13,5 mA at 2 Mbps
Sensitivity	-82 dBm at 2 Mbps -85 dBm at 1 Mbps -94 dBm at 250 kbps
Operating temperature	-40 to +85 ℃
Dimensions	2.9 x 1.5 x 1.2 cm

3. Proposed System

The proposed system divided into two main parts. These are transmitter and receiver. The transmitting part consist of sensors, Arduino UNO, and nRF24L01 RF Module. Receiving consist of Arduino UNO, nRF24L01 RF Module, and a laptop. The block diagram shown in Figure 6.

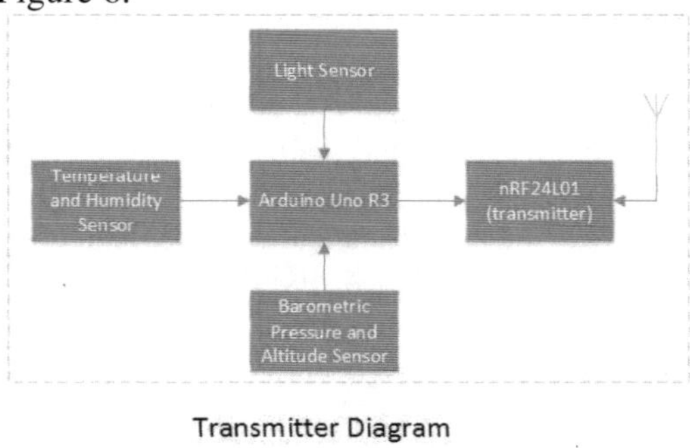

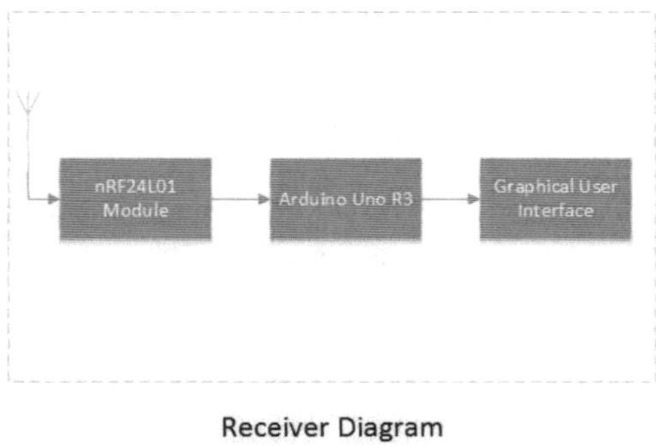

Figure 6. Transmitter and receiver block diagram

3.1. Transmitter System

Proposed transmitter system consists of DHT 11 sensor, BMP 180, light dependent resistor, Arduino UNO R3 and nRF24L01 module. Transmitter circuit design shown in figure 8. The modules and microcontroller used of the transmitter can be summarized as follows :

DHT 11

This sensor is used to measure current temperature and humidity.

BMP 180

Sensor is used to measure barometric pressure and altidude.

Light Dependent Resistor

This sensor is used for measuring the light intensity falling on the sensor.

Arduino Uno R3

is the main component that is centrally connected to all components.

nRF24L01

The RF module is used to transmit data that has been processed by Arduino Uno R3.

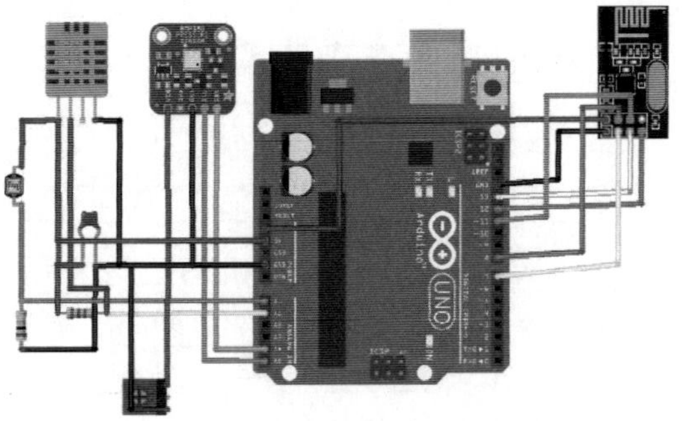

Figure 7. Transmitter circuit design

The transmitter system wiring table could be seen on Table 4.

Table 4. Transmitter System Spesification.

NRF24L01	**ARDUINO UNO R3**
VCC	VCC
GND	GND
CE	7
CSN	8
MISO	ICSP (MISO)
MOSI	ICSP (MOSI)
SCK	ICSP (SCK)
DHT11	**ARDUINO UNO R3**
VCC	VCC
GND	GND
DATA	3
NC	-
LDR	**ARDUINO UNO R3**
VCC	VCC
GND	GND
DATA	A0
BMP180	**ARDUINO UNO R3**
VIN	VCC
GND	GND
SCL	A5
SCA	A4

3.2. Receiver System

Proposed receiver system consists of Arduino UNO R3, nRF24L01 module and a personal computer (PC). Receiver circuit design shown in figure 9. The nRF24L01 module is used for receive sent data from the transmitter, the received data then processed by the Arduino Uno R3 and the data is displayed on a laptop using a graphical user interface.

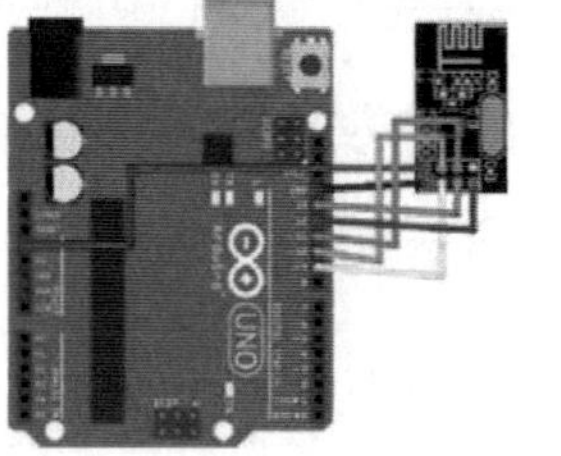

Figure 8. Receiver circuit design

The receiver system wiring table could be seen on Table 5.

Table 5. Receiver System Spesification.

NRF24L01	ARDUINO UNO R3
VCC	VCC
GND	GND
CE	7
CSN	8
MISO	ICSP (MISO)
MOSI	ICSP (MOSI)
SCK	ICSP (SCK)

3.3. Graphical User Interface

The graphical user interface was developed using C#. The software allows received data to be displayed in a user-friendly platform. This could simplify and made the data acquisition process. Beside of that, some of basic function of the GUI are to provide connection between receiver system and the PC, and to log all received data to a text file.

The GUI refresh in one second interval, updating to the latest received data on a real time update. Thus, more accurate weather monitoring could be achieved.

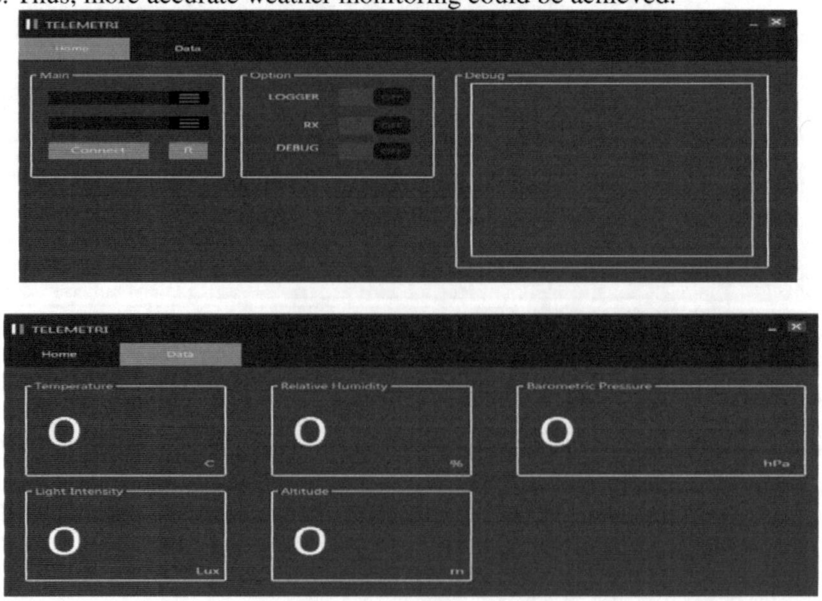

Figure 9. GUI display

4. Experimental Result and Discussion

Before starting the experiment a full system testing was done to ensure minimum error and no fatal malfunction in both transmitter and receiver system. The testing of each sensor was carried out successfully. Resulting in minimum error in sensor readings as shown on Table 7 and Table 9.

The experiment is carried out by powering the transmitter system using a portable power source and the receiver system connected to a PC. The GUI was launched and established a connection to the receiver system and begin to log data. Then distance and condition between transmitter system and receiver system varied until the GUI displayed a constant value which represent loss of signal or maximum transmitting range.

4.1. Sensors Test and Calibration

Temperature and lux sensor testing was carried out by comparing sensors reading value with respectively termometer, and lux meter.

Atmospheric pressure and altitude was not calibrated due to limitation of the required instruments.

Table 6. Temperature sensor test and calibration data.

Trial	Temperature sensor reading (C)	Termometer reading (C)
1.	28.00	27.80
2.	30.20	30.30
3.	33.40	33.50
4.	35.30	35.00
5.	25.80	25.40

Table 7. Temperature sensor errors.

Trial	Error	% Error
1.	-0.20	0.71
2.	0.10	0.33
3.	0.10	0.29
4.	-0.30	0.84
5.	-0.40	1,55

From Table 1 and Table 2 it can be concluded that used temperature sensor was able to sense temperature in centigrade unit which error is less than 2%.

Table 8. Lux sensor test and calibration data.

Trial	Lux sensor reading (lux)	Lux meter reading (lux)
1.	82.2	83.8
2.	84.5	83.5
3.	84.0	83.6
4.	83.7	84.4
5.	84.8	84.2

Table 9. Lux sensor errors

Trial	Error	% Error
1.	1.6	1.94
2.	-1	1.18
3.	-0.4	0.47
4.	0.7	0.83
5.	-0.6	0.70

From Table 3 and Table 4 it can be concluded that used lux sensor was able to sense light intensity in lux unit which error is less than 2%.

4.2. Telemetry System Result

Each experiment attempt was done in a different time and location. First attemp was done at second floor of A8 building in Engineering faculty of Surabaya State University. The results obtained are summarized in Table 5 and Table 6,

Table 10. Experiment attempt 1. Transmitting range.

LOS Distance (m)	T (C), RH (%), Lux (lux), Atmospheric pressure (hPa), Altitude (m)
1	27.4, 35, 890, 1007.97, 44.05
2	27.1, 38, 953, 1007.94, 44.31
4	27.2, 38, 958, 1007.99, 44.05
8	27.2, 38, 970, 1007.99, 44.05
10	27.1, 38, 1002, 1008.01, 43.30
12	27.1, 38, 1007, 1007.97, 43.89
14	27.1, 38, 1005, 1007.99, 43.97
16	27.1, 38, 987, 1008.04, 43.47
18	27.1, 38, 999, 1007.97, 43.89
20	27.1, 38, 1009, 1007.99, 43.72
22	27.1, 38, 1012, 1008.09, 43.80
24	27.1, 38, 1013, 1008.03, 43.39
26	27.1, 40, 1013, 1008.01, 43.72
28	27.1, 40, 1012, 1007.99, 43.89
30	27.1, 40, 1011, 1008.00, 43.63
32	27.1, 40, 1007, 1008.01, 43.72
34	27.1, 40, 1007, 1008.01, 43.72
36	27.1, 40, 1007, 1008.01, 43.72

Table 11. Experiment attempt 2. Transmitting range.

LOS Distance (m)	T (C), RH (%), Lux (lux), Atmospheric pressure (hPa), Altitude (m)
1	30.4, 33, 935, 1005.66, 63.39
2	30.4, 30, 935, 1005.65, 63.47
4	30.5, 33, 919, 1005.65, 63.47
8	30.6, 33, 928, 1005.64, 63.55
10	30.7, 34, 915, 1005.63, 63.64
12	30.8, 33, 927, 1005.64, 63.55
14	30.8, 33, 927, 1005.64, 63.55
16	30.7, 34, 929, 1005.62, 63.72
18	30.7, 33, 930, 1005.56, 64.22
20	30.7, 34, 929, 1005.61, 63.81
22	30.6, 26, 928, 1005.60, 63.72
24	30.6, 33, 926, 1005.60, 63.89
26	30.5, 18, 924, 1005.62, 63.72
28	30.5, 33, 924, 1005.58, 64.06
30	30.5, 34, 932, 1005.65, 63.47
32	30.5, 34, 932, 1005.65, 63.47
34	30.5, 34, 932, 1005.65, 63.47

Table 10 and 11 denotes received sensory data in a varied line of sight distance. Sensory data was not stay ini the same value due to the hysteresis effect of the used sensors, except lux variable. Light intensity was varied to make determination of the successful data transmission easier. Sensor hysteresis effect can be seen on Figure 10

It can be observed from Table 1 that the maximum transmission range is 32 meters, since the sensory value is constant. Meanwhile from Table 11, maximum transmission range is 30 meters. The difference in transmission maximum range between two experiment attempt could mainly due to different condition of the environment which was done at or another factors that were not considered.

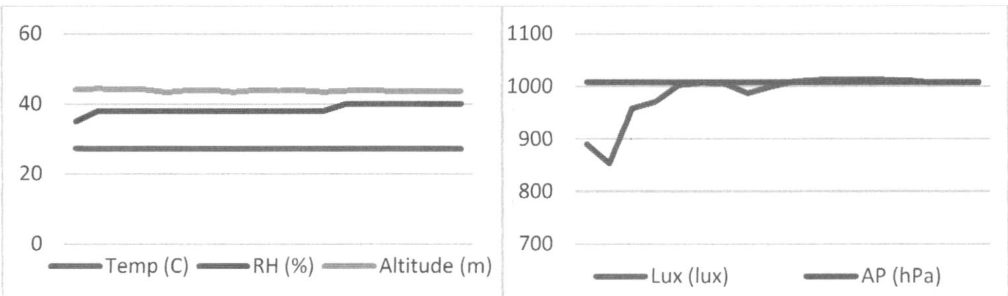

Figure 10. Sensors hysteresis effect value change

5. Conclusion

The main goal of this research is to develop and implement a weather monitoring telemetry system using nRF24L01+ 2.4 GHz RF Module. From the result obtained throughout the test and experiment carried out, it can be concluded: 1)The developed system work well, this can be

IOP Conf. Series: Materials Science and Engineering **336** (2018) 012024 doi:10.1088/1757-899X/336/1/012024

known from the system's capability to transmit dan receive weather data via a radio waves at a certain maximum distance. 2) nRF24L01+ maximum transmitting range is 30-32 meter line of sight, this value may differ from another conducted research due to differences in implemented programs, environment condition, hardware quality, and other factors that affecting radio wave

References

[1] Saefullah, Asep, Abas Sunarya, dan Deinsyah Fakhrizal. 2014. "Prototype Weather Station Berbasis Arduino Yun". Jurnal Volume 8, Nomor 2, *Creative Communication and Innovative Technology,* (57-65).

[2] Noordin, Kamarul Arifin, Chow Chee Onn, and Mohammad Faizal Ismail. 2006. "A Low-Cost Microcontroller-based Weather Monitoring System". Journal Vol. 5 (1). *CMU Journal*, (33-38).

[3] Olatomiwa, Lanre Joseph, and Umoru Sam Adikwu. 2012. "Design and Construction of a Low Cost Digital Weather Station". Journal Vol. 16 (2). *AU Journal,* (125-132)

[4] Susmitha, P and Sowmyabala, G. "Design and Implementation of Weather Monitoring and Control System", International Journal of Computer Aplication, Vol. 97 – No 3. July 2014.

[5] Akhilesh Chawla, Tejas Bangera, Chinmay Kolwalkar, Mahalaxmi Bhat. 2015. "Bluetooth Based Weather Station", International Journal of Engineering Trends and Technology (IJETT) volume 28 number 2 October 2015.

[6] Amber Katyal, Ravi Yadav, Manoj Pandey,"Wireless Arduino Based Weather Station". *International Journal of Advanced Research in Computer and Communication Engineering,*

[7] Karthik Krishnamurti, Suraj Thapa, Lokesh Kothari, Arun Prakash, " Arduino Based Weather Monitoring System", International Journal of Engineering Research and General Science Volume 3, Issue 2, March-April 2015.

[8] S. S. Sonavane, V. Kumar, B. P. Patil, "MSP430 and nRF24L01 based Wireless Sensor Network Design with Adaptive Power Control", ICGST-CNIR Journal, Volume 8, Issue 2, January 2009.

IOP Conf. Series: Materials Science and Engineering **336** (2018) 012025 doi:10.1088/1757-899X/336/1/012025

Student Team Achievement Divisions: Its Effect on Electrical Motor Installation Knowledge Competence

Ahmad Hanafi[*], **Ismet Basuki**

Graduate Program, Universitas Negeri Surabaya. Indonesia.

[*]ahmadhanafi@mhs.unesa.ac.id

Abstract. Student team achievement division (STAD) was an active learning strategy with the small group inside of the classroom members. The students would work in small heterogeneous groups (of five to six members) and help one another to comprehend the material given. To achieve the objectives of the study, this research aims to know the effect of STAD on competence of electrical motor installation. The objective of the student competence was knowledge competence. The data was collected from 30 students. the participants were the students of second class at electrical installation techniques, SMKN 1 Pungging Indonesia. The design of empirical test in this research was one shot case study. The result of knowledge test would be compared by criteria for minimum competence, which was 75. Knowledge competence was analyzed with one sample t test technique. From the analysis got average 84.93, which meant average of student competence had reached criteria for minimum competence. From that analyze, It could be concluded that STAD was effective on electrical motor installation knowledge competence. STAD could grow student motivation to learn better than other models. But, in the application of cooperative learning teacher should prepare carefully before the learning process to avoid problems that could arise during group learning such as students who were less active in the groups. The problem could be resolved by away the teachers took around to check each group. It was felt could minimize the problems.
Keyword: effectiveness, criteria for minimum competence, knowledge competence.

1. Introduction

Depend on the opening of Indonesian Republic 1945 Constitution was declared that one of the purposes is educate national life. Because of that every Indonesian people are entitled to earn a good education quality as conformity with the interestedness and galantines without regardless of social, racial, ethnic, religious, and gender status. To realize the purpose of national education, so it is established a standard of graduation competence. Based on Article 35 of Constitution Number 20 Year 2003, it is structured graduate ability that includes attitude, knowledge, and skill of students that must be fulfilled or achieved from an educational unit at the level of education at the level of education. Based on the Curriculum of 2013 emphasizes the four aspects of the competence of learners that include: spiritual attitudes, social attitudes, knowledge and skills that reflect the behaviour and mindset of learners. On the implementation of the 2013 curriculum of the learning process that is cantered on the learners, so that the role of educators in this curriculum is only as a facilitator in the learning process. Grouping students to work together becomes more and more acceptable to be applied by teachers. Students have a degree of potential, historical background, as well as future expectations are different. Because of the differences, Students can educate cooperative learning is consciously creating successive loving each other interactions so learning resources for students not only teachers and textbooks but also fellow students. So on the learning process; teacher is only as a facilitator.

IOP Conf. Series: Materials Science and Engineering **336** (2018) 012025 doi:10.1088/1757-899X/336/1/012025

STAD was developed by Robert Slavin et all from University of John Hopskin. This method is the simplest cooperative learning method. Educators use cooperative learning as a teaching method in all grade levels, in all curriculum areas, and for many different purposes, but all uses centred on the goal of getting students to understand and learn the material presented. Interest in cooperative learning gathered momentum in the early 1980s with the publication of the first meta-analysis involving 122 studies on the effects of cooperative, competitive, and individualistic goal structures on students' achievement and productivity in a sample of North American schools [4]. The results showed that cooperation was more effective than interpersonal competition and individualistic efforts; cooperation with intergroup competition was also superior to interpersonal competition and individualistic efforts; and there were no significant differences between interpersonal competitive and individualistic efforts. Moreover, these results were consistent across all subject areas (language arts, reading, mathematics, science, social studies and physical education), for all age groups, and for all tasks involving conceptual understanding, problem solving, categorizing, and reasoning. Cooperative learning on Felder & Brent (2004) explained that cooperative learning is a subset of collaborative learning in which students work in teams on structured learning tasks under conditions that meet five criteria. 1) Positive interdependence, Team members must rely on one another to accomplish goal. 2) Individual accountability, members held accountable for doing their share of the work and mastering all material. 3) Face-to-face interaction, some or all work done by members working together. 4) Appropriate use of interpersonal skills. 5) Team members practice and receive instruction in leadership, decision-making, communication, and conflict management.

Regular self-assessment of group functioning. Teams periodically reflect on what they are doing well as a team, what they could improve, and what (if anything) they will do differently in the future. Cooperative learning is not students sitting around a table studying together and group projects with one or two students doing all the work. Cooperative learning is a cooperative effort where interest and involvement can be generated, where all members are given tasks to work as a team. In cooperative learning, it involves small group of students (usually five to six members) of the different ability level working as a group for interaction to occur [9]. The effectiveness of cooperative learning principles and techniques in building a motivating, supportive learning environment is well known [6]. Cooperative learning is seen beneficial especially in a diverse classroom environment where students differ from religion, culture, race, etc. [2]. The main idea in all the cooperative learning approaches is that students work and learn together actively in small groups to accomplish a common goal in a mutually helpful manner. Cooperative learning is necessary in any teaching-learning situation, because this particular strategy "can foster educational excellence for all children regardless of race, class, or gender, and can provide students and teachers with the experience and expectations of active participation in controlling and changing the spheres of their lives [12]. So the researcher can argue that the Cooperative learning is a student-centered, instructor-facilitated instructional strategy in which a small team of students is responsible for its own learning and the learning of all team members. This way of grouping is believed to motivate the members of each group to get high scores.

Student Team Achievement Division (STAD) is a cooperative-learning strategy in which small groups of learners with different levels of ability work together to accomplish a shared learning goal [7]. The teacher presents a lesson, and then students work together within their teams to make sure that all team members have mastered the lesson. Finally, all students take individual quizzes on the material, at which time they may not help one another. Students' quiz scores are compared to their own past averages, and points are awarded on the basis of the degree to which students meet or exceed their own earlier performance [11]. STAD is considered a good strategy because as one the cooperative learning techniques, it can raise students' motivation in learning by exchanging and sharing information, reinforcing each other, giving feedback and having the responsibility for their tasks in group work [3]. STAD is one of the techniques of cooperative learning in which students are divided into groups of 4-5 members. Based on this technique, the teacher presents the lesson and then gives the students in each group a test based on the content of the instruction. The average of the students' score in each group is the score of STAD group. Accordingly, the STAD groups will be classified to good team, great team, and super team with respect to their scores achieved at the end of each session [12]. In STAD, learners work in small heterogeneous groups (of five to six members) and help one another to comprehend the given material. Individual quizzes are given at the end of the week and the best group is rewarded on the basis of individual improvement. The reward is given in different forms; their names may be written on the bulletin board or they may be given certificates at the end of every week [4]. The steps of STAD as type of cooperative learning as follows 1) Establish a group whose members are 4-6 people in a heterogeneous; 2) The teacher presents a lesson; 3) The teacher gives the task to the group to be done by members of the group. The members know to explain to other members until all members of the group understand; 4) The teacher gives a quiz/questions to all students at the time of answering the quiz should not help each other; 5) Provide evaluation; 6) Conclusion.

Based on the above descriptions can be concluded that Student Team Achievement Division is an active learning strategy with the small group inside of the classroom members. The students will be work in small heterogeneous groups (of five to six members) and help one another to comprehend the given material. To achieve the objectives of the study following null hypotheses were tested using a STAD; the average of student competence can reach above criteria for minimum competence.

2. Method

The design of empirical test in this research is one shot case study. The design of the empirical test design is shown in the following picture.

X O

Where:
X: Experimental treatment (use of STAD)
O: Post test score of knowledge competence

3. Respondent

The participating in this research is selected use simple random sampling technique. So the participating are the students of second class at electrical installation techniques. SMKN 1 Pungging Indonesia. There are 30 students. The research will be done at the second semester of school year 2016/2017.

4. Result

There are two kinds of analysis on this research. They are using descriptive analysis and using one simple t test. From the descriptive analysis, it can be known that N=30 with minimum score is 76 and maximum score is 96. Electrical motor installation average (mean) is 84.9333 with standard deviation is 7.25370. Variance value is 52.616. Skewness value is .259 and kurtosis value is -1.303.

Before competence's score is analyzed using one simple t test. This research has assuming that "by applying STAD, students can reach knowledge competence. The research result will be analyzed using a statistically test. That is one sample t test. Before using statistically test, the data must complete a requirement test. The requirement test is distributed normality using Komogorov Smirnov test. That is because the result is a single data. The result of distribution normality analyzed using IBM SPSS Statistics software. The result of distribution normality distribution normality with N=30 get value z = .152 and significance .076. From those description can be concluded for accepted H_0. Which means data is distributed normally. Because the distribution data is normal, so data can be analyzed using one sample t test. The test will be analyzed using IBM *SPSS Statistics software.* Before data that is analyzed using one sample t test. This research has statistical hypothesis 1) H_0: < 75 (students result average of knowledge competence by applying STAD less than 75) and 2) H_1: ≥ 75 (students result average of knowledge competence by applying STAD more or equal than 75). From that hypothesis we can make a conclusion for a result. The result of one sample t test is shown in the following table.

Table 1. Result of One Sample T Test

One-Sample Test

	Test Value = 75			
	t	df	Sig. (2-tailed)	Mean Difference
value	7,501	29	,000	9,93333

Depend on that Table 1, one sample t test get the significance .000. There is a different between average of electrical motor installation competence result (84.9333) and criteria of minimum competence (75). It means the average of student competence has reached criteria for minimum competence.

5. Conclusion

Base on the analysis. It can be known that the average of student commence has reached minimum competence criteria. So, It can be concluded that STAD is effective on electrical motor installation knowledge competence.

5.1. Recommendation

Teachers are encouraged to cooperative learning for some topic in teaching an electrical problem solving. However, the learning process has to be prepared as good as possible in advance. It is to avold problems in the learning using STAD or another strategy. Test the effectiveness of the other cooperative leaning strategies in teaching electrical or other related subject. But, In the application of cooperative learning should the teacher should prepare carefully before the learning process. This is to avoid problems that can arise during group learning such as students who are less active in groups or others.

REFERENCES

[1] Felder, R. M., & Brent 2004 Cooperative learning. National Effective Teaching Institute.

[2] Glomo, D. T. & Norzales 2015 *Student team achievement divisions (STAD):* Its effect on the academic performance of EFL learnens. *American Research Journal of English and Literature. 1-2378-9026.*

[3] Hanum, L 2015 A teacher's experience in teaching with student team-achievement divisions (STAD) technique. *Intern. J. of Inst. 8(2)*

[4] Ishtiaq, M., Ali, Z., & Salem. M. 2013 The effects of student teams achievement division (STAD) on motivation of Saudi EFL adult learners. *Intern. J. of Lang. Ed. and App. Ling.(IJLEAL).*

[5] Johnson, D. W., Maruyuma. G., Johnson. R., & Nelson. D., 1981 Effects of cooperative, competitive, and individualistic goal structures on achievement: A metaanalysis. Psychological Bulletin, **89**, pp. 47-62.

[6] Kupczynski, L., Mundy, M. A., Goswami. J., & Meling, V. 2012 Cooperative learning in distance learning: A meixed methods study. *Intern. J. of Inst.. pp.1308-1470.*

[7] Majoka, M. I., Dad, M. H., & Mahmood, T. 2010 Student team achievement divisions (STAD) as an active learning stategy: Empirical evidence from mathematics classroom. *J. of Ed. and Soc.. 2078-032X.*

[8] National Ministry of Education. 2003 National system education. Jakarta.

[9] Ocampo, R. O., & Ocampo, R. B. 2015 Effectitiveness of students' team achievement division on students' attitude toward physics. *Asia Pacific Journal of Multidisciplinary Research.* **3***(4). Pp. 112-117.*

[10] Republic of Indonesia 1945 Constitution of Indonesian Republic 1945. Jakarta

[11] Saniei, A. (2015). The contribution of student teams-achievement divisions (STAD) to teaching English collocations to intermediate learners. *Intern. J.of Lang. Learning and App. Ling. World (IJLLALW).* pp. 2289-3245.

[12] Tiantong, M., & Teemuansai, S. 2013. Student Team Achievement Divisions (STAD) technique through the moodle to enhance learning achievement. *Canadian Center of Science and Education..*

[13] Wyk, M. M. 2013. The effects of the STAD-cooperative learning method on student achievement, attitude and motivation in economics education. *J Soc Sci,pp. 261-270..*

IOP Conf. Series: Materials Science and Engineering **336** (2018) 012026 doi:10.1088/1757-899X/336/1/012026

Learning Media Application Based On Microcontroller Chip Technology in Early Age

Permata Ika Hidayati

Universitas Kanjuruhan Malang, Indonesia

permatahidayati@unikama.ac.id

Abstract. In Early childhood cognitive intelligence need right rncdia learning that can help a child's cognitive intelligence quickly. The purpose of this study to design a learning media in the form of a puppet can used to introduce human anatomy during early childhood. This educational doll utilizing *voice recognition* technology from EasyVR module to receive commands from the user to introduce body parts on a doll, is used as an indicator TED. In addition to providing the introduction of human anatomy, this doll $_{\text{jut.}}$ a user can give a shout out to mainly play previously stored voice module *sound recorder.* Results obtained from this study is that this educational dolls can detect more than voice and spoken commands that can be random detected. Distance concrete of this doll in detecting the sound is up to a distance of 2.5 meters.

1.Introduction

Education is an important asset for the progress of a nation; therefore every citizen must follow many levels of education both level of education of children ages. early childhood, basic education, secondary education and in the field of education of a child from birth require appropriate services in the fulfilment of educational needs, understanding of the characteristics of children according to their growth and development will be very helpful in adjusting the learning process for children with age, needs, and condition, respectively, both intelegencially, emotionally and socially. In early childhood development, parents experience many challenges in their child's education process. Cognitive knowledge such as the introduction alphabets and numbers, children also need to start recognizing part of their bodies. Parents usually train their children by saying the name and designate one member of the body to the late children can recognize what the name of limb designated, because the learning process of children mainly under the age of three years started off with seeing, hearing, and utter words. with imitate what was heard by him. In addition to learning, giving it time to play for children is also very important in the learning and development of their emotional, so it would be nice if the parent gives the right toy and educate their children, so that the children can learn while playing. A doll is a kind of toys with various kinds of shapes begins to humans, animals, and a fictional character. In general, dolls are used as children's toys and sometimes also used as a collection as well as a friend for doll fans. Of course, the existence of this one toy is very close to human life, especially children. Therefore, there is an opportunity to develop the great benefits of the doll one of them by insert the value of education.

To be able to design a doll. who can help parents in providing education to. his children, about the introduction of the anatomy of the human body. and how to pronounce the part of the body. Correctly, then required the role of technology in it, namely with the help of Voice Recognition technology. This Voice Recognition technology enables a device to recognize and understand the words spoken by the user. With using microcontroller ATmega328 as the central data processing and module Voice Recognition EasyVR as a sound sensor that is not only able to detect the presence of noise, but also be able to recognize words or sentences, doll education expected to be able to recognize words spoken by the parents and the child, for example when the parents of the child say the eyes ", then the LED light as a pointer to the puppet eye will light. In addition to teaching recognize the human anatomy, this doll is also said to teach children how to introduce themselves well, ie when parents say" hello ", then stuffed. Education will be answered by playing a sound." hello"stored in the module Sound Recorder.

2. Support Theory

2.1 Definition Media Learning

Medium term is derived from the Latin word is the plural of "medium", which literally means an intermediary or introduction [I]. Media is every something that can distribute information on resources to the recipient of the information learning process is also essentially a communication process, so that the media used in learning called learning media. Learning media is part of the learning resource merup will be a combination of software (learning materials) and hardware (learning tool). From the above understanding, in general it can be said that the substance of learning media is a form of the channel, which is used to distribute messages, information or material subject to the recipient of the message a.tau learners can also be said that the learning media is type of various components in the environment in which learning environment can stimulate learners to learn.

2.2 Microcontroller

Microcontroller CT is the main controller electronics devices today, including the robot course. [2]. Microcontroller famous and easy to get in Indonesia today is 89S51, ATMega AVR 8535, ATMega 16 ATMega 32, and ATMega 128. Another famous brands, such as PIC 16F877 and Basic Stamp microcontroller 2. Include a processor core, memory (a small amount of RAM, program memory, or both), and input and output equipment. In other words, the microcontroller is a digital electronic device that has inputs and outputs as well as control with a program that can be written and erased in a special way, the workings of the actual microcontroller to read and write data. Microcontroller is in the computer *chip* used to control electronic devices, which push and effectively costs. As literally be called "little control" where an electronic system which previously was often require components supporting components such as TTL and CMOS IC can be reduced minimized and centralized and controlled by the microcontroller [3]. ATmega328 is output from Atmel microcontroller that has a RISC architecture *(Reduce Instruction Set Computer)* wherein each execution process data faster than CISC architectures *(Completed Instruction Set Computer)* [4].

IOP Conf. Series: Materials Science and Engineering **336** (2018) 012026 doi:10.1088/1757-899X/336/1/012026

2.3 Voice Recognition

Voice Recognition Technology is a speech recognition technology that use human voice as an input signal for later identified by computer system. The technology is the development of interaction between human and computer equipment to minimize output *device* such as *a mouse, keyboard* and other *technological* equipment [5], technology allows a device to recognize and understand the words spoken by way of digitizing the word and match the digital signal with a specific pattern stored in a device. Word spoken, words transformed into a digital signal by converting sound waves into a set of numbers which are then adapted to the specific code for identify these words. Device Voice Recognition requires a sample of the actual words spoken from the user. The word sample will be digitized and then inserted in the computer. Then used as the database in the next match spoken words. Most of its speech recognition device is still dependent on the loudspeakers. This device can only recognize spoken words from one or two people and can only recognize separate words, i.e. words in his delivery there are pauses between words. Only a small part of devices using this technology that is not dependent on Loudspeaker. These devices can recognize spoken words and also many people can recognize continuous words, or words in their delivery. There are no pauses between words [6].

2.4 EasyVR

EasyVR as shown in Figure 1 is a multi-function recognition module that can be used in many control applications that require the detection of not only say yes noise but conversations [7], can be used *host* EasyVR module with UART interface, powered on. DC 5V working voltage 3.3V-, such as PIC and Arduinno board. This module is ideal for EasyVR k diverse applications such as home automation.

Figure 1. Module EasyVR

Compared with other sensor module sound like a sound sensor LM386, this EasyVR module not only detect the presence. sound waves, but also can detect words. or sentences spoken by the user by means of digitizing the word and match the digital signal with a pattern. certain stored in a device.

3. System Design

3.1 Design of Hardware and Software

The design of this educational puppet divided into two parts: the design of hardware and software. Hardware design consists of a *box* component assembly installation *drui* LED on the body of the doll. As for the software design using programming languages used to translate the input received by the sensor. and then will generate output LED lights turn on and play voice on the *sound recorder*.

3.1.1 Hardware Design

The dolls are used in designing this educational doll-shaped stuffed is a human which have dimensions Some of the electronic components used in the design of the system of education in order dolls education designed IIU able to receive voice commands from the user. The components used are as follows:

1. EasyVR

Used as a sensor to detect voice command of user, where in the sensor supports up to 32 *user-defined* voice commands.

2. ATMega328 Microcontroller

Used as a data processing center. This microcontroller 110 pin total of 14 pins, 6 of which PWM (*Pulse Width Modulation) output,* and 32 KB *Flash memory, T3 Sound Recorder SD182.0*

Figure 2. The microcontroller ATmega328

Module is used as a voice recorder or sound *recording module,* this module is also used to play a sound that has been stored previously.

Figure 3. Sound Recording Module IS 1820

4. LED

LED lights here is used as a pointer doll body parts ordered.

5. Lithium Battery Polymer

Used as voltage source for the microcontroller, *Voice Recognition module, sound recorder,* and an LED. The battery used has a 1000mAh power with a voltage of 11.1 volts, for components such as microcontrollers, modules *Voice Recognition, Sound Recorder,* and the battery will be placed on a box component is placed inside the body of the doll.

3.1.2 Software Design

The software in this system using the application as a programming center is using the Arduino IDE software with the Arduino programming language is the C language is easy to use functions that are simple so that it can learned easily. Application serves as place of making the source code which is then uploaded to the microcontroller is present on the system. So the system can run as expected.

1. The initial process begins by initializing the system as EasyVR module, LED pin, and pin Sound Recorder.
2. Then make sure the EasyVR module condition is already connected,
3. If you are connected, the Sound module Recorder turned on, while all LED light stays off,

IOP Conf. Series: Materials Science and Engineering **336** (2018) 012026 doi:10.1088/1757-899X/336/1/012026

4. For conversation, if EasyVR module detects the word "Hello", then Sound Recorder module will play a sound "Hello what, are you", And if EasyNTR detect phrase "Who is your name", Then the Sound module Record play voice "My name is Alphabete".
5. For the introduction of parts of the body, if the module detects words namely EasyVR "Nose".
   "Mouth", "Ear", "Head". "Neck", "Body". "Hand", "Legs", then the LED lights on the part of the word detected bodies will light.

4. Testing

After assembling the tools and system configuration, the next step is performing tests to analyze and determine whether the system can run as expected.

4.1 EasyVR Working Tests

Tests were conducted to determine how working voltage of the sensor noise when receiving voice commands. Working voltage of the sensor is 8 volts EasyVR sound coming from Lipo battery contained in the system, such as image input voltage testing using a measuring instrument AVO meter in Figure 4 below.

Figure 4. Measurement *input* voltage EasyNIR

4.2 Testing Commands Sound

First, testing command voice done by testing the response of the LED and sound recorder by saying a command phrase always. sequentially, namely: Hello, what's your name, eyes, nose, mouth, ears, head, neck, torso, arms, and legs. To begin testing this educational doll, first to do is turn on the system by pressing the switch that is on the edge of the box components. After the indicator lights up command receiver, then uttered the first trigger is the word "doll", in order to further the system can receive commands in the group as in Figure 5 below.

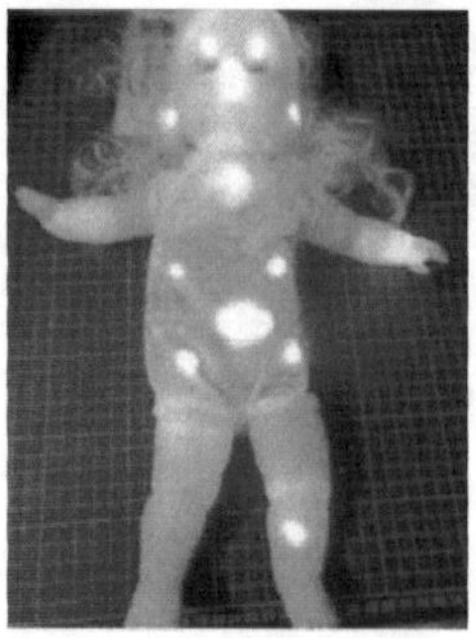

Figure 5. Voters response Dolls

IOP Conf. Series: Materials Science and Engineering **336** (2018) 012026 doi:10.1088/1757-899X/336/1/012026

If all the LED light as shown above, then the system can receive commands in the group, which is the first LED response when it detects the word "eye" In the picture, the LED portion of the eye responds without repetition pronunciation of command. Further testing LEDs response for the word "nose". From the test results all the voice commands, all command specified is already onshore (giving the corresponding output, but there word commands that require 1 to 3 repetitions to provide outputs, it also due to a lack of clarity when pronouncing words or phrases such order. To test a second voice command is done by testing the response of the LED and Sound Recorder to pronounce the word or phrase random order. The test results are shown in Table 1 below.

Table 1. Results of Tests Recurrence Randomized

NO	Command	Response	Repetition
1	Neck	LED ON	1
2	Eye	LED ON	0
3	Feet	LED ON	0
4	What is your name	Sound Recorder ON	
5	Body	LED ON	1
6	Nose	LED ON	2
7	Head	LED ON	1
S	Mouth	LED ON	0
9	Hand	LED ON	1 ,
10	Ear	LED ON	-
11	Hello	Sound Recorder ON	0

From the test results a second voice command, it can be seen that the system is capable of receiving commands

4.3 Testing Distance Sumbe r Sounds

This test is used to evaluate the response of the system based, distance of the sound source when pronouncing the word command, in this test using one word commands that eye, and this test is conducted in a room with difficult conditions. Results of the testing are presented in Table 2 below.

Table 2. Result Recognition of Sound Source Distance

NO	Distance	Assessment				
		I.	2	3	4	5
1	0.5 meters	0	0	0	O	0
2	1 meters	0	*O*	*0*	0	0
3	1.5 meters	0	O	O	O	0
4	2 meters	O	O	O	O	O
5	2.5 meters	0	0	0	0	O
5	3 meters	0	X	X	0	X
-	3,3 meters	O	X	X	X	0
g	4 meters	X	X	X	X	X

Description: 0 = Success X = Fail

From the results of tests performed, it is known effective distance education dolls in detecting a command word that is up to a distance of 2.5 meters, 2.5 meters More of sensor sensitivity is reduced and if a larger distanceof 3.5 meters to 4 meters sensor can not be detects spoken word commands.

4.4 Voice Testing Different

Testing was conducted to determine whether the system can detect more than one voice. In this testing use one word command is the word hello and testing is done with 3 types of different sound, the voice of men, women, and children. From the results of tests performed, the voice of women and children more failure, It caused by tone of voice in women and children. High child, while at the time of recording the voice command using voice men who have a low tone of voice

4.5 Battery Testing

Tests on these batteries are done to determine how long this educational dolls can be used. In peng, a battery test is divided into two types testing, namely:

1. No-load battery test (standby doll). In the no-load battery testing writer use digital measuring instrument for measuring current and voltage input to the system microcontroller connected to the overall system.Testing the battery power is intended to determine time in units of the battery is used up. The results of testing these batteries can result in no-load measurement system when (the doll in standby) is 70mA or 0.07A so if included in the formula be:

Time (t) = Current 1 battery load current Time = 1000 mA / 70mA
Time (t) = 14.2 hours

Figure 6. No-load battery testing

Testing the battery to the load (puppet reading voice commands).

Natural battery testing with a load of author use digital measuring instrument for measuring the flow and input voltage microcontroller system connected to the overall system. Testing the battery power is intended to determine how much time in other units until the battery is used up. Hasil of this battery test can result in measurement when the system load (puppet reading voice command) that is equal to 310mA or 0,31A so if included in the formula becomes:

Time = Current load current Battery 1 time = 1 000mA / 310mA
Time (t) = 3.2 hours

IOP Conf. Series: Materials Science and Engineering **336** (2018) 012026 doi:10.1088/1757-899X/336/1/012026

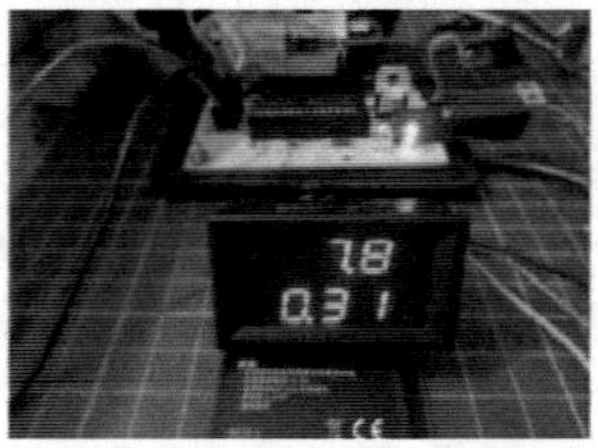

Figure 7. Battery testing with load

5. Conclusions

Based on the research result have been performed, it can be concluded like the following:

1. Designing draft educational doll *voice recognition* rbasis be done with a few step namely, designing component *box,* mounting the LED on the body of the doll, and determining which words or sentences acceptable whole system.
2. Sound recorders and LED module on the body of the doll can respond properly EasyVR sensor detects voice commands up to a distance of 2.5 meters, but sometimes it can take 1 to 3 times the repetition pronunciation order to provide a response.

6. References

[1] Muhson, Ali.2010. Pengembangan Media Pembelajaran Berbasis Teknologi. Jurnal Pendidikan Akuntansi Indonesia.
[2] Budiharto, 2010. Widodo Robotika Teori dan Implementasi. Yogyakarta ANDI.
[3] Ginting, Bernike, Natalia. 2012. Penggerak Antena Modem USB Tiga Dimensi Berbasis Mikrokomputer Menggunakan Arduino Uno. Universitas Sumatra Utara
[4] Dinata, Yuwono Marta. 2015. Arduino itu Mudah. Surabaya : PT Elex media Komputindo.
[5] Herdianto. 2014. Penerapan Speech Recognition Library untuk Query Basis Data. Jurnal Ilmiah Mahasiswa UBD
[6] Ardi, Jatra Kunaia., Nurussa'adah, Mochammad, Rif an. 2014. Implementasi Sistem Recognition pada Robot Pemindah Objek sebagai Sistem Navigasi. Universitas Brawijaya
[7] Birdayansyah, Radi., Soedjawanto, Noer., Zcbua, Osea. 2015. Pengendalian Kecepatan Motor DC Menggunakan Perintah Suara Berbasis Mikrokontroler Arduino. Jurnal Rekayasa dan Teknologi Elektro
[8] Ardi, Jatra Kunaia., Nurussa'adah, Mochammad. 2014. Rif an Implementation System Object Recognition on Robot Shifters as Navigation Systems. Brawijaya University

IOP Conf. Series: Materials Science and Engineering **336** (2018) 012027 doi:10.1088/1757-899X/336/1/012027

Applying Student Team Achievement Divisions (STAD) Model on Material of Basic Programme Branch Control Structure to Increase Activity and Student Result

Aulia Akhrian Syahidi1*, Arifin Noor Asyikin2, Asy'ari3
[1]Master of Computer Science, Faculty of Computer Science, Brawijaya University,
[2]Department of Electrical Engineering, State Polytechnic of Banjarmasin,
[3]Department of Information Technology Education, STKIP PGRI Banjarmasin,

* aakhriansyahidi@student.ub.ac.id

Abstract. Based on my experience of teaching the material of branch control structure, it is found that the condition of the students is less active causing the low activity of the students on the attitude assessment during the learning process on the material of the branch control structure i.e. 2 students 6.45% percentage of good activity and 29 students percentage 93.55% enough and less activity. Then from the low activity resulted in low student learning outcomes based on a daily re-examination of branch control material, only 8 students 26% percentage reached KKM and 23 students 74% percent did not reach KKM. The purpose of this research is to increase the activity and learning outcomes of students of class X TKJ B SMK Muhammadiyah 1 Banjarmasin after applying STAD type cooperative learning model on the material of branch control structure. The research method used is Classroom Action Research. The study was conducted two cycles with six meetings. The subjects of this study were students of class X TKJ B with a total of 31 students consisting of 23 men and 8 women. The object of this study is the activity and student learning outcomes. Data collection techniques used are test and observation techniques. Data analysis technique used is a percentage and mean. The results of this study indicate that: an increase in activity and learning outcomes of students on the basic programming learning material branch control structure after applying STAD type cooperative learning model.

1. Introduction

1.1. Background Information and Problem Statement

Based on my experience teaching the material of branch control structure, students are less active in responding to the material delivery of the branch control structure provided by the teacher. Students are less able to understand the subject matter and students are not enthusiastic about the delivery of material done by the teacher or the implementation of the task. Students feel the task given by the teacher is a heavy burden. Often the tasks assigned by the teacher cannot be solved well. In fact, teachers tend to be the center of all learning resources; it should be in the application of Curriculum 2013 students who should be more active. This condition leads to low student activity on attitude assessment during learning process on branch control material that is 2 students with the percentage of 6.45% good activity and 29 students with the percentage of 93.55% activity enough and less. The next reason is the result of student learning which is reviewed on the daily test result of branch control material, only 8 students with the percentage of 26% that reach the minimum completeness criteria and 23 students with 74% percent do not reach the minimum mastery criteria. Therefore it is necessary to change the way of learning, there are several factors that affect student learning outcomes, one of them is caused by teachers who tend to use the learning model that does not vary, has not been much-motivating students' activities to interact, and learning also tends to be still focused on the teacher both in terms of delivery material or settlement tasks. This situation makes the students cannot free to express what occurred in his mind, the emergence of mistrust in conveying ideas, reluctant to ask,

afraid of being wrong, embarrassed to ask, so that the learning process seemed to be a student reinforcement to develop and can lead to student saturation, material branch control structure.
A student who is in saturation, his wits system cannot work as expected in processing information items, so that learning achievement may decrease. Learning model that tends to minimize the involvement of students in the learning process resulted in the students become less active, less independent, did not have the courage to express their own opinions and also students are less able to think creatively so that student achievement is low. To attract liveliness and interest in student learning then the teacher must use learning model other than conventional learning model. Therefore, teachers are required to choose appropriate models of learning that are active, creative, effective, innovative, and effective. So in the learning process students can actively understand the material and their learning achievement increases. Vocational High School is one of the educational institutions that have the responsibility to create human resources (HR) that has the skills, abilities, and skills so that graduates can develop performance while in the workplace. In accordance with the Regulation of the Ministry of Education and Culture No. 70 of 2013 on the Basic Framework and Structure of the Vocational High School Curriculum, the objective of the Curriculum 2013 is to prepare Indonesian people to have the ability to live as individuals and citizens who are faithful, productive, creative, innovative and effective able to contribute to the life of society, nation, state, and world civilization. One of the vocational schools in Banjarmasin City that still apply the Curriculum 2013 is SMK Muhammadiyah 1 Banjarmasin. Currently, SMK Muhammadiyah 1 Banjarmasin has three skill programs, one of which is Computer and Network Engineering. In the program Computer and Network Engineering expertise, there are relatively new productive subjects in the Curriculum 2013 is Basic Programming. The Basic Programming Course is contained in the curriculum structure of Computer and Network Engineering skills program. These subjects belong to the C1 group of specialists in the area of expertise. The purpose of these Basic Programming subjects is as basic understanding for learners in a programming language which later becomes the basis for other subjects as well as train students' logic and creativity.

1.2. Basic Programming Subject

The program is an algorithm written in computer language. The programming language is the computer language used in writing programs. The programmer is the person who created the computer program. Programming is the activity of designing and writing programs. In programming there is the activity of writing program code, this activity is called coding [11]. The subject of Basic Programming in Vocational High Schools is a productive subject in the structure of the Computer and Network Engineering (TKJ) programming curriculum, the subject of specialization C1 is the area of fundamental expertise. Basic Programming teaches and trains students to understand and apply programming languages whose functions can also be applied in other programming languages, as a reinforcement of logic, and student creativity. The supporting application used in Basic Programming learning is Turbo Pascal 7.0 which uses the famous Pascal programming language with structured programming.

1.2.1. Student activity according to Diedrich, can be classified as follows: Visual activities, Oral activities, *Listening activities, Writing activities, Drawing activities, Motor activities, Mental activities, and Emotional activities.*

1.2.2. Learning outcomes are the changes that occur in students, both those involving cognitive, effective, and psychomotor aspects as a result of learning activities [22].

1.2.3. Student Team Achievement Divisions (STAD)

According to Slavin (The steps to be taken in applying STAD type cooperative learning model consists of six stages, as follows: the First step of delivery of goals and motivation Second step of group division Teacher's third presentation step Fourth step of team learning activities teamwork) with LKS, teamwork is STAD's most important image The fifth step of the quiz (evaluation) The sixth step of the team achievement award, then awarding the certificate or other prize for the group's success [16].

1.3. The Branch Control Structure
The branching control structure of programming generally uses IF (if), THEN (then), and ELSE (besides) keywords. There are several types of branch control structures based on the number of conditions encountered, below is the type of branching control structure that is as follows: Branching one condition, Branching two conditions, Branching more than two conditions, and Nested branching. The condition itself is a Boolean type expression, meaning it can only be true or false.

2. Research Methodology
2.1. Research Design
References [1] illustrate that: Classroom Action Research consists of four steps in one cycle and its repetition, presented in Figure 1.

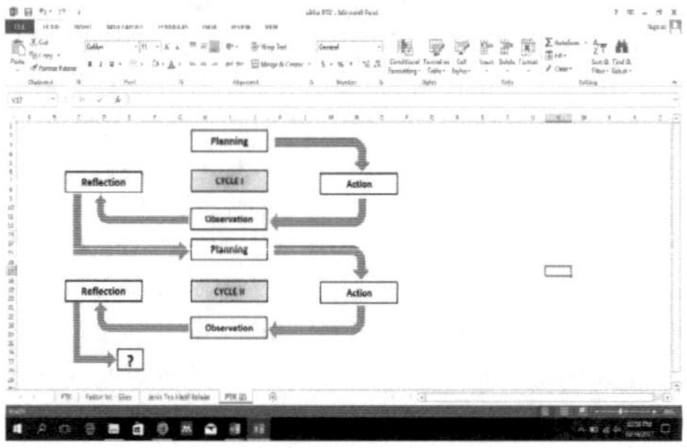

Figure 1. Class Action Cycle Research Chart

This Classroom Action Research will be carried out with the following stages: Cycle I preparation of the plan as follows: Prepare the implementation schedule of cycle I action., prepare the Learning Implementation Plan (RPP) of the subject of the branch control structure at each meeting, preparing the student activity sheets cycle I, preparing the ultimate evaluation sheet evaluation cycle instrument is an objective test, preparing the observation sheet instrument to observe the student activity in the learning process takes place. Action: Implementation of class actions carried out as many as three meetings. The first and second meetings are held in teaching and learning process with 2 x 35 minute time allocation, while for third meeting evaluation is done with 75 minutes time allocation. Teaching and learning process is done in accordance with the RPP that has been made at the stage of preparation of the plan. For evaluation, students are given objective tests in the form of multiple choice questions implemented at the end of the cycle. Observation: Observations made during the learning process took place conducted by two observers to observe student learning activities. Reflection: Reflection is done to examine what has been produced or what has not been completed at the stage of action based on the evaluation results at the end of the cycle and the results of observation. Cycle II: The stages performed in cycle II follow the stages in cycle I. In this second cycle, the preparation of plans prepared based on the results of reflection in cycle I. Activities undertaken in cycle II is intended to improve or improve the implementation of learning in cycle I.

2.2. Research Instruments
In the process of evaluating the learning or assessment process and learning outcomes, teachers often use measuring instruments in the form of certain instruments, both test, and non-test. This instrument has a function and role that is very important in order to know the effectiveness of the learning process in school [2]. Test: a good test instrument criterion needs to be tested in the following way: Validity, Level of Difficulty, Distinguishing Power, and Reliability. Non-test instruments can be used if we

want to know the quality of processes and products of a job as well as matters pertaining to effective domains, such as attitudes, interests, talents, and motivations. Each measured dimension and aspect requires different tools or instruments. In principle, whenever conducting a learning evaluation, we can use test and non-test techniques. Learning outcomes can be in the form of theoretical knowledge, skills, and attitudes. Theoretical knowledge can be measured using test techniques. Skills can be measured using action tests. The changes in attitude and growth of children in psychology can only be measured by non-test techniques, such as observation, interviews, attitude scales, and others [2]. To obtain data on learning activity of student in the learning process and implementation of STAD type cooperative learning model, hence in this research used the non-test instrument of observation type with a tool used is observation sheet/guidance. Before the non-test instrument, the observation sheet is used in advance the validity of the content through the experts' judgment is the assessment performed by the experts.

3. Data Collection Techniques

Observation is often interpreted as a narrow asset, i.e. pay attention to something by using the eyes. In the sense of psychology, observation is the activity of loading attention to an object by using all the senses. Thus, observation can be done through sight, smell, hearing, touch, and taste [1]. The test is a series of questions or exercises as well as other tools used to measure skills, knowledge of intelligence, abilities or talents possessed by individuals or groups. In this research, the test is done to know the result of student learning after learning model of cooperative learning type STAD applied [1]. Based on the number of learners, the test results of learning there are two types of group tests and individual tests. Group test is a group test. The teacher will deal with a group of learners. The individual test is a test conducted individually. The teacher will deal with a learner. To obtain data on student learning result in a final evaluation of cycle and implementation of STAD type cooperative learning model, this research use objective type test instrument in the form of multiple choice problem. Problems are made in accordance with the procedure of making multiple choice questions. Problems used also have done the validity of the content and validity of constructs in the form validity test, the level of difficulty, distinguishing power, and reliability.

4. Research Results and Discussion

4.1. Implementation of STAD

Implementation of the learning process by applying STAD type cooperative learning model, shown in the Figure 2a-d.

a. Presentation **b.** Teamwork **c.** Quiz **d.** Achievement

Figure 2. Implementation of the learning process by applying STAD type cooperative learning model

STAD type cooperative learning is one type of cooperative learning model using small groups with the number of members every four to five students in a heterogeneous way. In the beginning, we deliver learning objectives, materials, group activities, quizzes, and group awards. The main idea behind STAD is to encourage students to encourage and help each other to master the skills that teachers teach. If students want a group to get a reward, they should help their group's friends learn the lesson. They should encourage group mates to do their best, show the norms that learning is important, valuable, and fun. Based on the observations made during the research conducted, the things that make students active in basic programming learning material branch control structure, among others: Submission of objectives and motivation, in this step the researchers convey the lesson objectives to be achieved on learning STAD and motivate students to learn actively and creatively. Motivation in the form of invitations to learn and the benefits of learning the material and at the end of the meeting will be awarded a certificate. Division of the group, in this step the students are divided into seven groups, each group consisting of four to five students prioritizing class heterogeneity in academic achievement and gender. So it gives the stimulus for each student to work together, encourage each other, help each other to collect scores, and discuss them to become the best group. Presentation, in this step the researcher submits the subject matter by first explaining the lesson objectives to be achieved at the meeting and the importance of the material is learned. In the learning process, the researcher is assisted by Turbo Pascal 7.0, learning media based on PowerPoint and Flash, and ties it to real problems that occur in everyday life. Also explained about the skills and abilities those are expected to be mastered by students, and the work to be done as well as ways of doing it. Teaching activities in the team (teamwork), in these step students, learn in groups that have been formed. The researcher prepares the LKS as a guideline for group work, so all members master and each contributes. This LKS aims to stimulate students to understand the materials learned, in constructivist theory, students learn to build their own knowledge. As long as the team works, researchers are assisted by observers making observations, providing guidance, encouragement, and assistance when needed. Teamwork is the most important image of STAD. Individual quiz, in this step the researcher, evaluates the learning outcomes through the quiz about the material being studied and also evaluates the presence of the work of each group. Students are given seats individually and are not allowed to cooperate. This is done to ensure that individual students are accountable to themselves in understanding the material. Researchers set a score for each problem. Individual quiz questioning is given to lure students to understand the material and to know the students' mastery of the material they have learned during the lesson. The team achievement award, in this step after the quiz implementation, the researcher examines the student work and is given a number ranging from 0-100. Furthermore, the award of a certificate or other prizes for the success of the group in accordance with the scores of group development obtained. Comparison of inter-group rewards at each cycle is presented in Table 1.

Table 1. Comparison of Predicates of Inter-Group Appreciation in Each Cycle

Cycle	Meeting	Team Predicate						
		A	B	C	D	E	F	G
I	1	Super Team	Super Team	Super Team	Super Team	Good Team	Super Team	Super Team
I	2	Super Team	Great Team	Super Team	Great Team	Great Team	Super Team	Super Team
II	1	Super Team	Super Team	Super Team	Super Team	Super Team	Super Team	Super Team
II	2	Super Team	Super Team	Super Team	Super Team	Super Team	Super Team	Super Team

4.2. Student Activity

By applying STAD type cooperative learning model to basic programming subject material of branch control structure, student activity becomes increasing. A comparison of the increased student activity in cycle I and cycle II is presented in Table 2.

Table 2. Comparison of Student Activity Enhancement in Cycle I and Cycle II

Activity Indicators	Statement	Statement Number	Average Cycle I (%)	K	Average Cycle II (%)	K

Visual Activities	Students pay attention to demonstration experiments of others' work	1	96.77	AB	100	AB
	Students pay attention to drawings	2	96.77	AB	100	AB
Oral Activities	Students express opinions on discussion groups	3	59.68	K	82.26	B
	Students ask the discussion group	4	48.39	K	74.19	B
Listening Activities	Students listen to the teacher's description	5	100	AB	100	AB
	Students discuss with the teacher about the material that has not been understood	6	46.77	K	67.74	C
Motor Activities	Students are experimenting	7	69.35	C	82.26	B
	Students solve the problem	8	69.35	C	80.65	B
Emotional Activities	Students are interested in studying	9	96.77	AB	98.39	AB
	Students are eager to attend lessons	10	95.16	AB	100	AB
Average Cycle I and II			77.90	B	88.55	B

Based on Table 2 it can be seen that the average of student activity on statement number 1, 2, 3, 4, 6, 7, 8, 9, 10 has increased from cycle I to cycle II. At the first meeting, inactivity is considered reasonable because students have not been able to adapt to the applied learning situations and conditions. But at the next meeting, the students are getting used to the situation and condition and understand the rules/learning norms applied to STAD type cooperative learning model. So, all students experience increased activity at each meeting. The increase occurs because the students are already familiar and passionate about the application of the STAD model, so that the discussion and teamwork run smoothly, the students have started to feel confident to express their opinions, the students have started daring to ask, either between groups or with teachers, students are very passionate in doing experiments, students are very passionate in solving problems, students are very interested and eager to follow the lesson. However, from the observed activity, there is one activity that does not increase that is in statement number 5, this is because the score of activity of cycle I maximum is 100% so that in cycle II also have activity score 100%. The increase is also based on the advantage of applying the STAD model as follows: Students work together to achieve the goals by enforcing group norms, Students actively assist and motivate the spirit to succeed together, actively acting as peer group to further improve Group success, interaction among students in line with their increasing ability in opinion, enhancing individual abilities, enhancing group skills, not competitive, not taking vengeance [18]. All the advantages have been felt when applying STAD model to increase student activity. Average of student activity in cycle I with average percentage 77.90% with good qualification have fulfilled indicator of the success of research, but because hold on theory from Kemmis and Taggart hence researcher continue cycle II. The average activity of cycle II with a percentage of 88.55% with good qualification has also met the success indicators in this study.

4.3. Student Learning Outcomes

By applying STAD type cooperative learning model on the basic programming subjects of branch control material, student learning outcomes become increasing. Comparison of improvement of student learning outcomes at the end of the cycle I evaluation and final evaluation of cycle II is presented in Table 3.

Table 3. Comparison of Improved Student Learning Outcomes in the Evaluation of End of Cycle I and Cycle II

No.	Value	Alphabet	Predicate	Cycle I		Cycle II	
				F	%	F	%
1	97 – 100	A	Expert	-	-	1	3.23
2	88 – 96	A-	Expert	22	70.97	20	64.52
3	80 – 87	B+	Dominate	6	19.35	9	29.03
4	72 – 79	B	Dominate	2	6.45	1	3.23
5	63 – 71	B-	Dominate	-	-	-	-
6	55 – 62	C+	Less Mastery	1	3.23	-	-
7	47 – 54	C	Less Mastery	-	-	-	-
8	38 – 46	C-	Less Mastery	-	-	-	-
9	30 – 37	D+	Not Mastering	-	-	-	-
10	0 – 29	D	Not Mastering	-	-	-	-
The number of students				31	100	31	100
Average Student Results Based on Individual Completeness				89.25		90.32	
Student Learning Outcomes Based on Classical Mastery				97%		100%	

Based on Table 3 it can be seen that there is an increase in both the frequency, the average percentage of student learning outcomes based on individual completeness, and the average percentage of student learning outcomes based on individual completeness in each cycle. For students who scored between 97-100 with the letter A category Expert in the first cycle there has not been a single person and on the 2nd cycle increased marked by 1 person obtaining a value between 97-100 with the letter A category Expert with a percentage of 3.23%. Students who scored between 88-96 with the A- category Expert in cycle I amounted to 22 people with percentage 70.9% and in cycle II amounted to 20 people with a percentage of 64.52% decreased the number of students because 1 person increased to the value 97-100 and 1 person decreased to the value of 80-87. Students who scored between 80-87 with the letter B+ category Dominate in cycle I amounted to 6 people with percentage of 19.35% and in cycle II amounted to 9 people with a percentage of 29.03% increase in the number of students as much as 3 people because 1 person of value between 88-96 and 2 people of the value 72-79. Students who scored between 72 and 79 is categorized by letter B. Dominate in cycle I amounted to 2 people with percentage of 6.45% and in cycle II amounted to 1 person with percentage 3.23%, there was a decrease in the number of students because 2 people increased to the value between 80-87 and 1 person increase from the value between 55-62. Student learning outcomes consisting of classical completeness and individual completeness increased in each cycle. The first cycle of classical completeness with a percentage of 97% with an average score based on the individual completeness is 89.25 with the letter A- the predicate is Expert have fulfilled the indicator of success of the research, but because the researcher is shaped by the theory of Kemmis and Taggart in [2] proceed to cycle II. In the second cycle of classical completeness with 100% percent with the average value based on the individual completeness is 90.32 with the letter A- predicate Expert also has met the success indicator in this research.

5. Conclusion

Based on the results of research and discussion in Classroom Action Research conducted, it can be concluded as follows: Student learning activities increase in each cycle that is on the first cycle with the average percentage of 77.90%, and on the second cycle with an average percentage of 88.55%. This shows that by applying STAD type cooperative learning model can increase student activity on branching control material. Student learning outcomes consisting of mastery of classical and individual learning occurs in every cycle that is in the cycle I classical completeness with a percentage of 97% with an average value based on individual completeness is 89.25, and on cycle II with the classical completeness 100% percentage with average value based on individual completeness is 90.32. This shows that by applying STAD type cooperative learning model can improve student learning outcomes on the material of branch control structure.

5.1. Future Work

This research is limited to the basic programming subjects of branch control structural material, so there is a need for further research by applying STAD type cooperative learning model on other subject or material, and also as study material for the next researcher.

References

[1] Arikunto, Suharsimi 2014 *Research Procedures A Practice Approach*. Jakarta: PT. Rineka Cipta.
[2] Arifin, Zainal. 2014. *Evaluation of Learning*. Bandung: PT. Youth Rosdakarya.
[3] Budiyono. 2003. *Educational Research Methodology*. Surakarta: UNS Press.
[4] Dimyati and Mudjiono. 2013. *Learning and Learning*. Jakarta: PT. Rineka Cipta.
[5] Hidayati, Cana. 2012. *Non-Test Instrument Validation in Mathematical Research*. Yogyakarta: FMIPAUNY.
[6] Isjoni. 2014. *Cooperative Learning Develops Group Learning Ability*. Bandung: Alfabeta.
[7] Jihad, Asep, and Haris, Abdul. 2012. *Evaluation of Learning*. Yogyakarta: Multi Pressindo.
[8] Kunandar. 2012. *Easy Step Classroom Action Research as Teacher Professional Development*. Jakarta: PT. RajaGrafindo Persada.
[9] Kunandar. 2013. *Authentic Assessment (Assessment of Student Learning Outcomes Based on Curriculum 2013) A Practical Approach Accompanied by Example*. Jakarta: PT. RajaGrafindo Persada.
[10] Mulyasa. 2013. *Classroom Action Research Practices*. Bandung: PT. Youth Rosdakarya.
[11] Munir, Rinaldi. 2011. *Algorithms and Programming in Pascal and C Revised Edition*. Bandung: Informatics.
[12] Muslich, Masnur. 2009. *Implementing PTK (Classroom Action Research) is Easy Classroom Action Research*. Jakarta: PT. Earth Script.
[13] Regulation of the Minister of Education and Culture of the Republic of Indonesia Number 70 the Year 2013 on the Basic Framework and Structure of Vocational High School Curriculum / Madrasah Aliyah Vocational.
[14] Riduwan. 2005. *Learn Easy Research for Teachers-Employees and Beginner Researchers*. Bandung: Alfabeta.
[15] Riduwan and Sunarto. 2015. *Introduction to Statistics for Education, Social, Economic, Communication, and Business Research*. Bandung: CV. Alfabeta.
[16] Rusman. 2013. *Learning Models Develop Teacher Professionalism Second Edition*. Jakarta: PT. RajaGrafindo Persada.
[17] Sardiman A.M. 2014. *Interaction and Teaching Motivation Motivation*. Jakarta: PT. RajaGrafindo Persada.
[18] Shoimin, Aris. 2014. *68 Model of Innovative Learning in Curriculum 2013*. Yogyakarta: Ar-Ruzz Media.
[19] Slavin, Robert E. 2005. *Cooperative Learning Theory, Research, and Practice*. Translated by: Allymand Bacon. Bandung: Nusa Media.
[20] Sudijono, Anas. 2014. *Introduction to Education Statistics*. Jakarta: PT. RajaGrafindo Persada.
[21] Sugiyono. 2015. *How To Easily Organize: Thesis, Thesis, and Dissertation*. Bandung: CV. Alfabeta.
[22] Susanto, Ahmad. 2015. *Learning and Learning Theory at Elementary School*. Jakarta: Kencana Prenada Media Group.
[23] Sutikno, M. Sobry. 2013. *Learning and Learning Creative Efforts in Achieving Successful Learning*. Lombok: Holistica.
[24] Tiantong, M., & Teemuangsai, S. 2013. *Student team achievement divisions (STAD) technique through the moodle to enhance learning achievement*. International Education Studies, 6(4), 85–92.
[25] Trianto. 2009. *Designing Innovative-Progressive Learning Model*. Jakarta: Kencana Prenada Media Group.
[26] Trianto. 2012. *Designing Innovative-Progressive Learning Models*. Jakarta: Kencana Prenada Media Group.
[27] Widiyanto, Ardes. 2014. *Basic Programming Class X*. Bogor: Yudhishthira.
[28] X. Li, and Luo, L. 2010. *Probe into STAD cooperative learning based on moodle*. 2nd International Workshop on Education Technology and Computer Science, ETCS 2010, 2, 424–427.
[29] Zhao, J., McConnell, D., Li, K., and Akahori, K. 2002. *STAD modeling method for computer supported group learning*. Proceedings - International Conference on Computers in Education, ICCE 2002, 344–348
[30] Zhou, H., Wu, Z. F., and Wang, A. J. 2009. *A probe into the evaluation strategies to facilitate students' interaction and autonomous learning in CET*. Proceedings of the 1st International Workshop on Education Technology and Computer Science, ETCS 2009, 3, 418–421.

IOP Conf. Series: Materials Science and Engineering **336** (2018) 012028 doi:10.1088/1757-899X/336/1/012028

Development Of Ultrasonic Testing Based On Delphi Program As A Learning Media In The Welding Material Study Of Detection And Welding Disables In The Environment Of Vocational Education

Bagus Dwi Cahyono*, Chandra Ainur

Graduate Program, Universitas Negeri Surabaya

*baguscahyono16070895013@mhs.unesa.ac.id

Abstract: The development of science and technology has a direct impact on the preparation of qualified workers, including the preparation of vocational high school graduates. Law Number 20 the Year 2003 on National Education System explains that the purpose of vocational education is to prepare learners to be ready to work in certain fields. One of the learning materials in Vocational High School is welding and detecting welding defects. Introduction of welding and detecting welding defects, one way that can be done is by ultrasonic testing will be very difficult if only capitalize the book only. Therefore this study aims to adopt ultrasonic testing in a computer system. This system is called Delphi Program-based Ultrasonic Testing Expert System. This system is used to determine the classification and type of welding defects of the welded defect indicator knew. In addition to the system, there is a brief explanation of the notion of ultrasonic testing, calibration procedures and inspection procedures ultrasonic testing. In this system, ultrasonic input data testing that shows defects entered into the computer manually. This system is built using Delphi 7 software and Into Set Up Compiler as an installer. The method used in this research is Research and Development (R & D), with the following stages: (1) preliminary research; (2) manufacture of software design; (3) materials collection; (4) early product development; (5) validation of instructional media experts; (6) product analysis and revision; (8) media trials in learning; And (9) result of end product of instructional media. The result of the research shows that: (1) the result of feasibility test according to ultrasonic material testing expert that the system is feasible to be used as instructional media in welding material subject and welding defect detection in vocational education environment, because it contains an explanation about detection method of welding defect using method Ultrasonic testing in detail; (2) feasibility test results according to media experts, that this system has a very attractive visual, user friendly, compatible with windows and Linux and media size that is not too large; And (3) result of test by using data of indication of welding defect in PT PAL Surabaya, obtained classification data of welding defect in accordance with calculation of welding defect classification.
Keywords: Expert System, Ultrasonic Testing, Classification of Welded Disability, Delphi

1. Introduction

Currently, the development of science and technology has grown very rapidly. The rapid development of science and technology requires the preparation of qualified workers, including the preparation of vocational school graduates. Based on Law Number 20 the Year 2003 regarding National Education System explains that the purpose of vocational education is to prepare learners to be ready to work in certain fields. One of the learning materials in Vocational High School is welding and detecting welding defects. Introduction of welding and detecting welding defects, one way that can be done is by ultrasonic testing will be very difficult if only capitalize the book only. Because learning just by using a book will be bored and difficult o understand. Therefore this study aims to adopt ultrasonic testing in a computer system. This system is called Delphi Program-based Ultrasonic Testing Expert System. This system is used to determine the classification and type of welding defects of the welded defect indicator knew. In addition to the system, there is a brief explanation of the notion of ultrasonic testing, calibration procedures and inspection procedures ultrasonic testing. In this system, ultrasonic input data testing that shows defects entered into the computer manually. This system is built using Delphi 7 software and Inno Set Up Compiler as the installer. To limit this research to more focus on the formulation of the problem, several restrictions are required as follows:

1. Experts system is only used for the evaluation of the ultrasonic testing of welded specimens (carbon steel and iron ingots);
2. The process of entering data into the ultrasonic flaw testing computer is done manually;
3. Ultrasonic techniques and standard procedures in accordance with the standards of ASME Sec. V Non-Destructive Examination;
4. Criteria for acceptance of defect types by standard AWS D1.1 in Cyclically loaded no tubular connection and Statically loaded no tubular connection;
5. The calibration procedure used in accordance with the calibration procedure UT Flaw Detector with brand SIUI CTS-9005;
6. Provide appropriate solutions of the type of defects found, and efforts to overcome the defect appears.

1.1. Expert System
Expert systems are a computer-based application that is used to resolve problems in accordance with the thinking of experts. Experts are people who have special expertise that can solve problems that can not be solved by people in general [6].
The architecture of a typical expert system is shown in Figure 1 [10].

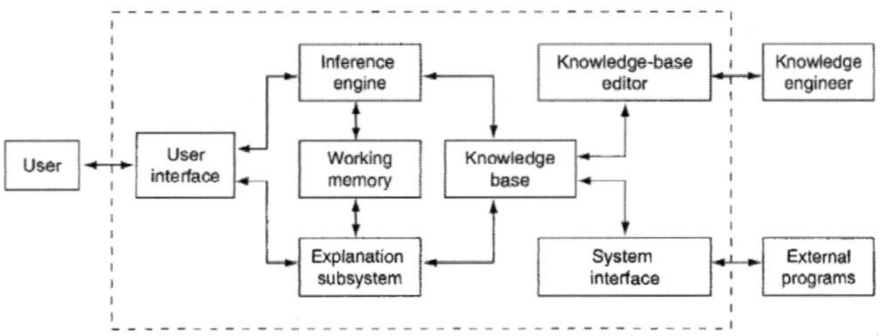

Figure 1. Architecture of Expert System

2. Method of Reasoning
The reasoning is also called inference method is the process to generate information from the facts that are known or assumed. In general, there are two methods of reasoning, trace-forward (forward chaining and backward chaining) [10].
Forward Chaining begins by placing the initial data in the form of symptoms as a premise in working memory. In the working memory, inference engine to match the initial data with the facts contained in the knowledge base, choose one rule and determine a conclusion.

2.1. Ultrasonic
The basic principle of this test is to transmit ultrasonic waves into the material being examined. The ultrasonic wave will propagate into the material and will be reflected when the direction of propagation found an air cavity. The reflected wave will be welcomed back by the probe (transducer) in the form of a pulse on the CRT screen which is a pulse or pulses disabled Backwall [2].

3. Experimental details
The method used in this research is Research and Development (R & D), with the following stages: (1) preliminary research; (2) manufacture of software design; (3) materials collection; (4) early product

development; (5) validation of instructional media experts; (6) product analysis and revision; (8) media trials in learning; And (9) result of end product of instructional media.

3.1. Input Requirement
There are two kinds of input requirements. The first material data form of materials, welding technique and material dimensions (thick material). The second data indicative of a form of ultrasonic defect, the location indication defect, a flaw in the screen display pulse UT flaw detector, the zero reference level, the level indication, sound path value, value and visual indication rating disabilities.

3.2. Process Requirement

3.2.1. Inference engines
To the needs of the input data as output data as input into output using the technique of reasoning trace Forward (Forward Chaining), the input is processed through three stages:
Phase I Design Creating a Program Flowchart In making the flowchart authors limit the materials to be used as a display as well as information that can be accepted by users. At the design, stage is also the input (data ultrasonic material and data) and output (the presence and type of weld defects, material acceptance, the solution to the existing defects) were collected and grouped according to the criteria. Phase II Development the criteria that have been grouped, made into a database which is then processed into a program. Phase III Testing At the last stage is a program that will process the input is matched to the output according to the criteria.

3.2.2. Output Requirement
To the requirement of output data showing the conclusion of the classification and type of defects in the welds are obtained, the cause of the disability and appropriate solutions. These conclusions were obtained through the execution of the program by entering the inputs that have been stored in the database.

3.3. Requirement of Software and Hardware
The software used in developing an expert system program is the Embarcadero Delphi 7 AlphaLite its components and software Inno compiler set up. The hardware needs to follow the requirements of the software.

3.4. Formulation Knowledge Base and Inference Machines
Material Data and Ultrasonic Data are grouped, arranged into a decision table and decision trees.

3.5. The design of the Main Window Interface

The design of the main window interface including: interface design of database page, interface design of evaluation defect classification page, interface design of evaluation defect classification page, and the design of the interface type of disability evaluation page

4. Program Algorithm
Algorithm program classification expert system for evaluation of weld defects and defect type is shown in the following figure2.

IOP Conf. Series: Materials Science and Engineering **336** (2018) 012028 doi:10.1088/1757-899X/336/1/012028

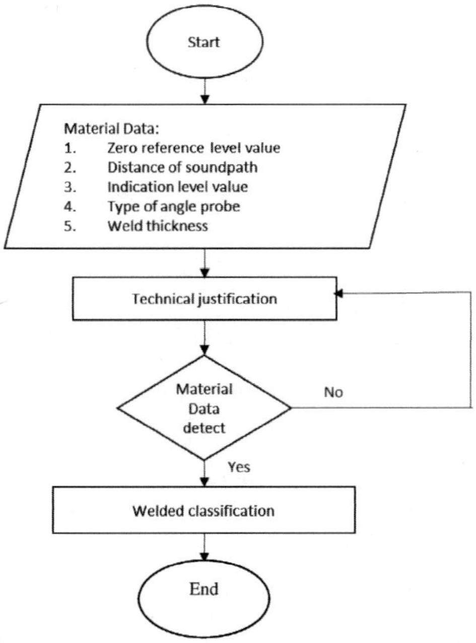

Figure 2. Flowchart of classification defect evaluation

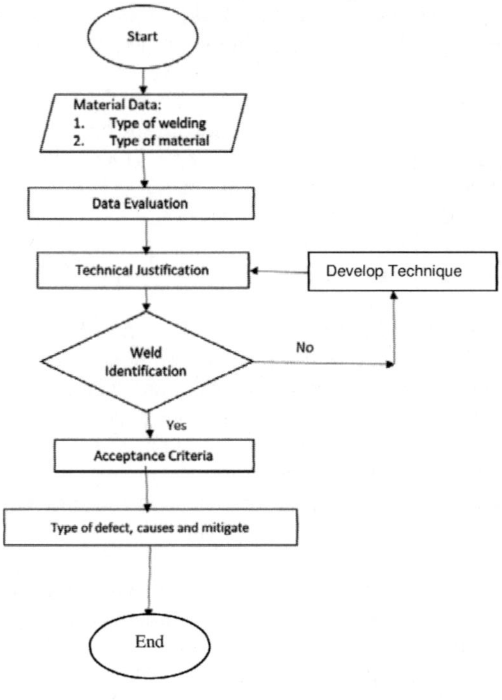

Figure 3. Flowchart of defect type evaluation

5. Results And Discussion

5.1. Implementation Program

Implementation is the stage where the program is ready to operate on a real stage, so it will be known whether the programme can actually produce the desired goal. This stage is done by running the program.

5.1.1. Testing of UT Basic Principles

In this test testing the form view of login in this form, the user is prompted to enter a username (User Name) and password to get into the expert system program detecting weld defects using ultrasonic methods with certain username and password.

After entering the username and password correctly then the user will go to the home page of an expert system program. On the home page form when we press the menu there will be several sub-menus are:

1. Choice: Useful for the next program option to choose.
2. Basis Data Cacat: the page is useful to go into an explanation disabled
3. Exit: To exit from the program expert systems.

On the basic principles UT group box, there are two buttons that are the database that serves to get to the database. A page where users can retrieve data associated with the UT in the form of a word (format. RTF). The data is already available in the Database folder, which folder should be placed in My Documents that can be opened directly.When the user presses the Open a File button, the user will go to the My Documents, then we select the Database folder and then the user must select the folder description of the UT is cool. As at this time the author took the test database on the basic principles of UT.

On this page there are several buttons which are:

1. Open: allows you to open a database in the form of a word (format. RTF).
2. Save: allows you to save the database in the form of a word (format. RTF) if there is a change.
3. Edit Font: allow you to edit the font size and type of letters used.
4. Pages setting: begun used to set the page before printing.
5. Printer setting: allows you to choose a printer that is used.
6. Print: allows you to print (print) database.

To return to the home page then the user must press the Return to Menu Submenu. As for the exit of the program, the user must press the Exit submenu on the Menu. To get help with this program the user must press the Help menu. At the main page the user selects the Start button on the Basic Principles UT group box then the user will go to the options page explanation of the UT. At the time, the selected menu shows the submenu option and the Return to Menu. Submenu option serves to select a group of UT explanation that users want. These options are About UT, Device UT and Couplant, Calibration Device, and Discontinuity. Furthermore, the user can just choose which one ultrasonic explanation required by him by pressing the related explanation.

At the time the user selects the option of Basic Principal Ultrasonic on the dialog box about UT, it will be displayed page explanation of the basic principles of ultrasonic. On this page, there are buttons to go up to a further explanation. And there is a menu that contains submenu Return to return to the options and sub-menu Return to Menu to return to the homepage. On this page, there is also an example of the calculations used to calculate the wave of thick material based on the formula contained in the explanation.

The next time the user will perform ultrasonic inspections in the field, need to perform calibration of UT equipment. In this program, there has been a description of the calibration procedure UT

IOP Conf. Series: Materials Science and Engineering **336** (2018) 012028 doi:10.1088/1757-899X/336/1/012028

equipment Sec.V. ASME standards Where the calibration procedure is divided into 2 kinds of probe normal calibration procedures and calibration of the probe angle. Apart from these two calibrations prior to the inspection of the calibration procedure needs to be done DAC (Distance Amplitude Correction), so that when the implementation of the inspection data obtained sound path proper distance. Normal probe calibration using the normal probe and reference blocks V1. While the calibration of the probe angle is divided into four steps:

1. Examination of the index point
2. Examination angle
3. Calibration mileage
4. Angle probe calibration using reference blocks V2

After performing the calibration procedure the user can perform an ultrasonic inspection. Ultrasonic inspection procedures contained in this expert system is based on the implementation of ultrasonic inspection procedures made by PT Multi Spec Sinergindo used in the implementation of UT inspection by PT PAL Indonesia.

5.1.2. Testing of the Weld Defect Evaluation

Testing of the weld defect evaluation is divided into two, namely the evaluation of the weld defect classification to determine whether these defects can be accepted or rejected. the second evaluation is the weld defect types.

In the test below is a test to determine the classification of weld defects:

In the testing of the weld defect classification is the user entering the data material, data and data of ultrasonic flaw indications are:

1. Material data:

The data included material is thick welds (mm)

2. Indicative of defects data:

Data included indications of defects is the indication level, zero level reference, distance sound path

3. Ultrasonic data:

Data is entered ultrasonic probe angle used during the implementation of inspection.

By entering the zero reference level, the level and range indication will be obtained sound path attenuation value and indication rating. The results obtained using the formula.

c=((sp/25)-1) x 2

d=a-b-c

With :

d = Indication rating, a = Indication level, b = Zero reference level, c = Attenuation factor, sp= Sound path distance

The rating indication should be round to the nearest number.

After knowing the value of the indication rating by looking at Table 1 for statically Loaded Non-Tubular Connection and Table 2 for Cyclically Loaded Non tubular Connection, it will be known to the classification of weld defects. Classification will be known whether the defective weld is accepted or rejected. To test this UT inspection test data used in PT PAL when I do the Job Training (KP) in PT PAL.

Oyong Extension Platform Inspection
Date: July 27th, 2011
Criteria : AWS D 1.1
Drawing No.: 9810-90-02-201
Material Type: Carbon Steel (CS)
Place: Plat Construction workshop II (Oyong Project)
Surface Condition: smooth
Surface preparation: As-Welded
Flaw Detector: SIUI CTS-9005

Calibration Block: IIW Block V1, V2
DB Reff (b) : 81,5
Couplant : CMC + Air
Probe : CB MWS 700 4MHz 8x9
Test results on the platform on July 27th, 2011 as shown at Table 1.

Table 1. Data Observations On Platform 1

No	JOINT NO	T (mm)	L (mm)	Acceptance Criteria			Sound Path (mm)	Length of discontinuity (mm)	Depth (mm)	Result	Error Type
				Indication Level(a)	Attenuation factor (c)	Indication Rating (d)					
1	J.100	8+10	190	87.5	0	6	23	15	5-7	Acc	-
2	J.122	8	150	82.5	0	-1	20	120	5-7	Rep	PJ
3	J.128	8	120	82.5	0	-1	20	120	5-7	Rep	PJ
4	J.129	8	120	82.5	0	-1	20	120	5-7	Rep	PJ
5	J.130	8+20	120	82.5	0	-1	20	130	5-7	Rep	PJ
6	J.130A	8+20	120	80.5	0	-1	20	130	5-7	Rep	PJ

After knowing the data is the data is entered into the expert system program to determine the indication rating and classification of weld defects. By entering the data obtained from ultrasonic inspection into the defect classification evaluation form, the obtained results of the comparison between the indication rating is calculated using a formula and a calculation program as in the table below.

A.1 Date: July 27th,2011
Criteria: AWS D 1.1
Drawing No. : 9810-90-02-201
Material Type: Carbon Steel (CS)
Place: Plat Construction workshop II (Oyong Project)
DB Reff (b) : 81,5
Probe: probe: CB MWS 700 4MHz 8x9

Differences in the indication rating are calculated using formulas and programs as in Table 2.

Table 2. Difference indication rating value from formula and program

No	JOINT NO	T (mm)	Acceptance Criteria					Error presentation (mm)	Length of discontinuity (mm)	Depth (mm)	Clasification	Result	Error Type
			Indication Level(a)	Attenuation factor (c)	Sound Path (mm)	Indication Rating (formula)	Indication Rating (Program)						
1	J.100	8+10	87.5	0	23	6	6	0	15	5-7	B	Acc	-
2	J.122	8	80.5	0	20	-1	-1	0	120	5-7	A	Rep	PJ
3	J.128	8	80.5	0	20	-1	-1	0	120	5-7	A	Rep	PJ
4	J.129	8	80.5	0	20	-1	-1	0	120	6-7	A	Rep	PJ
5	J.130	8+10	80.5	0	20	-1	-1	0	130	5-7	A	Rep	PJ
6	J.130 A	8+10	50.5	0	20	-1	-1	0	130	5-7	A	Rep	PJ

From table 2 can be known differences in the indication rating formulas and calculations in the program did not experience the difference (error percentage = 0%). Thus it can be said to be an expert system program has been successful in calculating the value of the rating and classification indication correctly. By knowing the value indication, it will be known classification rating. So that the classification will be known whether these defects can be accepted (Acc) or rejected (Rep). UT inspection on the implementation of Table 2 is used to statically Loaded No tubular Connection. This is because of the welded joints on Platform Extension Oyong a static connection. This evaluation can only be known the classification of weld defects that would be useful to know the acceptance of weld defects. As for knowing the type of welded defects then the user can go into the form of defects that contain the question dialog box. The question dialog box contains questions relating to the indication of welded defects that users must answer for welded defects. After knowing the type of welded defects then the user can know the explanation about the type of defect in the form explanation type welded defects. In this form, we find it difficult to obtain data related to indication of welded defects that can be identified using ultrasonic methods. Because during this ultrasonic inspection is only used to determine the depth of defects and classification in the determination of acceptance of welded defects. So the question in this dialog box only contains the indication data of welded defects that can be identified using ultrasonic method in the form of ultrasonic display, the effect of changing the position of the transducer to the height of amplitude and visual appearance.

5.2. Validation

From Media Expert: according to experts in the Expert system using Delphi said it expects the results of testing on Friday the 13th July 2012 This system has a very attractive visual, user friendly, compatible with windows and linuc and media size is small. But, the database type weld defects remains to be stored in My Documents that can be opened by an expert system that will be a little troublesome users.

From Material Expert of Ultrasonic: According to experts in the field of NDT said it expects the results of testing on Friday the 13th July 2012, an expert system for evaluation of weld defects using ultrasonic method to function properly. The existence of defect classification evaluation can help to determine the classification of weld defects acceptance. In addition, this expert system contains an explanation of its procedures and calibration of ultrasonic inspection ultrasonic. With it will be helpful for the implementation of inspections in the field. But, this expert system database contains twenty-eight types of disabilities and limited to the welding area, so the expert system cannot be used to identify and evaluate other defects in the weld which is a representation of material outside the weld area.

6. Conclusions

The result of the research shows that: (1) the result of feasibility test according to ultrasonic material testing expert that the system is feasible to be used as instructional media in welding material subject and welding defect detection in vocational education environment, because it contains an explanation about detection method of welding defect using method Ultrasonic testing in detail; (2) feasibility test results according to media experts, that this system has a very attractive visual, user friendly, compatible with windows and linuc and media size that is not too large; And (3) result of test by using data of indication of welding defect in PT PAL Surabaya, obtained classification data of welding defect in accordance with calculation of welding defect classification.

References and Notes

[1] Arif 2001 NDT Training Program Ultrasonic Method (Dual Language), ASNT Continuing Education In Nondestructive Testing. PT Multi Spec Sinergindo (MSS).
[3] ASME 2001 ASME Sec.V Non-destructive Examination. Bandung: Balai Besar Penelitian dan Pengembangan Industri Bahan dan Barang Teknik (B4T).
[4] AWS 2002 AWS D1.1. Bandung: Balai Besar Penelitian dan Pengembangan Industri Bahan dan Barang Teknik (B4T).
[5] Bechtel Power Corporation 2003 Welding Inspection Handbook. San Francisco: Bechtel Power Corporation.
[6] Fadli, Arie. 2010. Sistem Pakar Dasar.pdf. IlmuKomputer.Com
[7] Kusrini. 2008. Aplikasi Sistem Pakar. Yogyakarta: Andhi Offset.
[8] Mardjanto, Djoko. 2012. Private Communication.
[9] Pusat Pendidikan dan Pelatihan. 2002. NDT Umum. Jakarta: Badan Tenaga Nuklir Nasional.
[10] Shang, Yi. 2005. Artificial Intelegent. Missouri; Academic Press, University of Missouri.
[11] Supriyono.2012.Private Communication.
[12] Tri Atmoyo, Bagus.2011. Sistem Pakar Untuk Manual Perbaikan Citra Untuk Evaluasi Radiografi Industri.Yogyakarta: STTN BATAN.

IOP Conf. Series: Materials Science and Engineering **336** (2018) 012028 doi:10.1088/1757-899X/336/1/012028

[13] Wiryosumarto,Harsono & Okumura,Toshie. 1985. Teknologi Pengelasan Logam. Jakarta: PT.Pradnya Paramita.

IOP Conf. Series: Materials Science and Engineering **336** (2018) 012029 doi:10.1088/1757-899X/336/1/012029

Curriculum Development Based On INQF and Business/Industries Sector for Improvement Competency of Basic Pattern Making Students at Vocational High School

Yoni Nilasari, Dasining*
Graduate Program, Universitas Negeri Surabaya

*dasininglaksmi@gmail.com

Abstract. In this era of globalization, every human resource is faced with a competitive climate that will have a major impact on the development of the business and industrial sector. Therefore it is deemed necessary to research the development of curriculum based on INQF and the business/industries sector in order to improve the competence of Sewing Technique for Vocational High School Students of fashion clothing program. The development of curricula based on INQF and the business/industries is an activity to produce a curriculum that suits the needs of the business and industries sector. The formulation of the problem in this research are: (1) what is the curriculum based on INQF and the business/industries sector?; (2) how is the process and procedure of curriculum development of fashion program profession based on INQF and the business/industries sector?; And (3) how the result of the curriculum of fashion expertise based on INQF and the business/industries sector. The aims of research are: (1) explain what is meant by curriculum based on INQF and business/industries sector; (2) to know the process and procedure of curriculum development of fashion program profession based on INQF and the business/industries sectors ; And (3) to know result the curriculum of clothing expertise based on INQF and the business/industries sector. The research method chosen in developing curriculum based on INQFand business/industry sector is using by 4-D model from Thiagarajan, which includes: (1) define; (2) design; (3) development; And (4) disseminate. Step 4, not done but in this study. The result of the research shows that: (1) the curriculum based on INQF and the business/industries sector is the curriculum created by applying the principles and procedures of the Indonesian National Qualification Framework (INQF) that will improve the quality of graduates of Vocational High School level 2, and establish cooperation with Business/industries as a guest teacher (counselor) in the learning process; (2) process and procedure of curriculum development of fashion program profession based on INQF and business/industries sector is process and procedure of curriculum development of fashion program profession based on INQF and business/industries sector there are several stages: feasibility study and requirement, preparation of initial concept of curriculum planning based on INQF and the business/industries sector in the field of fashion, as well as the development of a plan to implement the curriculum based on INQF and the business/industries sector in the field of fashion, this development will produce a curriculum of fashion proficiency program in the form of learning competency of sewing technology where the implementer of learning (counselor) Is a guest teacher from business/industries sector. (3) the learning device validity aspect earns an average score of 3.5 with very valid criteria and the practicality aspect of the device obtains an average score of 3.3 with practical criteria.

1. Introduction

In this era of globalization every human resource is faced with a competitive climate that will have a major impact on the development of the business world and the world of industry will find ways to maintain their own existence, by always improving competent Human Resources (HR) Which occurred in this century resulted in changes in the overall life of society, not least the education sector. In this sector, education should be able to address and anticipate the development of labor market liberalization and development of science-based society. Therefore, innovations in various educational methods and models should also be developed (UNESCO: 2006). Looking at the development situation and demands on the world of work or business/industry, then vocational school graduates should be able to print a skilled and competent workforce. To create competent graduates, schools must be able to provide the best for their student's one of them, through learning with an innovative curriculum and in accordance with INQF. Therefore it is deemed necessary to research the development of curriculum based on INQF and the business/industry in order to improve entrepreneurship competence for SMK Students of fashion clothing programming. The expected result of this research is a learning tool for curriculum development based on KKNI and business/industry become feasible with the valid and practical element [1].

The formulation of the problem in this research are: (1) what is the curriculum based on INQF and the business/industries sector?; (2) how is the process and procedure of curriculum development of fashion program profession based on INQF and the business/industries sector?; And (3) how the result of the curriculum of fashion expertise based on INQF and the business/industries sector. In its development, the curriculum involves various parties, especially those who directly or indirectly have an interest in the existence of a designed education, ranging from educational experts, study experts, teachers, students, education officials, practitioners or role models or Other members of the community.

According to Hidayat, said six principles of curriculum development is as follows: principles of relevance, it is the closeness of the relationship. When associated with education with the community then must have a close relationship so that the educational results obtained will be useful for the life of learners in society. The principle of flexibility, the developed curriculum must have room for freedom of action. In this regard, it is related to the flexibility in choosing educational programs and flexibility in the development of learning programs. Efficiency principle, this principle relates to the effort, cost, time and energy used in the learning process can produce optimal learning process and result. Thus, curriculum development should be efficient. Principles of effectiveness It is the extent to which curriculum planning can be achieved in accordance with predetermined desires. The effectiveness of the curriculum is related to the teaching process of the educator, and the learning process of the learner.

The principle of continuity, this principle in curriculum development indicates a link between education level, type and educational program and field of study.

A purpose-oriented principle, the principle asserts that the goals are the direction for the development of other components in curriculum development. For that purpose, the curriculum should be clear, meaning that the curriculum objectives must be clearly understood by the curriculum implementers to be elaborated into other, more specific and operational objectives. Curriculum objectives should also be unity, meaning that cover various aspects of education. [2]

According to Hamalik, the curricula development consists of six stages. Phase 1 is a feasibility study and needs, phase 2 is the preparation of the initial concept of curriculum planning, stage 3 is the development of plans to implement the curriculum, stage 4 is the implementation of curriculum trials in the field, stage 5 is the implementation of the curriculum, and stage 6 is the implementation of assessment and monitoring the curriculum, and stage 7 is time Frame repairs and adjustments.

Stage 1: Feasibility study and needs the curriculum developer conducts the program needs analysis activities and formulates the basics of consideration for the development of the curriculum. For that, the developer needs to do documentation studies and/or field studies. Stage 2: Preparation of the initial concept curriculum. This initial concept is formulated based on the formulation of ability, and then formulate the objectives, content, learning strategies in accordance with the pattern of the systemic curriculum. Stage 3: Develop a plan to implement the curriculum. The preparation of this plan includes the preparation of syllabi, the development of lesson materials and other material resources. Stage 4: Implementation of the curriculum in the field. The field curriculum testing is intended to determine the level of reliability, its possible implementation and success, the constraints and problems that arise and the supporting factors available, and others related to the curriculum implementation. Stage 5: Implementation of the curriculum There are 2 activities that need to be done, our dissemination activities, namely the implementation of the curriculum within the broader scope of the sample and implementation of a comprehensive curriculum covering all educational units at the same level. Stage 6: Implementation of curriculum assessment and monitoring. During the curriculum implementation, it is necessary to conduct a review and monitoring regarding curriculum design and the results of curriculum implementation and its impact. Stage 7: Implementation of improvements and adjustments. Based on the assessment and monitoring of the curriculum, accurate data and information can be used as a material for conducting the curriculum if necessary or adjusting the curriculum to the circumstances. Improvements were made to several aspects of the curriculum [3].

Regulation of the President of the Republic of Indonesia Number 8 Year 2012 regarding the Indonesian National Qualification Framework which explains that the level of Vocational qualification is at level 2 which has the following competencies: 1) Ability to carry out a specific task, using commonly used tools, information and working procedures , And demonstrate performance with measurable quality, under the direct supervision of his / her supervisor, 2) Have basic operational knowledge and factual knowledge of specific work areas, to be able to choose available solutions to common problems, 3) Responsible for selfemployment and Given the responsibility of guiding others [1].

According to Yunus, business/industry is one of the important elements in the world of employment, so the need for adjustment between business/industry with the world of education as a source of labor income and employers appropriate with industries sector needed [4]. According to Megasari in his research, the role and function of the curriculum are: (1) as the structural path of the psar as well as the absorption of human resources and the users of scientific and technological knowledge; (2) to help develop relevant curriculum with the SMK; (3) to help provide experienced workforce as guest teachers in order to improve the quality and skills of educators in accordance with the development of science and technology; (4) to provide information on the development of science and technology in business world and work world; (5) provide technical advice in related fields; (6) to give input to SMK about job opportunity concerning SMK study program of kind, type and specification; (7) provide facilities of good means, facilities and to support the accuracy of teaching in accordance with industrial procedures; (8) provide suggestions on improving the quality of SMK graduates in accordance with the needs of the world of work, so that students who have completed vocational education can be ready to work in accordance with their respective expertise field with all the power it has [5].

2. Method

This research takes place in SMKN 3 Kediri because this school is one of the schools that often use guest teachers from the business/industry in the process of learning. The method used in this research is the research and development, using the Four-D models. According to Thiagarajan, here are four stages of research and development using the Four-D models that define, design, develops, and disseminate. 4D development model Consists of four main stages: Define (Define), Design (Design), Develop (Development) and Disseminate (Spreading) [6].

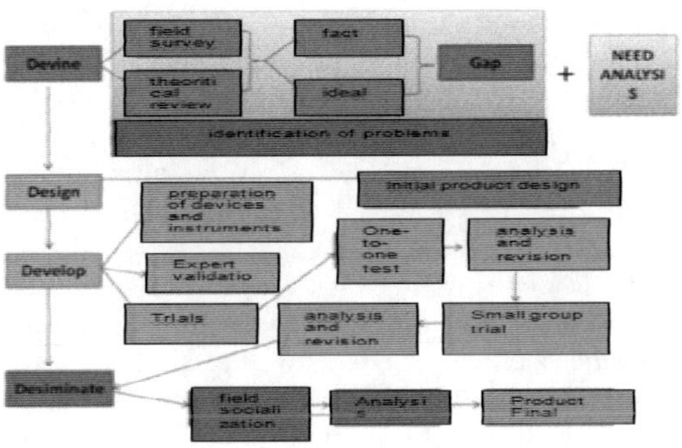

Figure 1. 4 D model of development

IOP Conf. Series: Materials Science and Engineering **336** (2018) 012029 doi:10.1088/1757-899X/336/1/012029

Development and trial stage: The purpose of the define phase is to determine and establish learning objectives. This initial stage will be analyzed the needs of learning tools (RPP, LKS, and LP) in accordance with the curriculum based on INQF and the business/industry then set the learning objectives and teaching materials and constraints that will be developed. Stage define or definition has five steps which the analysis of the front end (front-end analysis), analysis of learning (learner analysis), analysis of the task (task analysis), analyzes the concept (concept analysis), set a goal (Specifying instructional objectives).

The purpose of the design phase is to design a learning device (lesson plans, worksheets, and test). This phase can begin with the preparation of test grids, media selection, format selection, and initial design. Four steps in the design phase (design) are constructing the criterion-referenced tests (establishing the criteria and grating tests), media selection (selection of media), the format selection (selection format), the initial design (the initial design). The purpose of step develop (development) is to modify the type of design of learning materials through the validation and expert assessment (expert appraisal). The empirical test phase is to test the revised learning tool to get consistent and effective results. To declare a learning tool that is compiled valid or not is to interpret the criteria of the assessment of 1 guest teacher from the business/industry (fashion businessman "yayuk' modes"), 1 person deputy of the head of the curriculum and 3 teachers productive clothing, as follows.

Table 1. Criteria of validity coeffitien of tets instrument

Range	Assesmen criteria
3,4 s/d 4	very valid
2,6 s/d 3,3	Valid
1,7 s/d 2,5	less Valid
3,4 s/d 4	invalid

The analysis performed is the result of the research with the average score (x) of the validator in accordance with the validity criterion.

$$X = \frac{\Sigma xi}{n}$$

, information:

X = Average value of validation results
Σxi = Number of **answers** observer
n = number of observer

3. Results And Discussion

3.1. Meant by curriculum based on INQF and business/industries sector

According to According to Allais 2011, the national framework or the level of qualification/type of learning qualification at a certain level of education and sector that has been agreed by the government to improve the quality of graduates of an educational institution [7]. According to Toma 2013, the national vocational qualification framework is the entire educational qualification structure that contains all the qualifications of vocational education to improve the learning achievement or improve student achievement [8].

According to Serbian Ministry of Education 2013 or Kemendikbud in Serbia in 2013, the national vocational qualification framework is a qualifying instrument that organizes students in vocational schools such as labor market/industry as well as applying long-life learning [9]. According to Stanciu, the framework of the national qualification is the introduction stage of learning that is bleeding the world of work as well as adjusted to the needs of the world of an industry will be labor market and society [10].

According to Gudeva, the national qualification framework should be elaborate, comparable and appropriate for application at the vocational secondary education level, in order to create graduates, with the capacity to work, the level and the learning. [11] Thus, all graduates can enter the workforce in accordance with their capacity. The business world/industry is one of the important elements in the world of manpower ... so it is necessary to adjust the business/industry with the education world as a source of a competent workforce and in accordance with the needs of the world [4]. According to Notonegoro, cooperation between vocational schools or SMK with the business world/industry is an effort to print students (graduates) into the appropriate workforce needed by the industry. Schools that make guest teachers from the business world/industry should have practical equipment in accordance with industry procedures in order to support the learning process. [12] This is very important to do because with the complete equipment then the guest teacher will easy to carry out the learning.

According to Indriaturahmi, et al, education field / vocational skills programs in collaboration with business/industries sector also must apply INQF and has the infrastructure, financing and adequate teaching staff so that the potential of graduates according to the needs of the workforce. [13] Based on the study above that can be interpreted is a curriculum-based INQF and business / industry is a curriculum that is made by applying the principles and procedures Indonesia National Qualifications Framework (INQF) that will improve the quality of graduates of Vocational High School at level 2, and establish cooperation With the business / industry as a guest teacher (counselor) and using equipment in accordance with industry standards in the learning process.

3.2. Process and procedure of curriculum development of fashion program profession based on INQF and the business/industries sectors.

Stage 1: Feasibility study and needs the curriculum developer conducts the program needs analysis activities and formulates the basics of consideration for the development of the curriculum. For that, the developer needs to do documentation studies and/or field studies. Stage 2: Preparation of the initial concept of curriculum planning. This initial concept is formulated based on the formulation of ability, then formulate the objectives, content, learning strategies in accordance with the pattern of a systemic curriculum. Stage 3: Development and plan to implement the curriculum. The preparation of this plan includes the preparation of syllabic, the development of lesson materials and other material resources.

Curriculum development in these n empirically using only three stages to stage 3 curriculum developments and implement the plan covers the development of appropriate learning tools KKNI and business/industries sector.

3.3. Results-based curriculum development and INQF business/industry based on analysis of 4D

Stage Define

From the analysis of existing devices obtained data that learning tools consist of the syllabus, RPP, and the question sheet, still not oriented stimulus of desired learning outcomes. The test is only done at the end of learning materials, which only prioritize aspects of knowledge, and test questions never in test valid test.

The material presented is good and covers the current developments; the material uses examples that can be understood by the students. The language used by the teacher is appropriate and easy to understand by the students, teachers is attractive in the learning process. From observed result can be concluded that learning device can be accepted with some improvement. a) Define Stage, From the analysis of existing devices obtained data that learning tools consist of a syllabus, RPP, and the question sheet, still not oriented stimulus of desired learning outcomes. Test only has done at the end of learning materials, which only prioritize aspects of knowledge, and the test never In test valid test. The material presented is good and covers the current developments; the material uses examples that can be understood by the students. The language used by the teacher is appropriate and easy to understand by the students, teachers activities in the learning process from observed result can be concluded that learning device can be accepted with some improvement. b) Design stage, based on the availability of facilities and infrastructure that adjust to the learning materials to be delivered then used sewing machine industry (high speed). Learning model used in this learning process using direct learning model. So every step in the RPP will be developed using syntax in the direct learning model that will be developed in accordance with the curriculum based on INQF and the business/industries sector. The stages of preparing the test plan begin with the analysis of student tasks, concept analysis, and learning objectives so that the grid can be developed into a matter of test. Question made include aspects of attitude, knowledge, and skills. c) Develop stage, In this development stage, learning tools are validated by the instrument. The result of the instrument assessment is calculated using the average formula. With the following results:

Table 2. Device validations results

No	Materials	average	category
1	Syllabus	4	Valid
2	RPP	3,0	Valid
3	LKS	3,3	Valid
4	LP and keys	3,7	very valid
Average rate		3,5	very valid

Table 3. Device practicality results

No	Aspec	averange	categori
1	Preliminary	3.6	Very practical
2	Core	3.0	practical
3	Cover	3.3	practical
4	Class management	3.3	practical
Average overall		3,3	parktis

Based on descriptive analysis by categorizing valid learning tools based on the development of INQF-based curriculum and business world/industry that syllabus shows average score 4 is in very valid category, RPP shows average score of 3.0 is invalid category, LKS shows average score 3,3 are In the valid category as well as LP and key of LP shows the score of 3.7 are in very valid category, so the overall average shows the 3.5 score is in very valid category. Based on descriptive analysis by categorizing the practicality of learning devices based on the development of INQF-based curriculum and business world / industry that the opening shows average score 3.6 is in the category of very practical, the content shows the average score of 3.0 is in the practical category, the cover shows the average score of 3.3 is In the category of practical as well as class management showed 3.3 scores are in the practical category, so the overall average shows a score of 3.3 in the practical category. d) The disseminate stage. At this stage of deployment is not done but there is in this study.

4. Conclusion

The conclusions of this study are: Developing a curriculum-based KKNI and business / industry is a curriculum that is made by applying the principles and Indonesia National Qualifications Framework (INQF) procedures that will improve the quality of graduates of Vocational High School at level 2, and establish cooperation with business / industry as a guest teacher (supervising personnel) using equipment in accordance with industry standards in the learning process. The process and procedures for curriculum development program membership dressmaking based INQF and business/industry are processes and procedures of curriculum development program membership dressmaking based INQF and business/industry there are several stages of a feasibility study and needs, drafting the initial concept of curriculum planning Based on KKNI and the business world / industry in the field of fashion, as well as the development of a plan to implement the curriculum based on KKNI and the business world / industry in the field of fashion, this development will produce a curriculum of fashion proficiency program in the form of learning competency of sewing technology where the implementer of learning (counselor) a guest teacher from the business/industry. Results of devices based learning KKNI and World business/industry terms of validity learning device obtained an average score of 3.5 with very valid criteria, And the practicality aspect of the device obtained an average score of 3.3 with practical criteria. It can be concluded that the eligible device for use in learning because it is very valid and practical criteria.

References

[1] DPKP 2012 Kurikulum Kursus Dan Peltihan Tata Busana Jenjang 2 dan 3 Berbasis KKNI. Kemendikbud.

[2] Hidayat, Sholeh. 2013 *Pengembangan Kurikulum Baru.* Bandung: PT Remaja Rosdakarya.

[3] Hamalik, Umar 2007 *Manajemen Pengembangan Kurikulum. Bandung*: PT Remaja Rosda Karya.

[4] Yunus, 2006. Penyesuaian pendidikan SMK dengan Dunia kerja. Jakarta: Mahardika.

[5] Megasari, Dindy shinta. Tingkat Kesesuaian kurikulum SMK TKR Ditinjau Dari Kompetensi Yang Dibutuhkan Dunia Usaha / Dunia Industri. Universitas Negeri Surabaya. Thiagarajan, S.

[6] Semmel, D. S., & Semmel, M. I. 2013 *Instructional Development for Training Teacher Of Exceptional Children: A Sourcebook.* Bloomington: Centre for Innovation on Teaching The Handicapped.

[7] Steephanie 2011 What Is National Qualifications Framework Consideration From a Study of National Qualification Framework From 16 Countries. University of The Witwaterstand Education Policy Unit. Johanseburg, Shout Afrika**, 5**.

[8] Toma, Steliana 2013 Contribution Of The Design Of The Romanian National Qualification Framework. Buchares, Romanian **2** (122).

[9] Serbia Manistry of Education. NQF in Serbia, National Qualification System Level I to V. Republic of Serbia. 2013.

[10] Stanciu, Simona and Viorica Banciu. National Qualification System (NQS) in Romania And Validation of Formal And Non Formal Learning. Oradea, Romania. Vol.69 No.816-820. 2012

[11] Gudeva, Liljana Koleva, and friends. Implementations Of National Qualification Framework For Higher Education in Republic of Machedonia. Repubic Of Machedonia. Vol.46. No.2556-2560. 2012

[12] Notonegoro, Arief Yulianto. Model Kompetensi Lulusan Sekolah Menengah Kejuruan Berbasis Kompetensi Dunia Usaha Dan Dunia Industri. 2008

[13] Indriaturrahmi, and Sugiyanto. Peran Dunia Usaha dan Dunia Industri Dalam Penyelenggaraan SMK Berbasis Kearifan Lokal Di Kota Mataram. UNY. Vo.6. No.2. 2016

IOP Conf. Series: Materials Science and Engineering **336** (2018) 012030 doi:10.1088/1757-899X/336/1/012030

Developing Learning Tool of Control System Engineering Using Matrix Laboratory Software Oriented on Industrial Needs

Subuh Isnur Haryudo*, Achmad Imam Agung, Rifqi Firmansyah

Department of Electrical Engineering, Universitas Negeri Surabaya

*subuhisnur@unesa.ac.id

Abstract: The purpose of this research is to develop learning media of control technique using Matrix Laboratory software with industry requirement approach. Learning media serves as a tool for creating a better and effective teaching and learning situation because it can accelerate the learning process in order to enhance the quality of learning. Control Techniques using Matrix Laboratory software can enlarge the interest and attention of students, with real experience and can grow independent attitude. This research design refers to the use of research and development (R & D) methods that have been modified by multi-disciplinary team-based researchers. This research used Computer based learning method consisting of computer and Matrix Laboratory software which was integrated with props. Matrix Laboratory has the ability to visualize the theory and analysis of the Control System which is an integration of computing, visualization and programming which is easy to use. The result of this instructional media development is to use mathematical equations using Matrix Laboratory software on control system application with DC motor plant and PID (Proportional-Integral-Derivative). Considering that manufacturing in the field of Distributed Control systems (DCSs), Programmable Controllers (PLCs), and Microcontrollers (MCUs) use PID systems in production processes are widely used in industry.

1. Introduction

The development of science and technology is very fast. The competition between industries in the world will certainly cause the need for workers' ability to run the industrial process well. Thus, training or educational institution tries to improve the competence of graduate candidates demanded by the working world.

The comfort and speed system are highly demanded in the modern world where the system that works automatically will be more and more developed. The industrial automation field is a discipline that studies the device or system automation [18]. Learning automation theoretically is not enough because learning automation technology requires training probs (trainer) to support the theory obtained by students. Practically, vocational education is different from general education because it requires "up to date" equipment to train students to have a good life skill and be the competent graduates. The limitations of trainer learning media become a serious matter to be observed in vocational education. Using learning media, students can easily understand content of the lesson. As stated by Briggs [1] that learning media is the means to deliver learning content or materials, such as books, modules,

IOP Conf. Series: Materials Science and Engineering **336** (2018) 012030 doi:10.1088/1757-899X/336/1/012030

films, videos, trainers and so on. Then, the communication is printing and hearing-sight form, including hardware technology.

Planning an automation system in both large and small industrial scale will not be separated from an assumption how this system will run well viewed from the review behavior or system characteristics. The main characteristics that must be known is the electrical characteristics of the system such as start current surge, transient voltage profile to transient analysis when having system interruption. The ability to know the actual system condition will provide good and optimal planning results. The interpreting process or interpreting the system behavior is not an easy task because of dealing with the static and dynamic behavior system. Modeling and simulation should be performed in an interactive and real time. Using Matrix Laboratory software will also determine the accuracy of the model taken.

Regulatory system is the process of setting or controlling one or several quantities (variables, parameters) to obtain a price or within a certain range. The regulatory system is also a system in which the components are connected that forms system configuration. The control system regulates its own system or other systems to obtain response of the required system. According to [1], the student learning test shows the achievement of complete learning of all students was100%, after following the lectures using the engineering module on fundamentals of controlling system. Referring to the successful previous research, the writer developed learning media application of wider regulatory techniques. The regulatory techniques using Matrix Laboratory Software with Inquiry Based Learning device for Industrial Needs by giving examples of mathematical equations in controlling system applications using PID (Proportional-Integral-Derivative). Considering that manufacturing in the field of Distributed Control System (DCSs), Programmable Controllers (PLCs), and Microcontrollers (MCUs) use PID systems in production processes widely used in industry.

2. Research Method

The design of this study was experimental research using research and development (R & D) methods modified in the form of engineering methods of setting technique. The design of Regulatory Techniques Using Software Matrix Laboratory for industrial needs requires experts in education, engineering, and language. Thus, it takes a multi-disciplinary team-based approach, which the industry's manufacturing used PID systems in the production process. Figure 1 shown stage for using trainer in operating and setting the system using PID.

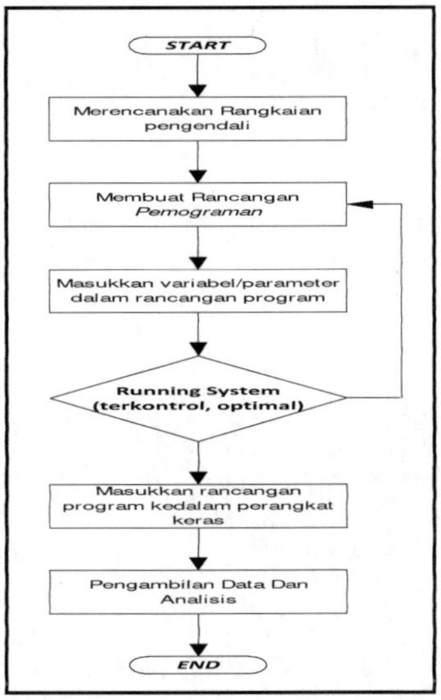

Figure 1. Flowchart of control system engineering design

PID control is famous for its structure simplicity, as well as the ease of tuning the control parameters, reliability, easy to fix, and more importantly PID control provides excellent control compared to other techniques that usually requires more effort and higher fund. Implementing logical theory is capable of creating a revolution in technology. The basic configuration of the PID control system can be seen in Figure 2.

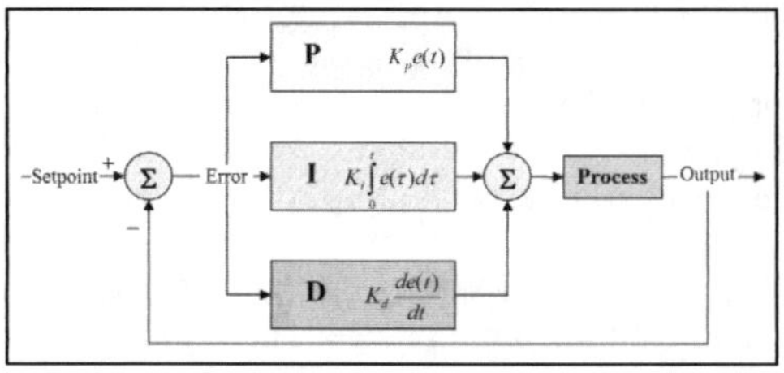

Figure 2. Configuration of PID Controller

This research was arranged in the form of learning device on trainer's regulatory technique with DC motor plant and PID Controller using Matrix Laboratory software with Arduino Uno Microcontroller.

3. Results and Discussion

This research has succeeded in producing learning device of PID-based system automation Trainer using Arduino equipped with ultrasound sensor. This trainer is a system to control DC motor speed. The speed of DC motor can be used to support the competence of regulatory techniques so that students can simulate the measurement and trial in an effective and efficient way.

This trainer consists of several components placed on the PCB board as a component of placement medium. Figure 3 shows the component specification design of the trainer setting system:

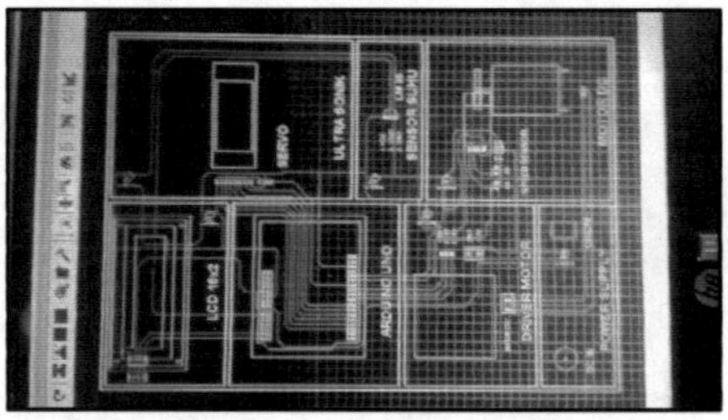

Figure 3. Design of Control System Engineering Trainer Board

The Trainer engineering method refers to industrial competence standards and is aimed to understand, operate, use and apply basic principles of basic regulatory techniques in electronic equipment including Arduino programming, LCD character programming, ultrasonic sensors using Arduino UNO microprocessors and Matrix Laboratory assisted design program on PID experimental circuit that controls DC motor.

Figure 4. Trainer of Control System Engineering

To ease the learning process, students must understand the making process of series and programming with Arduino. The design simulation is done by making programming design on PC / Laptop. Simulation is done to minimize the circuit error application into automatic control using *Arduino Uno*. Then, the results of simulation and programming design are transferred to Arduino Uno hardware. Some competences gained in the learning process using these regulatory technique trainer is the first lesson of programming to display Ultrasonic Sensor Data:

```
int trig= 13;
int echo= 12;
long durasi, jarak;
void setup() {
pinMode(trig, OUTPUT);
pinMode(echo, INPUT);
Serial.begin(9600);      }
void loop()
{digitalWrite(trig, LOW);
delayMicroseconds(8);
digitalWrite(trig, HIGH);
delayMicroseconds(8);
digitalWrite(trig, LOW);
delayMicroseconds(8);
durasi= pulseIn(echo, HIGH);
jarak= (durasi/2) / 29.1;  Serial.println(jarak :); }
```

The second program learning to show temperature data and PWM of motor:

```
void lm35(){
  suhu=analogRead(A0);
  suhu_celcius=(suhu*500)/1023; }
void setup() {
pinMode(trigPin1, OUTPUT);
pinMode(echoPin1, INPUT);
pinMode(trigPin2, OUTPUT);
pinMode(echoPin2, INPUT);
pinMode(trigPin3, OUTPUT);
pinMode(echoPin3, INPUT);
pinMode(in1, INPUT);
pinMode(in2, INPUT);
pinMode(en1, OUTPUT);
Serial.begin(9600);
lcd.begin(16,2);}
void PWM_out() {
 if(speed_motor>0)
 {
   PWM=(unsigned char)speed_motor;
   digitalWrite(in1,HIGH);  digitalWrite(in2,LOW);  analogWrite(en1,PWM);
//maju
 }
 Else  {
   speed_motor=(unsigned char)255+speed_motor;
   PWM=speed_motor;
   digitalWrite(in1,LOW);  digitalWrite(in2,HIGH);  analogWrite(en1,PWM);
//mundur
 }  }
```

```
void setup()
{
pinMode(trigPin1, OUTPUT);
pinMode(echoPin1, INPUT);
pinMode(trigPin2, OUTPUT);
pinMode(echoPin2, INPUT);
pinMode(trigPin3, OUTPUT);
pinMode(echoPin3, INPUT);
pinMode(in1, INPUT);
pinMode(in2, INPUT);
pinMode(en1, OUTPUT);
Serial.begin(9600);
lcd.begin(16,2);
}
void loop()
{
digitalWrite(11,LOW);
lcd.clear();
  lm35();
  us_pid();
    lcd.setCursor (0,0); lcd.print ("P:");
  lcd.setCursor (2,0); lcd.print (kp);
  lcd.setCursor (4,0); lcd.print ("I:");
  lcd.setCursor (6,0); lcd.print (ki);
  lcd.setCursor (8,0); lcd.print ("D:");
  lcd.setCursor (10,0); lcd.print (kd);
  lcd.setCursor (13,0); lcd.print (us3);
  lcd.setCursor (14,0); lcd.print (us2);
  lcd.setCursor (15,0); lcd.print (us1);
  lcd.setCursor (0,1); lcd.print ("E: ");
  lcd.setCursor (2,1); lcd.print (error);
  lcd.setCursor (5,1); lcd.print ("PWM: ");
  lcd.setCursor (9,1); lcd.print (PWM);
  lcd.setCursor(13,1);
  lcd.print(suhu_celcius);
  lcd.setCursor(15,1);
  lcd.write(0xdf);
  delay(200);
```

And the third program of PID program to ultrasonic sensor to control motor:

```
}
```

The programming learning is then compiled into the system, the result of the compilation can be displayed on the LCD screen as shown in Figure 5.

Figure 5. the view of LCD to display ultrasonic data and motor speed

Article I. The results of data analysis and equipment testing in Student Competency Competition (LKS) 2012-2016 showed that the developed module has reached the fourth stage of research development method that is dissemination method *(disseminate)*. The developed teaching materials should not be the latest technology or the appropriate technology but include the development of learning device, so that this trainer is well used to improve students' competence in automation system design to support the regulatory technique course.

4. Conclusion

Based on the results of data analysis and discussion, the researchers can take the following conclusions:

This developed medium is research result and monitoring research conduced by team for 4 years which continuously has tried to develop the applied product research entitled regulatory techniques using matrix laboratory software with inquiry based learning device for industrial needs that is expected to produce the compatible regulatory technique learning device for industrial needs.

5. References

[1] Arysad, Azha. 2013 *Media Pembelajaran.* Jakarta: Rajawali Press.

[2] Depdikbud 1993 *Konsep Sisem ganda pada sekolah menengah Kejuruan di Indonesia*,Jakarta, Depdikbud RI.

1998. Keputusan Menteri Pendidikan Menengah Dan Kebudayaan RI No. 323/U/1997 tentang *Pelaksanaan Pendidikan Sistem Ganda Pada Sekolah Menengah Kejuruan*. Jakarta; Direktorat Dikmenjur.

[3] Dit. Dikmenjur, 1996 *Kerangka Acuan Studi tentang Pelaksanaan Program Dukungan Pelatihan Industri (Industrial Training Support Program*). Jakarta: Proyek Pendidikan Teknologi dan Kejuruan.

[4] Daniel JS. 1996 The knowledge media. In: Mega-universities and knowledge media. *Technology strategies for higher education*. London: Kogan pp.101-135.

[5] Devitt P, Palmer E. 1999 Computer-aided learning: an overvalued educational resource Med Ed; 33: 136-139

[6] Elves AW, Ahmed M, Abrams P. 1997 *Computer-assisted learning*; experience at the Bristol Urological Institute in the teaching of urology. Br J Urol **80** (3) pp. 59-62

[7] Erickson, H.L 1998 *Concept Based Curriculum Instruction*: Teaching Beyond The Thousand Oaks: Corwin

[8] Finch, Curtis R. & john R Crunkilton 1983 *Curriculum Development in Vocational and Technical Education*, London: Allyn and Bacon, Inc.

[9] Greinert, WD., 1992 *The Dual System Of Vocational Training In The Federal Republic of Germany*; Structure and function. Postfach 1, 6101 6326 Eschborn, Federal Republic of Germany

[10] Mardapi, Djemari., 1999 "*Performance Based Evaluation*" Bahan Lokakarya Tentang Performance Based Evaluation di Program Pascasarjana UNY.

[11] Norton, R.E, 1985. *DACUM* handbook. Columbus: The National Centre For Research In Vocational The Ohio State University

IOP Conf. Series: Materials Science and Engineering **336** (2018) 012030 doi:10.1088/1757-899X/336/1/012030

[12] Sulifan, R., 1995 *The Competency Approach To Training*. JHPIEGO strategy Paper.
[13] Slamet PH., 1993 "*Pendekatan Alternatif Dalm Penyiapan Dan Pengembangan Tenaga Kerja Profesional Di Indonesia*" Disampaikan Pada Seminar Nasional Sistem Permagangan Dalam Pendidikan Dan Kejuruan Dan Penyiapan Tenaga Kerja, Malang
[14] Kallinowski F, Mehrabi A, Gluckstein C, Benner A, Lindinger M, Hashemi B, et al. *Computer-based traininga new method in surgical education and continuing education.* Chirurg 1997; 68: 433-438
[15] Rogers DA, Regehr G, Yeh KA, Howdieshell TR. 1998 *Computer-assisted learning versus a lecture and feedback seminar for teaching a basic surgical technical skill.* Am J Surg **175** pp. 508-10
[16] Son Kuswadi, Eru Puspita 2005 *Modul Praktikum Interaktif Teknik Digital.* Modul praktikum Politeknik Elektronika Negeri Surabaya – ITS, Surabaya
[17] Son Kuswadi, Eru Puspita 2006 *Modul Praktikum Computer Assisted untuk Pengendalian Kecepatan Motor DC*, Modul praktikum Politeknik Elektronika Negeri Surabaya – ITS, Surabaya
[18] Wicaksono, Handy 2009 *Programmable Logic Controller*. Graha Ilmu.

IOP Conf. Series: Materials Science and Engineering **336** (2018) 012031 doi:10.1088/1757-899X/336/1/012031

Adjusted Framework of M-Learning in Blended Learning System for Mathematics Study Field of Junior High School Level VII

Lipur Sugiyanta*, Moch. Sukardjo

Study Program of Education for Informatics and Computer Engineering/ Faculty of Engineering, State University of Jakarta

*lipurs@unj.ac.id

Abstract. The 2013 curriculum requires teachers to be more productive, creative, and innovative in encouraging students to be more independent by strengthening attitudes, skills and knowledge. Teachers are given the options to create lesson plan according to the environment and conditions of their students. At the junior level, Core Competence (KI) and Basic Competence (KD) have been completely designed. In addition, there had already guidebooks, both for teacher manuals (Master's Books) and for learners (Student Books). The lesson plan and guidebooks which already exist are intended only for learning in the classroom/in-school. Many alternative classrooms and alternatives learning models opened up using educational technology. The advance of educational technology opened opportunity for combination of class interaction using mobile learning applications. Mobile learning has rapidly evolved in education for the last ten years and many initiatives have been conducted worldwide. However, few of these efforts have produced any lasting outcomes. It is evident that mobile education applications are complex and hence, will not become sustainable. Long-term sustainability remains a risk. Long-term sustainability usually was resulted from continuous adaptation to changing conditions [4]. Frameworks are therefore required to avoid sustainability pitfalls. The implementation should start from simple environment then gradually become complex through adaptation steps. Therefore, our paper developed the framework of mobile learning (m-learning) adaptation for grade 7th (junior high school). The environment setup was blended mobile learning (not full mobile learning) and emphasize on Algebra. The research is done by R&D method (research and development). Results of the framework includes requirements and adaptation steps. The adjusted m-learning framework is designed to be a guidance for teachers to adopt m-learning to support blended learning environments. During mock-up prototype, the adjusted framework demonstrates how to make successful implementation of early blended mobile learning through framework. The Social area is in focus of adaptation because participation is important to improve the sustainability. From the short practice of mock-up prototype, blended mobile learning can be an effective pedagogical model in supporting students in inquiry-based learning.

1. Introduction

Education in the future is more likely in the form of education and training by applying the distance education (distance learning). The lesson materials developed tend to combine conventional learning with blended learning. [1] There is no consensus on a single definition for blended learning. Some references define blended learning differently. Learning that combines conventional learning (face to face) with learning based on information technology is known as Blended Learning. [2]

Blended Learning does not replace the conventional learning in the classroom (face to face), but it reinforces the learning model through the educational technology. Blended Learning Learning focuses primarily on students. Students must be independent at certain times. The learning atmosphere requires students to play a more active role in their learning. Students are encouraged to obtain enrichment materials with their own efforts and initiatives. Through Blended Learning the learning system becomes more flexible and not rigid.

In general, blended learning is implemented in the form of: 1) integration / integration of traditional learning with application-based approach (on-line web and mobile application); 2) a combination of media and equipment (eg smartphones) used in learning environments, and 3) a combination of teaching and learning approaches regardless of the technology used. In this study, blended learning is interpreted as the integration of face-to-face learning and learning methods using smartphone devices. The term used in this study is Blended Mobile Learning (blended m-learning). Blended Mobile Learning (m-learning) consists of several tools; learning materials, software, instructional planning, learning media, and smartphone devices. The learning process does not stop in the classroom, but students can continue it at home and even start learning from home.

Mobile learning itself has rapidly developed in the last ten years and many initiatives have been conducted worldwide. However, few of these researches have produced any lasting outcomes. It is evident that mobile applications are complex and, hence will not become sustainable. Long-term sustainability results from continuous adaptation to changing conditions [3]. Frameworks are therefore required to avoid sustainability pitfalls. The framework defines the life-cycle of a mobile learning initiative and identifies the importance of emphasizing the concepts of scalability and sustainability during the development processes.

The former result research [4] defines the life-cycle of a mobile learning initiative and identifies the importance of emphasizing the concepts of scalability and sustainability during the development process. The resulted framework could reduce the complexity of mobile learning and its settings, and a means to improve the outcomes of mobile learning initiatives in terms of long lasting usable results. This framework was analyzed from eight scientific publications that have been presented at different international conferences. Five of the them explore the field of mobile learning and its practice while the other three publications present the central ideas that serve as the basis for the proposed framework, how it has been developed, and the motivations behind its creation.

This framework will be examined in junior high school 7th grade in Mathematics at SMPN 77 Jakarta Pusat. 7th grade of SMPN 77 has implemented 2013 curriculum and its learning has been using a scientific approach. However, the implementation of mathematics with a scientific approach still difficult. This is caused by the participation of students in learning are poor.

Mathematics for junior high school (SMP/MTs) is directed to encourage students to find out knowledge from various sources, able to formulate problems, and solve simple problems. Mathematics aimed to have math skills. Math skills are part of the life skills that students must possess especially in the reasoning, communication, and problem solving encountered in students' daily lives. [4] Therefore, Blended Learning is potentially developed for Mathematics subjects. There are some materials which is essential in mathematics in 2013 curriculum. Algebra is essential to students, because students have not received this material in Elementary School. The material on algebra should be mastered by the students before the material of linear equations and linear inequality of one variable were delivered. The scope of algebra includes: variables, coefficients, constants, similar tribes to its operations (addition, subtraction, multiplication, and division).

2. Related Research

The "Framework for sustainable mobile learning in schools" that [5] was developed in the context of an Australian school. The school's efforts sought to introduce and sustain an mLearn programme using personal digital assistants (PDAs), and the school allocated resources to internally fund the programme. Sustainability of mobile technology integration in schools has thus been researched in other countries. However, a sustainability framework where teachers and district officials' perspectives were applied to refine a sustainability framework in the context of resource-constrained environments is novel. This framework worked as full mobile learning initiative.

[8] propose an analytic self-regulated learning (SRL) model of mobile learning as a conceptual framework for designing and analyzing mobile learning. SRL notion of self-regulation as the core (Figure 1).

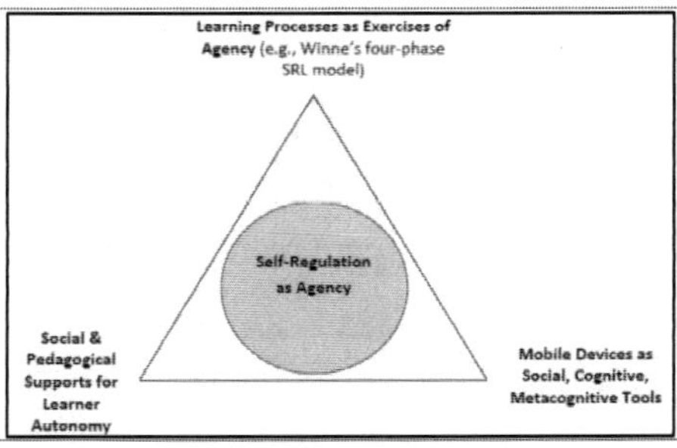

Figure 1. SRL model for Mobile Learning

The centre of the model referring to the students' characteristics that function as internal driving forces initiating and sustaining a self-regulated mobile learning process. The key personal factors include domain knowledge, prior experiences, motivation, and metacognitive awareness, epistemological beliefs, and so on. Mobile learning processes that are regarded as manifestation/exercises of agency (fundamentally composed of motivation and metacognition) can be understood, analyzed, and assessed from the theories and methodologies of SRL. Mobile learning activities are supposed to be mediated by mobile technologies and devices, which presumably function as social, cognitive, and metacognitive tools. The existing studies in mobile learning largely focus on the social and cognitive functions but ignore the metacognitive function.

In [1], the use of Blended Learning in Indonesian Language Learning at SD Insan Amanah was conducted in the classroom. Observations made on student motivation and activity and also students' learning ability. Learning begins with apperception activities by inviting all students to access (logging in first) the website http://elearning.insanamanah.sch.id and then students begin to open the lesson material and work on. The student's enthusiastic starts from the beginning of the lesson. Students' passion is also visible when online questions is marked with right answer.

Sustainability to improve motivation and ability to learn in SD Insan Amanah depend from these supporting factors. These are:

1. great student enthusiasm for learning with Blended Learning;
2. the percentage of students which has Internet access is increasing;
3. many schools have facilities that support Internet access in computer lab;
4. online assignment makes students more independent;
5. learning process will become more innovative and interactive with no time limit so that students feel enjoy learning compared when learning is done conventionally; and
6. good cooperation and responsiveness from the principal and other teachers.

This was blended learning alternative using website.

The framework [4] consists of a life-cycle that divides an initiative into four stages: Idea, Trial, Project, and Release. The purpose and key activities of each stage are summarized in Table 1. Each stage is divided into four Areas of concern: Technology, Learning, Social, and Organization. These represent the four major groups of stakeholders as well as the four main sources of risks. The four areas are linked together and depend on each other. A change to one will affect the others. In order to

IOP Conf. Series: Materials Science and Engineering **336** (2018) 012031 doi:10.1088/1757-899X/336/1/012031

reason about these propagating changes, the concept of Equilibrium is used to reason about the state of a stage. If a stage is in equilibrium, all changes have propagated and the effects of this propagation have been dealt with (or at least considered). Equilibrium is reached for a stage when there is no longer a need for reaction and adjustment.

Table 1. Result of Adjusted Development Framework

Stage	Purpose	Outcome	Key Activities
1 Ideas	Establish soundness of the idea	Plan of how to go ahead	Investigate technology
	Establish technical platform		Investigate feasibility Surveys
2 Trial	Test the idea	Information about what works and what does not	Produce learning materials Produce learning tools
	Elaborate the learning	Considerations made on what needs to change to move ahead	Offer the learning material
	Small scale testing		Measure how well it is received
3 Project	Speed launch/display	Expand beyond initiators	Similar to trial but larger scale
	Large scale testing	Information on how the material is received, both in terms of students and teachers experience	Report to funding organization
	Formalized in terms of resources and outcome		
	Establish social interplay		
4 Release	Hand over to target organization	Implemented and in use	Integrate into the organization (training)
	Remove reliance on initiators		Establish facilities (servers, studios, etc)

In order to reduce the complexity introduced by the areas of concern and the propagating changes, the concept of Focus is used. During each stage of the evolution, one (or at most a few) area of concern is in focus. The area in focus represents the area where the development is currently the most active, and where the direct changes will occur. The other areas are only changed through propagated effects of a change to the area in focus. Focus thus provides a means to reduce complexity and a way to introduce a sequential work-flow to the framework. Figure 2 depicts the framework and the major concepts.

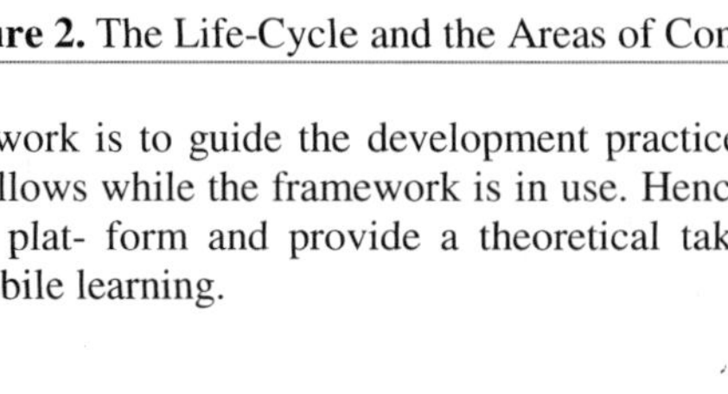

Figure 2. The Life-Cycle and the Areas of Concern

The major role of the framework is to guide the development practices not to tie down the models, methods, and tools, which follows while the framework is in use. Hence, the framework is there to be a work and communication plat- form and provide a theoretical take on the information systems development occurring in mobile learning.

3. Method and Tools

To design sustainable blended mobile learning classroom in practices, we took research approach to address simple problems in real classroom of Mathematics 7th grade in collaboration with teachers. As researchers from the university, we serve as meso-level actors to recontextualize pedagogic discourse [8] and help the school students and teachers interpret curricula of blended mobile learning and actualize them into teaching and learning practices.

The iterative cycle of research involving four linked phases, illustrated in Figure 3 [9]. Field studies and survey research have been planned, sebagai data awal, in turn the gathering of the empirical data of student's mobile device. Analisis data awal tersebut help to increase the competence and self-help ability in the field of mobile learning.

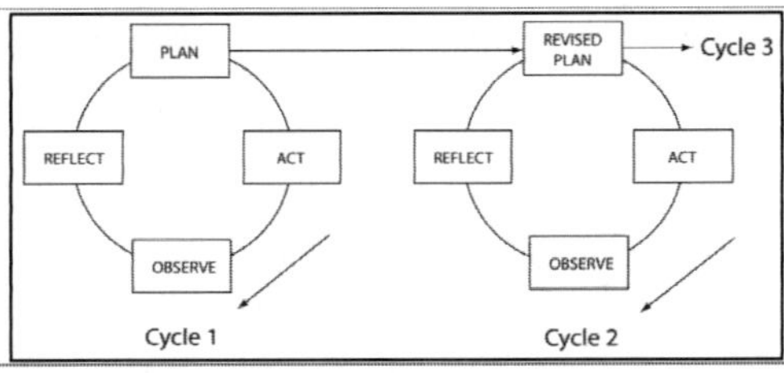

Figure 3. The Iterative Cycles of Action Research

Concerning the curricular commitment, Blended Mobile Learning use of the 5E Instructional Model [10]. This 5Es model consists of the following phases: engagement, exploration, explanation, elaboration, and evaluation. The model is used to sequence the learning activities in Mathematic 7th. Blended mobile learning is designed to provide opportunities for the student in the engagement phase of 5E using Mobile Learning Environment (MLE). In the Engage phase, the teacher motivates the inquiry by doing some video demonstration and posing some questions. These experiences engage the students with an event or question.

We designed four Lesson Plan (RPP: Rencana Pelaksanaan Pembelajaran) units in the first semester of Algebra emphasized on engagement and exploration. Before, during and after first cycle for RPP unit, we reflected upon the lessons and apply such understanding to inform the design of the next RPP unit (iterative as shown at Fig. 3). This was to facilitate the students' gradual changes in their habits towards learning seamlessly. For example, while the earlier activities involve the students expressing their understanding using MLE or capturing artifacts of the classroom and insert them to the RPP. [8]

The designed RPP was developed with the use of software apps on the Moodle MLE (Mobile Learning Environment) that runs on an Android operating system. The Moodle MLE enables teachers to create courses easily via its online learning management system and it enables students to easily personalize their learning experiences. Moodle MLE supports teachers in creating complete, coordinated, curriculum-based lessons that employ multiple media and applications (e.g., video, images, documents, animations, and the like) also multiple activities. All types of files can be uploaded and accessed through Moodle, but the student needs to have the correct software to be able to open them. There are 14 different types of activities in the standard Moodle, e.g., Assignments, Chat, Choice, Feedback, Forum, Quiz, and Survey.

The site of research is domained https://lipursukardjo.moodlecloud.com/. MoodleCloud supports the most recent versions of browsers (Google Chrome, Firefox, Safari, and Internet Explorer). To access MoodleCloud sites on mobile devices, students should install the Moodle Mobile app. The latest version of our Moodle Mobile app for Android (Moodle Mobile 3.3.2, release date 29 September 2017) can be obtained from the Google Play Store (recommended). Moodle Mobile was hosted on 27 July 2017. The research was conducted in SMPN 77 Jakarta Pusat.

IOP Conf. Series: Materials Science and Engineering **336** (2018) 012031 doi:10.1088/1757-899X/336/1/012031

4. Result and Analysis

The teachers' challenges faced in blended mobile learning assessed in school. During interviews, the teachers raised these concerns:

1. They are not sure if the designed lesson plans in the right way. Some teachers expressed doubts on the students' ability for doing self-directed learning.
2. Some teachers expressed that they needed help in developing and practicing questioning skills in the classroom.
3. Students' smartphones were varied.
4. Some regulations need to be reviewed.

The first response provide support to the teachers by giving personalized feedback after prototype lesson, and by helping the teachers to adapt the lessons for different ability students. The second response was for sharing in weekly meetings to discuss skills via analyzing recorded classroom sessions.

4.1. The Adaptation Scenario

The socio-technical system for mobile learning consists of three systems: the social, the personal and the technical [11]. Technology is the hardware and software used, Learning includes the pedagogy and teaching goals, while Social includes how students and teachers use and interact using the technology.

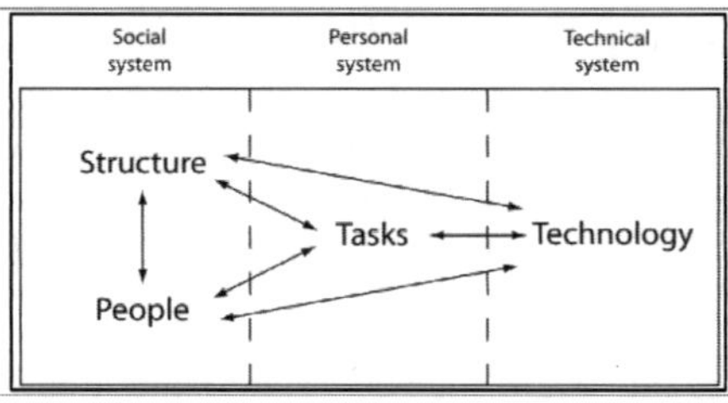

Figure 4. An Extended Socio-Technical System for Mobile Learning

The Social area is more focus because teachers and students' participation needed to improve the sustainability. The adaptation scenario was starting from a few users (teacher and students) with mock-up prototypes. At the SMPN 77 Jakarta, we have set 35 students (1 teachers) as mock-up prototype. Thus, in going to scale, the issue is learning process and infrastructure – how do we provide blended mobile learning supports and Internet access and make it more reliable and more stand-alone. The scale-up comprises these multiple dimensions:

1. Mobilized curricula (to lead students to self-directed learning and to bridging informal learning spaces),
2. Teacher facilitation skills,
3. Teacher readiness,
4. Student facilitation skills, and
5. Student readiness,
6. Technology infrastructure, e.g. WiFi and 3G/4G Connectivity; availability of mobile devices in 1:1.

From the short practice, the area of social is an effective pedagogical aspect of blended mobile learning model in supporting students with inquiry-based learning. Developed inquiry-based learning features equipped with learning media such as;

- a summary of the subject matter;
- learning videos created according to Engagement, Exploration and Explanation phase targets (drama, speech, poetry reading, etc.)
- animation;

IOP Conf. Series: Materials Science and Engineering **336** (2018) 012031 doi:10.1088/1757-899X/336/1/012031

- images from various sources on the internet to aid the learning process; and
- various interactive quiz questions with various models such as: multiple choice, short field, true false, matching, and TTS.

Video material is the most suitable/popular media for students. Students can replay on certain parts to see a more focused picture. This is difficult to achieve when video is delivered through media such as television. It is suitable to teach the material and faster to deliver messages than text media.

Sumber: https://matematohir.wordpress.com/2014/01/22/penerapan-konsep-aljabar-dalam-pemecahan-masalah/

Pak Idris mempunyai kebun apel berbentuk persegi, sedangkan Pak Tohir mempunyai kebun jeruk berbentuk persegi panjang. Ukuran panjang kebun jeruk Pak Tohir 20 m lebih dari panjang sisi kebun apel Pak Idris. Lebar kebun Pak Tohir 15 m kurang dari panjang sisi kebun apel Pak Idris. Jika diketahui kedua luas kebun Pak Idris dan Pak Tohir adalah sama, maka tentukan luas kebun apel Pak Idris?

Permasalahan yang terdapat pada kasus di atas dapat diselesaikan dengan model matematika yang dinyatakan dalam bentuk aljabar. Untuk memahami lebih lanjut mengenai bentuk aljabar, pelajari uraian bab ini dengan seksama.

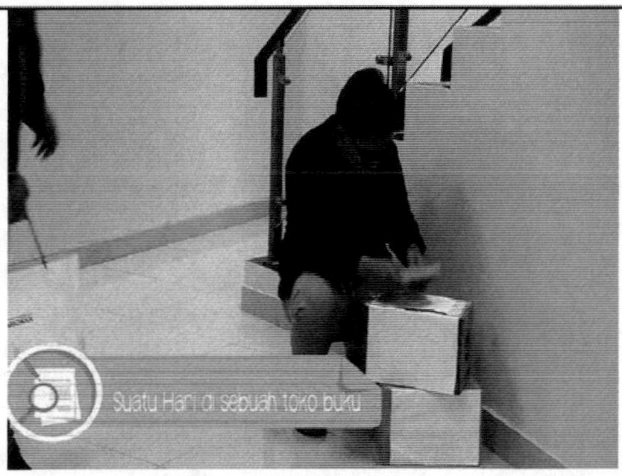

Story (Narrative)	The similar video of the story

Figure 5. Focus for adaptable mobile learning in the 7th grade of Mathematics

To able to take advantage mobile technologies, the revised lesson plan and the formative assessments of current learning process need to be adjusted.

Bu Halimah mempunyai sekeranjang apel. Bu Halimah ingin membagikan apel yang ia miliki tersebut kepada setiap orang yang ia temui. Setengah keranjang ditambah satu apel untuk orang pertama. Kemudian setengah dari sisanya ditambah satu, ia berikan kepada orang kedua yang ia temui. Selanjutnya, setengah dari sisanya ditambah satu, diberikan kepada orang ketiga yang ia temui. Sekarang, Bu Halimah hanya memiliki satu apel untuk ia makan sendiri. Tentukan banyak apel semula.

Sumber: Kemdikbud
Gambar 3.1 Ibu Halimah sedang membagikan sekeranjang apel

Kalian mungkin bisa memecahkan permasalahan tersebut dengan cara mencoba-coba dengan suatu bilangan. Namun berapa bilangan yang harus kalian coba, tidak jelas. Cara tersebut terlalu lama, tidak efektif, dan terkesan kebetulan.

Kalian bisa memecahkan persoalan tersebut dengan cara memisalkan banyak apel mula-mula dalam keranjang dengan suatu simbol. Lalu kalian bisa membuat bentuk matematisnya untuk memecahkan permasalahan tersebut. Bentuk tersebut selanjutnya disebut dengan bentuk aljabar, dan operasi yang digunakan untuk memecahkan disebut operasi aljabar. Untuk lebih mengenal tentang bentuk dan operasi aljabar, mari mengikuti pembahasan berikut.

Figure 6. Example inquiry material for 7th grade of Mathematics

4.2. The Adjusted Adaptation Framework

The former result research will be adjusted framework ([11]) and summarized in Table 2. The adjusted framework was created with a focus on Engagement activities, lesson plan, and its supporting materials.

Table 2. Result of Adjusted Development Framework

	Stage	Purpose	Outcome	Key Activities
1	Ideas	Establish soundness of the idea	Plan of how to go ahead	Investigate technology
		Establish technical platform		Investigate feasibility Surveys
				Investigate the migration/blended scenario
2	Trial	Test the idea	Information about what works and what does not	Produce learning materials Produce learning tools
		Elaborate the learning	Considerations made on what needs to change to move ahead	Offer the learning material
		Small scale testing		Measure how well it is received
3	Project	Speed launch/display	Expand beyond initiators	Similar to trial but larger scale
		Large scale testing	Information on how the material is received, both in terms of students and teachers experience	Report to funding organization
		Formalized in terms of resources and outcome		
		Establish social interplay		
4	Release	Hand over to target organization	Implemented and in use	Integrate into the organization (training)
		Remove reliance on initiators		Establish facilities (servers, studios, etc)

The migration scenario will take portion more stage 1, 2, and 3 (see table below).

Social	Entities involvements (teachers, students, parents).
Technological	Android based (based on majority smartphones used in school). Get used to mobile learning applications.
Pedagogical	Set the roles (teachers, parents, students, lesson plans, learning materials, assesments.

Figure 7. Focus for adaptable mobile learning in the 7th grade of Mathematics

In the revised lesson plan, the engagement activities help students to make connections with what they know and can do. During the engagement phase, the teacher can stimulate students' curiosity through:

- Create a need to know/create an interest
- Focus on a problem/ask questions

In the Evaluate phase, the teacher reviews the work done on MLE to detect students' developing conceptions of the Algebra. The four Lesson Plan (RPP: Rencana Pelaksanaan Pembelajaran) units in the first semester of Algebra emphasized on engagement and exploration. Before, during and after first cycle for RPP unit, we reflected upon the lessons and apply such understanding to inform the design of the next RPP unit (iterative as shown at Fig. 3). This was to facilitate the students' gradual changes in their habits towards learning seamlessly. For example, the earlier activities involve the students expressing their understanding using MLE should inserted to the RPP. [8] The designed RPP was developed with the use of software apps on the Moodle MLE (Mobile Learning Environment) that runs on an Android operating system.

C. Learning Activity Steps

1. Introduction Activities

No.	Activities	Duration (minutes)
1.	Conditioning the class	10
2.	Delivering competencies to be learned	
3.	Giving motivation	
4.	Delivering learning goals	

2. Core Activities

No.	Activities	Duration (minutes)
1.	Exploration The teacher divides the students into study groups	10
2.	Elaboration - Students in the group watched the video being aired - Students are distributed worksheets to hypothesize the issues under study - The teacher guides the students to use symbols that express many variables present in a box - Each student in the group fills the worksheet with teacher guidance to get the algebraic form according to the video - The experimental results are written on the provided worksheet. - Each Group makes a conclusion of the algebraic form.	35

Activities	Description of activities	Time Allocation
Pendahuluan	**Pre-Learning** Teachers provide training and simulation of mobile learning at previous meetings, where teachers have asked learners to bring Android Handphne with a minimum 5 inc screen. Training shaped introduction and how to use the app. Previous instructional instruction in accordance with attachment 1. **Conditioning the class** 1. Teachers invite learners to pray and check attendance 2. Teachers ask students if there are technical and non technical issues experienced by students while using Android app	20 minutes
	Delivering competencies to be learned 1. The teacher reiterates the basic competencies to be learned (Basic competencies are displayed on Android apps) 2. The teacher presents the learning experience for this meeting	
	Delivering learning goals 1. Teachers present learning indicators 2. Teachers convey concept maps to learners 3. The teacher informs the scope (attitudes, knowledge, and skills), and the peering technique to be used.	
	Giving motivation	

Figure 8. Example revised lesson plan with emphasize on the engagement activities.

5. Conclusion

As an enabler of new teaching learning activities supported by mobile technologies. The research implemented innovations of former research was adjusted for adoption of blended mobile learning by schools in formal learning of SMPN 77 7th grade. This adjusted framework demonstrates how to make successful implementation of early mobile learning sustainability through blended mobile learning adaptation framework. The Social area is in focus of adaptation adjusted framework because participation is important to improve the sustainability. From the short practice of mock-up prototype, blended mobile learning can be an effective pedagogical model in supporting students in inquiry-based learning.

Acknowledgments
We would like to express my sincere gratitude to my advisor DR. Agus Dudung, MPd for the continuous support of research, for his patience, motivation, enthusiasm, and immense knowledge. His guidance helped me in all the time of research and writing of this paper. My sincere thanks also go to principal of SMKN 39 Jakarta, teachers and students of 10th grade for offering us the opportunities in their school and leading us working on exciting research.

References
[1] S. Nurhayati 2017 Pemanfaatan Blended Learning dalam Pembelajaran Bahasa Indonesia di Sekolah Dasar in *Konferensi Nasional Sastra, Bahasa, dan Budaya (KS2B)* Malang

[2] I. J. Dewanto, 2006 Web Desain Metode Aplikasi dan Implementasi, Yogyakarta: Graha Ilmu

[3] M. Fullan 2004 Leadership & Sustainability: System Thinkers in Action, Ontario: Corwin.

[4] S. T. Dave Lawrence, 2007 Balanced Website Design (Optimising Aesthetics, Usability and Purpose), London: Springer,.

[5] W. Ng and H. Nicholas 2013 A framework for sustainable mobile learning in schools., *British Journal of Educational Technology,* **44**(5) p p. 695–715.

[6] T. Bell, A. Cockburn, A. Wingkvist and R. Green 2009 Podcasts as a supplement in tertiary education: An experiment with two Computer Science courses in *The International Conference on Mobile Learning Technologies and Applications (MoLTA'07)*, Auckland.

[7] A. Wingkvist and M. Ericsson 2009 A meta-model describing the development process of mobile learning," in *the 8th International Conference on WebBased Learning (ICWL 2009)*, Aachen

[8] C.-K. Looi and L.-H. Wong 2014 Implementing Mobile Learning Curricula in Schools: A Programme of Research from Innovation to Scaling," *Educational Technology & Society,* **17**(2) pp. 72-84

IOP Conf. Series: Materials Science and Engineering **336** (2018) 012031 doi:10.1088/1757-899X/336/1/012031

[9] S. Kemmis and M. Wilkinson 1998 Participatory action research and the study of practice," in *Action Research in Practice: Partnerships for Social Justice in Education*, London, Routledge pp. 21-36.

[10] R. B. e. all 2006 The BSCS 5E Instructional Model: Origins, Effectiveness, and Applications BSCS, Colorado

[11] A. C. Wingkvist 2009 Understanding Scalability and Sustainability in Mobile Learning: A Systems Development Framework. Thesis for the degree of Doctor of Philosophy," Intellecta Infolog, Göteborg, Sweden

[12] D. Juju 2006 Tip dan Trik Desain Web untuk Pemula, Jakarta: PT Elex Media Komputindo

[13] B. Permana 2004 Kiat Praktis Menjadi Desainer Web Profesional, Jakarta: PT Elex Media Komputindo.

[14] J. N. Robert Mack 1993 Usability inspection methods: report on a workshop held at CHI'92," ACM SIGCHI Bulletin **25** (1)

[15] J. Simarmata 2010 Rekayasa Web, Yogyakarta: Andi Offset

[16] F. Sulianta 2010 IT Ergonomics, Jakarta: PT Elex Media Komputindo

[17] A. H. Suyanto 2007 Web Design Theory And Practices, Yogyakarta: Andy

[18] A. P. D. S. Ahmet Mentes and A. P. D. Aykut H. Turan 2012 ASSESSING THE USABILITY OF UNIVERSITY WEBSITES: AN EMPIRICAL STUDY ON NAMIK KEMAL UNIVERSITY," *TOJET: The Turkish Online Journal of Educational Technology* **11** *(3)* , pp. 61-69

[19] L. Hasan and E. Abuelrub, "Common Usability Problems on Educational Websites," in *International Conference on Education and Education Technologies*, Greece, 2003.

[20] KEMENTERIAN PENDIDIKAN DAN KEBUDAYAAN 2016 SILABUS MATA PELAJARAN MATEMATIKA SEKOLAH MENENGAH PERTAMA/MADRASAH TSANAWIYAH (SMP/MTs), Jakarta: KEMENTERIAN PENDIDIKAN DAN KEBUDAYAAN

[21] A. Mariani 2014 Mencermati Materi Matematika Kelas VII Pada Kurikulum 2013 [Online]. Available: http://guraru.org/guru-berbagi/mencermati-materi-matematika-kelas-vii-pada-kurikulum-2013/. [Accessed 10 September 2017].

[22] A. Wingkvist and M. Ericsson 2009 Addressing sustainability for research initiatives in mobile learning through scalability," in *the 8th IASTED International Conference on Web-Based Education (WBE'09)*, Sharm El Sheikh

[23] C.-K. Looi and L.-H. Wong 2014 Implementing Mobile Learning Curricula in Schools: A Programme of Research from Innovation to Scaling," *Educational Technology & Society,* **17**(2), pp. 72-84

[24] L. Wong, C. Chai, C. Chin, Y. Hsieh and M. Liu 2012 Towards a seamless language learning framework mediated by the ubiquitous technology," *International Journal of Mobile Learning and Organisation,* **6**(2), pp. 156-171

[25] Y. C. Hsu and Y. H. Ching 2015 A review of models and frameworks for designing mobile learning experiences and environments *Canadian Journal of Learning and Technology,* **41**(3) pp. 1-22.

IOP Conf. Series: Materials Science and Engineering **336** (2018) 012032 doi:10.1088/1757-899X/336/1/012032

Measurement of Usability for Multimedia Interactive Learning Based on Website in Mathematics for SMK

Moch. Sukardjo*, Lipur Sugiyanta

Study Program of Education for Electrical Engineering/ Faculty of Engineering, State University of Jakarta

*msoekardjo@unj.ac.id

Abstract. Web usability, if evaluation done correctly, can significantly improve the quality of the website. Website containing multimedia for education shoud apply user interfaces that are both easy to learn and easy to use. Multimedia has big role in changing the mindset of a person in learning. Using multimedia, learners get easy to obtain information, adjust information and empower information. Therefore, multimedia is utilized by teachers in developing learning techniques to improve student learning outcomes. For students with self-directed learning, multimedia provides the ease and completeness of the courses in such a way that students can complete the learning independently both at school and at home without the guidance of teachers. The learning independence takes place in how students choose, absorb information, and follow the evaluation quickly and efficiently. The 2013 Curriculum 2013 for Vocational High School (SMK) requires teachers to create engaging teaching and learning activities that students enjoy in the classroom (also called invitation learning environment). The creation of learning activity environment is still problem for most teachers. Various researches reveal that teaching and learning activities will be more effective and easy when assisted by visual tools. Using multimedia, learning material can be presented more attractively that help students understand the material easily. The opposite is found in the learning activity environment who only rely on ordinary lectures. Usability is a quality level of multimedia with easy to learn, easy to use and encourages users to use it. The website Multimedia Interactive Learning for Mathematics SMK Class X is targeted object. Usability website in Multimedia Interactive Learning for Mathematics SMK Class X is important indicators to measure effectiveness, efficiency, and student satisfaction to access the functionality of website. This usability measurement should be done carefully before the design is implemented thoroughly. The only way to get test with high quality results is to start testing at the beginning of the design process and continuously testing each of the next steps. This research performs usability testing on of website by using WAMMI criterion (Website Analysis and Measurement Inventory) and will be focused on how convenience using the website application. Components of Attractiveness, Controllability, Efficiency, Helpfulness, and Learnability are applied. The website in Multimedia Interactive Learning for Mathematics SMK Class X can be in accordance with the purpose to be accepted by student to improve student learning outcomes. The results show that WAMMI method show the usability value of Multimedia Mathematics SMK Class X is about from 70% to 90%.

1. Introduction

There are many things to consider when building a new website for education. Basically, the site needs to be attractive enough that students want to use at it. It also needs to contain all of the materials that students need in order to help them achieve the objective for learning targets. One of the most important aspects of building a website is testing for usability. Internet users (students) are accustomed to being able to figure out how to use a website quickly. Most of them will not take the time to figure out a site that is not usable.

Web usability is how easy a website is to use. The main reason that usability is so important is because there are so many similar websites that people will go to the next site if the first one they visit is not usable. Beautiful website may not achieve its presence if users are unable to figure out how to navigate the site quickly. Web usability testing aims to determine whether a website is in accordance with the needs of the user (students) or not.

This research does usability testing on Website containing Multimedia Interactive Learning Based in Mathematics for SMK on by using field observation method is by observing users how they use the application. The website is not only a cost efficient and timely method to communicate with various students. That's vitality of usability issue for the websites. Usability is importance in terms of satisfying website users' needs and expectations. In this research, usability components tested consist of attractiveness, controllability, efficiency, helpfulness, and learnability. It also presents a description of two such methods developed by [1] and [2].

2. Related Research

The criteria used in the research of usability websites are quite diverse and vary from one journal to another. There are two papers that become reference in doing this research, i.e.,:

- "Assessing the Usability of University Websites: An Emperical Study on Namik Kemal University" [1]
- "Common Usability Problems on Educational Websites" [2]

[1] uses five usability criteria defined by WAMMI (Website Analysis and Measurement Inventory). These criteria can be seen in table 1.

Table 1. Criteria used in paper 1 "Assessing the Usability of University Websites: An Empirical Study on Namik Kemal University"

	Criteria	Definition
1	Attractiveness	Website have a visually pleasing appearance and can draw the user's attention to visit the web.
2	Controllability	The level of control that a user perceives when a person interacts with a website. The site with good controllability enable users to navigate the website easily and can do the things they want to do.
3	Efficiency	Users can achieve their target with a short visit without the need to use a lot of cognitive effort.
4	Helpfulness	Website has the structure and content that matches the expectations of users.
5	Learnability	Ease in the time and effort of users to learn the website.

That research aims to measure the usability of the Namık Kemal University (NKU) web site via the five factors of usability: attractiveness, controllability, helpfulness, efficiency and learnability. The

results reveal that the five usability factors are positively related with usability perception, i.e., attractiveness, helpfulness, efficiency, learnability, and controllability.
[2] uses five different criteria. The criteria used in the paper and its definition can be seen in able 2 below.

Table 2. Usability criteria for paper 2 "Common Usability Problems on Educational Websites"

	Criteria	Definition
1	Navigation	Assessing whether a site has the tools and links that facilitate user navigation on the site.
2	Architecture	Associated with the information structure where the site structure is divided into logical groups and each group contains related information.
3	Ease of use and Communication	Relates to the cognitive effort required to use a website.
4	Design	Visual Attraction website, use of good design and use of images, fonts, good color.
5	Content	Assessing whether a website already has the information the user needs.

The criteria in Table 2 comprehensively evaluated the usability of three large public university websites in Jordan, using the heuristic evaluation method. The results identified a list of 34 specific types of common usability problems that were found on the selected Jordanian university websites, and described the frequency number of these problems. The results of 34 problems that were uniquely identified in this research included: Misleading links, links don't open the destination pages, links cause disappearance of the menu, broken links, various types of inconsistency problems, inappropriate orientation of the page design, ineffective text format, broken images, inappropriate choice of color, empty page, inappropriate content, difficult interaction with a website, and lack of support to the Arabic language.
Futhermore, the common and specific usability problems that were uniquely identified in that research, could be used as guidelines for universities in Jordan, to investigate and improve their universities' websites, and therefore to achieve the advantages of usable educational websites.
Therefore, given the different criteria used by the two papers, further studies were conducted by investigate and improve their educational websites followed by comparing the definitions of the two groups. Table 3 shows the results of the comparison:

Table 3. Comparison of Criteria

	Paper 1	Paper 2
1	Attractiveness: Website have a visually pleasing appearance and can draw the user's attention to visit the web.	Design: Visual Attraction website, use of good design and use of images, fonts, good color.
2	Controllability: The level of control that a user perceives when a person interacts with a website. The site with good controllability enable users to navigate	Navigation: Assessing whether a site has the tools and links that facilitate user navigation on the site.

	the website easily and can do the things they want to do.	
3	Efficiency: Users can achieve their target with a short visit without the need to use a lot of cognitive effort.	Ease of use and Communication: Relates to the cognitive effort required to use a website.

Table 3. Comparison of Criteria

	Paper 1	Paper 2
4	Learnability: Ease in the time and effort of users to learn the website.	Architecture: Associated with the information structure where the site structure is divided into logical groups and each group contains related information.
5	Helpfulness: Website has the structure and content that matches the expectations of users.	Content: Assessing whether a website already has the information the user needs.

Based on the review results of the two papers above, it is known that both groups of criteria have the similar form of definition but criteria in Paper 1 is more extensive coverage than Paper 2, therefore in this research, criteria in the Paper 1 will be used.

3. Method and Tools

Set up a website usability testing is carefully construct activities to build a scenario wherein a person performs a list of tasks that someone who is using the website for the first time is likely to perform. This form will be used to evaluate a product (in this case a website) by testing is on users (students). Researcher observes and listens to the students who is performing the tasks while taking notes. Watching students perform common tasks on a website is a great way to test whether the site is usable because quick response will immediately be able to see whether they are able to perform the tasks and any difficulties they have while doing so.
The following is a brief description of the main usability testing methods that are used.

Table 4. Description of Criteria

	Main Criteria	Sub-Criteria	Definition
1	Attractiveness	View	Applied good resolution at least 640 x 480 Can be viewed using mobile with a resolution of 1280 x 800 (HD) to 320 x 240 (QVGA)
		Color	Color used is dominant cool color (pastel color or grey color)
		Font	Selected of fonts is readable (Arial 10-15 pt or Calibri 10-15 pt)
		Images	Images have good quality that is type PNG or type TIFF
		Video	Video is encoded with minimum resolution of 480p: 854x480
		Animation	Animation embedded is not auto play
2	Controllability	Design	Website visualization is easy to understand by users in the form of buttons and links with no different from the usual looking of Facebook.
		Menu navigation	Menus are grouped according to their similarity
3	Efficiency	Speed launch/display	Time needed to load/display the entire web in a short time from the beginning of the click (is ranging from 1-5 to 21-25 second)

Table 4. Description of Criteria

Main Criteria	Sub-Criteria	Definition

4	Learnability	Ease of use	Users can use the website without any constraints even though it has not opened for 1 month
5	Helpfulness	Content architecture	Information is divided into logical groups and each group contains updated regularly

The criteria above need to be quantitative to make student's perceptions on usability of website could

Table 5. Ranking of Criteria

	Main Criteria	Sub-Criteria	A	B	C	D	E
1	Attractiveness	View	Website use 1280x800 resolution	Website use 1024x768 resolution	Website use 1024x600 resolution	Website use 800x600 resolution	Website use 640x480 resolution
			Could open via mobile with 1280x720 resolution (HD)	Could open via mobile with 960x540 resolution (HD)	Could open via mobile with 800x600 resolution (SVGA)	Could open via mobile with 640x480 resolution (VGA)	Could open via mobile with 320x240 resolution (QVGA)
		Color	Dominant color is cool (green, blue, purple)	Dominant color is warm (red, yellow, orange)	Dominant color is neutral (chocolate, grey)	Dominant color is black and white	Dominant color is black
		Font	Arial size 10-15	Open Sans size 10-15	Times New Roman size 10-15	Verdana size 10-15	Calibri size 10-15
		Images	Format PNG (Portable Network Graphics)	Format JPEG (Join Photographic Expert Groups)	Format JPG	Format BMP	Format TIFF (Tagged Imaged Format File)
		Video	Format MKV, MOV	Format MP4	format WMV	Format FLV, SWF	Format AVI, MPEG, 3GP
		Animation	Format 3D	Format SWF, SWI	Format EXE	Format PPTX	Format GIF
2	Controlability	Design	Button underlined, Pointer hover	Button in individual frame, Pointer hover	Pointer hover	Pointer change size	No effect
		Menu navigation	Grouped according to head menu	Grouped	Some are grouped	Few are grouped	None
3	Efficiency	Speed launch/display	1-5 secs	6-10 secs	11-15 secs	16-20 secs	21-25 secs
4	Learnability	Ease of use	Users capable using the website after 30 days	Users capable using the website after 20 days	Users capable using the website after 15 days	Users capable using the website after 10 days	Users capable using the website after 7 days
5	Helpfulness	Content architecture	Updated regularly every 3 days	Updated regularly every 5 days	Updated regularly every week	Updated regularly every month	Updated regularly every 2 months

be analyzed. This quantitative fulfilment restriction is useful for each sub-criteria to have a maximum and a minimum constraints response that can be fulfilled. The limits used letters A-B-C-D-E, each representing the mark of the criteria. The letter "A" representing the best rating of fulfillment and the lowest rating represented by letter "E". Students will provide value in accordance with column that has been provided. The Table 5 below is the criteria and quantitative description of usability website used in the research.

The website become research object is named http://www.sukardjo-lipurs.com. The website usability testing was using the famous browser Google Chrome. In terms of this test, only 3 pages from the website will be tested. Almost all of usability criteria were evaluated through visual or eyesight not get

IOP Conf. Series: Materials Science and Engineering **336** (2018) 012032 doi:10.1088/1757-899X/336/1/012032

into the elements of the website source code. Students and 3 expert (teachers) are involved during this testing. Several criteria require a period to complete the usability testing. Researcher also perform testing using online tools to measure of a website portfolio. The chosen online tools are responsivepx.com; fontface. Ninja; and pingdom (https://tools.pingdom.com/). For helpfulness, these sub-criteria is verified with 16-item questionnaire was administered manually using PSSUQ (Post-Study Usability Questionnaire) template.
Website usability testing was conducted on 27 July 2017 until 10 September 2017. The research was conducted in SMKN 39 Jakarta Pusat.

4. Result and Analysis

The result of the website usability testing based on criteria showed in Table 6. below.

Table 6. Result of Testing

	Main Criteria	Sub-Criteria	**Result**
1	Attractiveness	View	**A**
			A
		Color	**A**
		Font	**B**
		Images	**A**
		Video	**A**
		Animation	**B**
2	Controllability	Design	**B**
		Menu navigation	**A**
3	Efficiency	Speed launch/display	**B**
4	Learnability	Ease of use	**A**
5	Helpfulness	Content architecture	**B**

4.1. Usability Website of Attractiveness

Website has been using the resolution of 1280 x 800 then the rating is "A". The screenshot view using responsivepx.com is given below. Website can be accessed via mobile (Smart Phone) with the best size 1280 x 720 (HD) hence the rating is "A".

Figure 1. Website view using http://responsivepx.com/?sukardjo-lipurs.com%2F#816x984&scrollbars

The “A” was marked on sub-criteria for color, because the dominant color in the homepage of the website is cool (blue and green). Psychologically, the cool color is able to attract the attention of users because of its beauty and cool mind for whom views the website.

Mark “B” on the sub-criteria Fonts because the selection of fonts from the website in accordance with the website in the academic world, (most of) using "Open Sans" font size 10-13pt. This font has a distinctive feature as a clear, easy to read and often used by students in academic activities. Verification of this font using software called “font face ninja”.

Figure 2. Font view using FontFace Ninja.

4.2. Usability Website of Controllability

Mark “B” on the design sub-criteria because the navigation on the website can be distinguished by the frame limit that as shown below figure 4.3. The key menu and navigation is easy to understand because it uses language commonly used by most websites and textbooks. Navigation buttons position is designed stay in place even though visitors move pages of the website.

Figure 3. Navigation button grouped in frame.

4.3. Usability Website of Efficiency

Mark “B” in the sub-criteria loading time obtained from the testers using online tools called pingdom (https://tools.pingdom.com/). This tool is useful to measure time required (seconds) for the entire website to load. The result of loading time website is 6,38 s with capacity of website size equal to 15,7 MB.

Figure 4. Navigation button grouped in frame.

4.4. Usability Website of Learnability

Learnability is defined as the capability of a website to enable the student to learn how to use it. This is a key usability feature that is often disregarded. In this case, the testing whether website is learnable is over a certain period (month or week), using the same students testing the interface multiple times at regular intervals. A more learnable website is one that reduces the time it takes to complete tasks as users spend more time with a website faster than others. This sub-criteria is marked with "A". Student can use the website even though not open for 30 days. That website rank relies on the familiarity and simplicity design.

4.5. Usability Website of Helpfulness

Helpfulness is degree to which students feel that the site enables them to solve their problems with finding information and navigating. Website is updated regularly every 5 days. This sub-criteria is marked with "B". Students find the website attractive, and also find it helpful.

This sub-criteria is verified with 16-item questionnaire was administered manually using PSSUQ (Post-Study Usability Questionnaire) template. The PSSUQ is a 16-item survey that measures users' perceived satisfaction with a product or system. Obtaining an overall satisfaction score is done by averaging the four sub-scales of System Quality (the average of items 1-6), Information Quality (the average of items 7-12), and Interface Quality (the average of items 13-16). The PSSUQ is highly reliable (0.94 of 1.00) and is entirely free. (https://chaione.com/blog/ux-research-standardizing-usability-questionnaires/)

The Post-Study Usability Questionnaire Version 3 — Strongly agree (1) … Strongly disagree (7)

		1	2	3	4	5	6	7	NA
1	Overall, I am satisfied with how easy it is to use this system.	O	O	O	O	O	O	O	O
2	It was simple to use this system.	O	O	O	O	O	O	O	O
3	I was able to complete the tasks and scenarios quickly using this system.	O	O	O	O	O	O	O	O
4	I felt comfortable using this system.	O	O	O	O	O	O	O	O
5	It was easy to learn to use this system.	O	O	O	O	O	O	O	O
6	I believe I could become productive quickly using this system.	O	O	O	O	O	O	O	O
7	The system gave error messages that clearly told me how to fix problems.	O	O	O	O	O	O	O	O
8	Whenever I made a mistake using the system, I could recover easily and quickly.	O	O	O	O	O	O	O	O
9	The information (such as online help, on-screen messages and other documentation) provided with this system was clear.	O	O	O	O	O	O	O	O
10	It was easy to find the information I needed.	O	O	O	O	O	O	O	O
11	The information was effective in helping me complete the tasks and scenarios.	O	O	O	O	O	O	O	O
12	The organization of information on the system screens was clear.	O	O	O	O	O	O	O	O
13	The interface* of this system was pleasant.	O	O	O	O	O	O	O	O
14	I liked using the interface of this system.	O	O	O	O	O	O	O	O
15	This system has all the functions and capabilities I expect it to have.	O	O	O	O	O	O	O	O
16	Overall, I am satisfied with this system.	O	O	O	O	O	O	O	O

**The "interface" includes those items that you use to interact with the system. For example, some components of the interface are the keyboard, the mouse, the microphone, and the screens (including their graphics and language).*

Figure 5. PSSUQ survey.

Using PSSUQ instrument, the questionnaire was distributed to 10 respondents chosen from the sample, ie students and teachers who have used the website online. From the answers, all questions were answered validly. Reply of the questionnaires produce an output shows in figure 5.

IOP Conf. Series: Materials Science and Engineering **336** (2018) 012032 doi:10.1088/1757-899X/336/1/012032

The Post-Study Usability Questionnaire	Strongly Agree					Strongly Disagree		Total Respond
	Selected Answer							
ITEM	1	2	3	4	5	6	7	
1	5	3	2	0	0	0	0	100 %
2	7	3	0	0	0	0	0	100 %
3	2	8	0	0	0	0	0	100 %
4	0	8	2	0	0	0	0	100 %
5	3	6	1	0	0	0	0	90 %
6	7	3	0	0	0	0	0	100 %
7	0	0	0	0	7	3	0	100 %
8	0	0	0	6	4	0	0	100 %
9	2	6	2	0	0	0	0	100 %
10	4	5	0	0	0	10	0	100 %
11	0	2	7	1	0	0	0	100 %
12	1	9	0	0	0	0	0	100 %
13	1	8	1	0	0	0	0	100 %
14	2	7	1	0	0	0	0	100 %
15	4	6	0	0	0	0	0	100 %
16	7	3	0	0	0	0	0	100 %

Figure 6. Quetionaires results of PSSUQ survey.

Data processing produces the users' interpretation of 70% satisfaction with the website. This satisfaction is based on the simplicity of the application (> 70%), ease of use (> 70%), application response to errors (> 70%), appropriate content (> 70%) and pleasant interface > 80%). The results of data processing are as follows:

	The Post-Study Usability Questionnaire	Result
1	Overall, I am satisfied with how easy it is to use this system	From 10 respondents, 5 people (50%) chose the item "1", 3 people (30%) chose item "2" and 1 person (20%) chose item "3". It can be said for item "1" generally satisfied enough.
2	It was simple to use this system	From 10 respondents, 7 people (70%) chose the item "1", 3 people (30%) chose the item "2". It can be said for item "2" generally the website used is quite simple.
3	I could complete the tasks and scenarios quickly using this system	From 10 respondents, 2 people (20%) chose item "1", 8 people (80%) chose item "2". It can be said for item "3" generally learning task can be quickly can be done after using this website.
4	I felt comfortable using this system	From 10 respondents, 8 people (80%) chose the item "2" and 2 people (20%) chose item "3". It can be said for item "4" the website is very convenient.
5	It was easy to learn to use this system	From 10 respondents, 30 people (30%) chose the item "1", 6 people (60%) chose item "2" and 1 person (10%) chose item "3". It can be said for item "5" generally very easy to use this website.
6	I believe I could become productive quickly using this system	From 10 respondents, 7 people (70%) chose the item "1" and 3 people (30%) chose item "2". It can be said for item "6" the website is very productive.
7	The system gave error messages that clearly told me how to fix problems	From 10 respondents, 7 people (70%) chose the item "5" and 3 people (30%) chose item "6". It can be said for item "7" generally website show a clear error message when there is a problem.
8	Whenever I made a mistake using the system, I could recover easily and quickly	Similar with question 7 that the majority of respondents stated whether appear a misuse, it is easy to fix. 6 people (60%) answered item "5" and 4 people (40%) answered answer "5". It can be said that if there are errors of use then it is easy to fix it.
9	The information (such as online help, on-screen messages and other documentation) provided with system was clear	From 10 respondents, 2 persons (20%) chose the item "1", 6 people (60%) chose the item "2" and 2 people (20%) chose item "3". It can be said for item "1" information submitted in website is clear.
10	It was easy for me to find the information I needed	From 10 respondents, 4 people (40%) chose the item "1", 50 people (50%) chose the item "2" and 1 person (10%) chose the item "6". It can be said for item "1" that 90% of respondents said it is very easy to get the required information.
11	The information was effective in helping me complete the tasks and scenarios	From 10 respondents, 2 persons (20%) chose the item "1", 7 people (70%) chose item "3" and 1 person (10%) chose item "4". It can be said for item "1" that 90% of participants say information is complete enough to help learning tasks.
12	The organization of information on the system screens was	From 10 respondents, 9 people (90%) chose the item "1", 1 person (10%)

The Post-Study Usability Questionnaire	Result
clear	selects item "2". It can be said for item "1" generally the organization of website on the screen quite clear.
13 The interface* of this system was pleasant	From 10 respondents, 1 person (10%) selects item "1", 8 people (80%) selects item "2" and 1 person (10%) selects item "3". It can be said for item "1" interface in this website is fun.
14 I liked using the interface of this system	From 10 respondents, 2 persons (20%) chose item "1", 7 people (70%) chose item "2" and 1 person (10%) chose item "3". It can be said for item "1" that the participants really like to use interface in this website program.
15 This system has all the functions and capabilities I expect it to have	From 10 respondents, 4 people (40%) chose the item "1", 6 people (60%) chose the item "2". It can be said for item "1" that program in website got function and capability as expected.
16 Overall, I am satisfied with this system	From 10 respondents, 7 people (70%) chose the item "1", 3 people (30%) chose the item "2". It can be said for item "1" that user feel satisfied with this website.

5. Conclusion

Web usability is how easy a website is to use. The website has fulfilled usability testing aimed to determine whether a website is in accordance with the needs of the user (students) or not. The website https://www.sukardjo-lipurs.com/ in usability testing achieved from 70% to 90% of the criteria (Attractiveness = almost A, Controllability = almost A, Efficiency = B, Helpfulness = B, Learn ability = A). Therefore, this website is qualified in order to be used in research to assess the effectiveness of self-learning using the website.

Acknowledgments

I would like to express my sincere gratitude to my advisor DR. Agus Dudung, MPd for the continuous support of research, for his patience, motivation, enthusiasm, and immense knowledge. His guidance helped me in all the time of research and writing of this paper. My sincere thanks also go to principal of SMKN 39 Jakarta, teachers and students of 10th grade for offering us the opportunities in their school and leading us working on exciting projects.

References

[1] A. P. D. S. Ahmet Mentes and A. P. D. Aykut H. Turan 2012 ASSESSING THE USABILITY OF UNIVERSITY WEBSITES: AN EMPIRICAL STUDY ON NAMIK KEMAL UNIVERSITY," *TOJET: The Turkish Online Journal of Educational Technology 11 (3)*, pp. 61-69.

[2] L. Hasan and E. Abuelrub 2003 Common Usability Problems on Educational Websites," in *International Conference on Education and Education Technologies*, Greece

[3] S. T. Dave Lawrence, 2007 Balanced Website Design (Optimising Aesthetics, Usability and Purpose), London: Springer

[4] I. J. Dewanto 2006 Web Desain Metode Aplikasi dan Implementasi, Yogyakarta: Graha Ilmu

[5] D. Juju 2006 Tip dan Trik Desain Web untuk Pemula, Jakarta: PT Elex Media Komputindo

[6] B. Permana 2004 Kiat Praktis Menjadi Desainer Web Profesional, Jakarta: PT Elex Media Komputindo,

[7] J. N. Robert Mack 1993 Usability inspection methods: report on a workshop held at CHI'92," ACM SIGCHI Bulletin **25** (1)

[8] J. Simarmata 2010 Rekayasa Web, Yogyakarta: Andi Offset

[9] F. Sulianta 2010 IT Ergonomics, Jakarta: PT Elex Media Komputindo

[10] A. H. Suyanto 2007 Web Design Theory And Practices, Yogyakarta: Andi

IOP Conf. Series: Materials Science and Engineering **336** (2018) 012033 doi:10.1088/1757-899X/336/1/012033

Development of Servo Motor Trainer for Basic Control System in Laboratory of Electrical Engineering Control System Faculty of Engineering Universitas Negeri Surabaya

Endryansyah*, Puput Wanarti Rusimamto, Adam Ridianto, Hariyadi Sugiarto
Department of Electrical Engineering, Universitas Negeri Surabaya. Indonesia

*endryansyah@gmail.com

Abstract. In the Department of Electrical Engineering FT Unesa, there are 3 majors: S1 Electrical Engineering Education, S1 Electrical Engineering, and D3 Electrical Engineering. Courses the Basic System Settings go to in the curriculum of the three programs. Team lecturer college of basic system settings seek learning innovation, focused on the development of trainer to student practicum at the laboratory of systems control. Trainer developed is a servo motor along with the lab module that contains a wide variety of theories about the servo motor and guide the practicum. This research type is development research using methods Research & development (R & D). In which the steps are applied in this study is as follows: pay attention to the potential and existing problems, gather information and study the literature, design the product, validate the design, revise the design, a limited trial. The results of the validation of learning device in the form of modules and trainer obtained as follows: score validation of learning device is 3,64; score validation lab module Servo Motor is 3,47; and questionnaire responses of students is 3,73. The result of the whole validation value is located in the interval >of 3.25 s/d 4 with the category of "Very Valid", so it can be concluded that all instruments have a level of validity "Very Valid" and worthy of use for further learning.

1. Introduction

In an effort to increase the quality of learning outcomes for students of Electrical Engineering Unesa in particular the department of Electrical Engineering Education, then the team lecturer college System's Basic Settings have been conducting research with the innovation learning model. That is by applying Cooperative Learning Model (MPK) is completely supported by the module correctly. The module in this study is a practical module motor servo consists of Modules Students, Module Lecturers and trainer. Module Student contains (1) the Map position of the module and the ability prerequisites; (2) an Introduction that includes a description, prerequisites, instructions for use of the modules, competencies and indicators; (3) learning Activities, in which each learning activity includes the competencies and indicators, a description of the material, sheet of student activity and formative test that is not accompanied by the answer and (4) Evaluation that includes cognitive, affective and psychomotor.

Assessment team of experts to the module in the form of a response team of experts on: (1) Purpose, which includes the final destination is relevant to the curriculum of Department of Electrical Engineering Unesa and Learning Plan Semester (RPS) courses in the Basic System Settings. (2) the Concept or the Substance, which includes the truth of the concept, structured competency-based, is the substance of the principal essential oils, are grouped in parts that are logical (the material of each learning activities relevant to the RPS), the material in accordance with the time available, related to the material of the previous (sequence), feasibility of tools and materials, the use of notation, symbols, and units, no bias of gender, ethnic, religious. (3) Format, which includes an agreed format, the numbering system is clear, the text and illustrations are balanced, the attractiveness module, (4) Language which includes the writing of the Indonesian Language that is good and true, the suitability of the language with the level of development of students, the phrase is written in an effective, efficient, and communicative, the definition clearly described and the terminology used appropriately, commands, instructions, and assignment are written clearly. (5) Illustrations, which includes help understanding the concept, lists the source, if the quote illustrations from other sources, illustrations presented in a clear, interesting, and easy to understand, not gender-biased, ethnic, religious. (6) Evaluation, which includes relevance to the learning objectives, equipped with a grating/table specifications, and guide the evaluation. (7) a Bibliography, which includes writing by default (referring to WHAT is or agreed upon), actually referred to in the text.

2. Theory Review

2.1 Cooperative Learning
According to Trianto, 2007, learning under the theory of constructivists is cooperative. Cooperative learning emerge from the concepts that the students will more easily find and understand difficult concepts if they discuss with friends. Students routinely work in groups to help each other solve issues that are complex. So the nature of the social and the use of peer group a key aspect of cooperative learning. According to Sugiyanto, 2010, cooperative learning (cooperative learning) is a learning approach that focuses on on the use of small groups of students to work together to maximize the learning conditions for achieving learning goals. According to Lie (in Sugiyanto, 2010), in cooperative learning there are several elements that must be known. These elements are: (1) the interdependence is positive, (2) face to face Interaction, (3) Accountability of the individual, (4) Skills to establish personal relationships.

2.2 Steps of Cooperative Learning
There are six main steps or stages in a lesson using cooperative learning. Phase 1, the lesson begins with the teacher convey the objective of the lesson and motivated students to learn. Phase 2, the teacher conveys information, often times the reading material rather than verbally. Phase 3, students are grouped into teams-team learning. Phase 4, teachers guide students work together to complete the task with them. Phase 5, the percentage of results end of working group or evaluation about what they have learned. Phase 6, the teacher gave the award to the efforts of groups and individuals.

2.3 Module
According to Mulyasa (2004:148) module is a learning process for a unit subject specific which are arranged in a systematic, operational and targeted to be used by the students, accompanied by guidance on its use to teachers. Learning with the module system has the following characteristics: (1) Each module must provide information and give instructions for implementation are clear about what should be done by the learners, how to do it, and a source of learning what should be used. (2) a Module is a learning individual, so seek to involve as much as possible the characteristics of the learners. (3) the learning Experience in the module is provided to help learners achieve the learning objectives effectively and seefesien possible, as well as allow the learners to do learning actively, not just read and hear, but more than that, the module provides the opportunity to role-play (role playing), simulation and discussion. (4) the learning Material is presented logically and systematically, so that learners will know when they start and when the end of a module, and does not pose the question about what should be done or studied. (5) Every module has a mechanism to measure the achievement of the learning objectives of the students, especially to provide feedback/response for the students in achieving mastery learning.

IOP Conf. Series: Materials Science and Engineering **336** (2018) 012033 doi:10.1088/1757-899X/336/1/012033

2.4 Trainer

According to Hasan's (2006: 3) the trainer is a set of equipment in the laboratory is used as a medium of education which is a combination between a working model and a mock-up. Trainer indicated to support the learning of students in applying the knowledge/concepts acquired on real objects. The Model mock up is a simplification of the arrangement of the constituent parts of a process or system that is more complicated. From these statements it can be concluded, the trainer is learning media in the form of a replica or a miniature of a device that simulated and applied in practical activities in the laboratory/workshop that help and facilitate the students understand and master the learning material being delivered. According to Suryani (2006:5) some of the excess media trainer as a learning media are as follows: (1) Not all systems can be presented in the mathematical model, simulation is the appropriate alternative. (2) Can experiment without any risk on a system that is real, with the simulation allows to perform experiments against the system without having to bear the risk of the system running. (3) the Simulation can estimate system performance under certain conditions and provide design alternatives to best fit with the desired specifications. (4) the Simulation allows to perform long-term studies delam a relatively short time. (5) Can use the input data varied. (6) is a concrete and more realistic in the bring up subject matter, if compared with verbal language. The Media trainer also has a weakness as a medium of learning, according to Suryani (2006:5), namely: (1) the Quality and analysis model depending on the maker model of learning. (2) Only estimate the characteristics of the system based on certain input. (3) show Only the perception of the senses the eye, the size is only limited can be viewed by a group of students.

2.5 Servo Motor

Servo Motor is a DC motor with feedback system is closed in which the position of the motor will be informed back to the control circuit in the servo motor. A servo Motor is essentially a motor with special qualifications in accordance with the application “sevosing” in the control technique, consists of a DC motor, a series gear, the control circuit and the potentiometer. Servo Motor is widely used on the R/C (remote control) cars, planes, helicopters, and ships, as well as the actuators of the robot and drive on the camera.

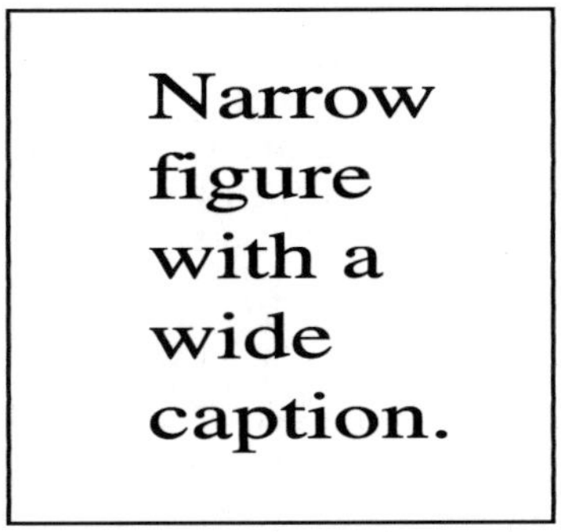

Figure 1. Servo Motor standar Hitec HS-311

As the name suggests, a servomotor is a servo. More specifically, is a closed loop servo that uses position feedback to control the movement and final position. Input control is some signal, either analogue or digital, representing the position commanded for the output shaft. Principle it works the motor is controlled with the signal modulisasi pulse width through the control cable. Pulse width control signal is given will determine the position of the angle of rotation of the shaft of the servo motor. Example, the width of the pulsation with a time of 1.5 ms (mill seconds) will rotate the motor shaft of the servo to an angle of 90. When the pulsation is shorter than 1.5 ms, then the shaft will re-spin to position 0 or to the left in a counter clockwise direction, whereas when the pulsation given longer than 1.5 ms, then the shaft of the servo motor will rotate to a position 180 can be seen in figure 2.

IOP Conf. Series: Materials Science and Engineering **336** (2018) 012033 doi:10.1088/1757-899X/336/1/012033

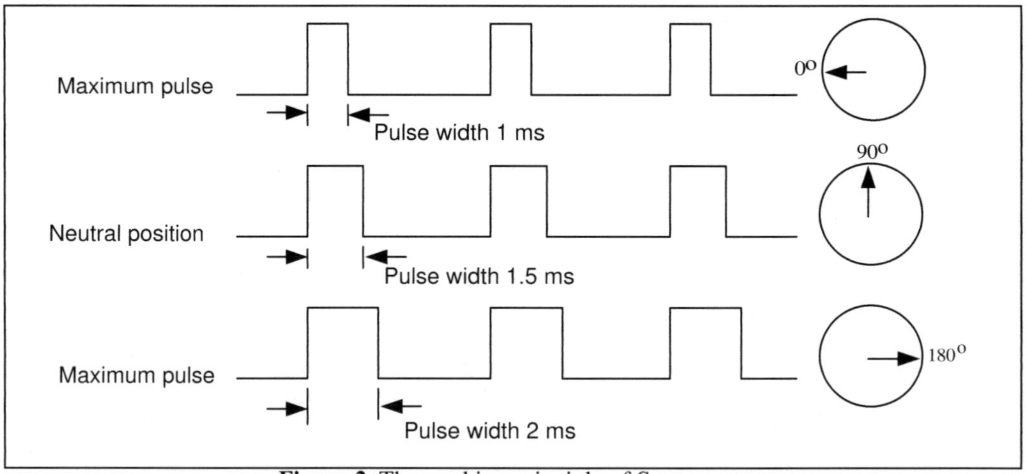

Figure 2. The working principle of Servo motor

When the width of the control pulse has been given, then the shaft of the servo motor that will move or spin toward a position that has been commanded and stopped in that position and will remain in that position. Such as if there is an external force to try to move, rotate or change the position by force, then the servo motor will resist and fight with the magnitude of the strength of the tors which has (rating torque servo). But the servo motor will not maintain its position forever, the signal width of the control pulse must be repeated every 20 ms once to remain instructed that the position of the motor shaft of the servo the remain or still hang on to his position.

3. Research Methods

This research type is development research using methods Research & development (R & D).This research is the product developed is a Software of Learning Practicum Servo Motor as a Supporting Lab for the Basic courses System Settings in the Department of Electrical Engineering, Faculty of Engineering, and Universitas Negeri Surabaya. According to Sugiyono (2015:409) there are ten steps of research R&D. However, in this study not used fully such measures, but modify the stages of his research only into six steps and is intended for the manufacture of products in the scope of small-scale and limited and not intended to mass produce its products. Thus, the six steps or the procedure of his research that have been modified are as follows. Stage 1 needs analysis learning device. Stage 2 Design the initial product. Stage 3 Validation of expert and revision. Phase 4 trial usage. Stage 5 Reporting and analysis. Stage 6 the results of the final product. Because the stages or the research procedures have been modified, then the procedure his research into like Figure 3 below.

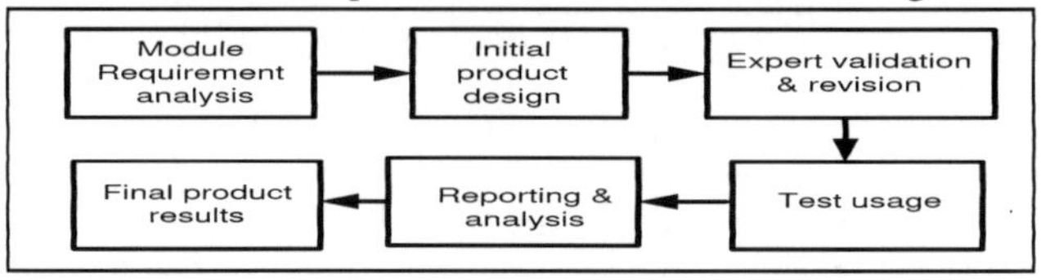

Figure 3. The modified step or procedure of the study

4. Validation of Expert and Revision

Validation of expert and revision is an activity that is intended to analyze and to test systematically the results of the product, namely the learning device. At the stage of validation performed two types of validation, i.e. the validation module and validation trainer. The validation module is intended to provide input and evaluation regarding the content of the material in the modules in the learning device customized with the material taken is about the servo motor, while the validation of the trainer is meant to provide input and evaluation regarding the principal instrument in the module that is used as a function of learning media lab. Input and evaluation based on the results of the validation of the 3 expert/validator composed of experts in the substance, media experts and experts in pedagogy from the faculty of Electrical Engineering, Universitas Negeri Surabaya.Such a result is expected to find flaws and shortcomings in the learning device. Then do the revision on the product which is prepared by fixing the weaknesses and shortcomings of the product until the product is really worth to test.

5. Results and Discussion

5.1 Module validation results

On the instrument the Validation of the Lab Module, there are 4 indicators assessed by the validator, i.e. the Physical Trainer, Discussion, Grammar, Size and Shape of the Letters. As for the recapitulation of the results of the calculation of the validation module is shown in Table 1.

Table 1. Recapitulation Of The Results Of The Validation Lab Module

Indicators	Assessment Criteria				∑ validator value
	1	2	3	4	
Physical Module					
The attractiveness of the face cover				3	12
Contains elements of the title				3	12
Contains the name of the author				3	12
Contains the name of the university				3	12
Load the logo of the university				3	12
The cover image is interesting				3	12
Discussion					
The material in accordance with the RPS			2	1	10
Image in accordance with the content of the material.			1	2	11
The text can be read clearly.			1	2	11
The instructions on the module help			2	1	10
The task of encouraging active student learning.			3		9
The task encourages students to think critically.			3		9
The level of truth the concept of the material in the module.			3		9
Text and image are interrelated			2	1	10
. Object image according to the material.			2	1	10
Object the image is clear or not blurred			3		9
Grammar					
Language in accordance with EYD.				3	12
The language is easy to understand.			1	2	11
Not many use the term regionalism				3	12
Not many using a foreign language				3	12
The language used in accordance with the intellectual			1	2	11

Indicators	Assessment Criteria				∑ validator value
	1	2	3	4	
development of students.					
Size and Letters					
The size of the font proportional and can be read.			1	2	11
The length and width of the modules have already considered.			1	2	11
The size of the letters according to the rules.			1	2	11
Do not use letter models				3	12
Total					237

Based on Table1, the result of validation of the module to the whole indicator on the module validation sheet obtained by the total score is 273, and to know the validity level of the module it is known that the average is calculated by the formula:

$$\text{Average score} = \frac{\Sigma \text{answer validator}}{\Sigma \text{ validator x } \Sigma \text{ item}} = \frac{273}{3 \text{ x } 25} = 3{,}64$$

Based on the calculation of the formula, the average is 3.64 and when viewed based on the interpretation table the score is at interval> 3.25 s / d 4 that is included in the criteria "Very Valid". Based on the results of the overall assessment can be concluded that the module "Very Valid" and its meaning is very applicable for research in the Department of Electrical Engineering State University of Surabaya.

5.1.1 Trainer Validation Results

In the Trainer Validation Instrument, there are 3 aspects that are evaluated with the total details of the 15 different aspects assessed by the validator namely Media Compatibility with Curriculum, Display and Media Quality, Media Compliance with Module Practicum. The recapitulation of Trainer validation results is shown in Table 2.

Table 2. Recapitulation of Trainer Validation Results

No	Rated aspect	Assessment				∑ validator value
		1	2	3	4	
Media Suitability with Curriculum						
1	Media trainer in accordance with the teaching materials delivered			2	1	10
2	The making of trainers helps the understanding of the material presented				3	12
Display and Media Quality						
1	Design trainer.			1	2	11
2	The suitability of the box trainer size with the circuit.			1	2	11
3	The layout of the circuit on the trainer.			2	1	10
4	Picture clarity / circuit scheme on the trainer (description of component symbol and part name of each circuit.			1	2	11
5	The writings on the trainer read clearly.			1	2	11
6	The accuracy of wiring-point layout (cabling place)			1	2	11
7	Length of the connector cable as needed.			1	2	11
8	Ease of connecting connectors on the trainer.			1	2	11
9	Easy access to each circuit.			2	1	10
10	Performance trainer.			3		9
11	Ease of operation / use of trainer			3		9
Media Conformity with Jobsheet						
1	Conformity of trainer with module material.			2	1	10
2	Practical activities with trainers in accordance with the module.			3		9

Based on Table 2 the results of Trainer validation recapitulation on all aspects of the trainer validation sheet obtained the total score is 156, and to know the level of validity of the learning device it is necessary to know the average is calculated by the formula:

$$\text{Average score} = \frac{\Sigma \text{answer validator}}{\Sigma \text{ validator x } \Sigma \text{ item}} = \frac{156}{3 \text{ x } 15} = 3{,}47$$

Based on the calculation of the formula, it is obtained averages of 3.47 and when viewed based on the interpretation table scores are in the interval> 3.25 s / d 4 that is included in the criteria "Very Valid". Based on the overall results of the assessment can be concluded that the trainer "Very Valid" and its meaning is very applicable for research at the Department of Electrical Engineering State University of Surabaya.

5.1.2 Limited Trial

After going through the stage of validation, then do a revision on the product as per the advice of the validator. Then it is tested limited to ten students of the Department of Electrical Engineering study program S1 TE class of 2014. Trial usage was done with the aim to determine the response of the students. So it is expected after a trial use of this learning device is able to determine the feasibility as Supporting Media Practicum for courses the Basic System Settings in the Department of Electrical Engineering, Faculty of Engineering, Universitas Negeri Surabaya, as well as the material is lifted and presented in accordance with what is required of the students of the Department of Electrical Engineering State University of Surabaya.

5.2 Results of Student Response

In the Student Response response validation instrument, there are 2 indicators assessed by the validator of the instruction, and the language. The recapitulation of Student Response validation calculation results is shown in Table 3.

Table 3 Recapitulation of Student Response Selection Validation Results.

Indicator	Assessment				$\sum$ validator value
	1	2	3	4	
Construction					
The statement is formulated briefly (not with 20 words) and clear.			2	8	38
The sentence is independent of the irrelevant statement of the object in question or the sentence is a necessary statement only.			3	7	37
The sentence is free of double negative statements.			1	9	39
The sentence is free from statements referring to the past.			4	6	36
The sentence does not contain elements that harm the other side.			3	7	37
The sentence is free from factual statements or can be interpreted as facts.			5	5	35
The sentence is free of statements can be interpreted more and free of statements that may be approved or vacated by almost all respondents.			3	7	37
Each statement contains only one complete idea.			4	6	36
Sentence does not contain any questions.			1	9	39
The sentence is free of uncertain statements like all, always, sometimes, none, never.			4	6	36
Do not use words a lot, just, just.			3	7	37
Use as necessary			3	7	37
Grammar					
Language problem should be communicative and in accordance with the level of education of students or respondents.			1	9	39
The matter should use standard Indonesian.			2	8	38
Problem not using local / taboo language.			2	8	38
Total					559

Based on Table 3, the result of the recapitulation of validation of student response questionnaire to the overall indicator on the validation sheet of student response questionnaire obtained total score is 559, and to know the level of validity and Questionnaire response need to know the average is calculated by the formula:

$$\text{Average score} = \frac{\Sigma\text{answer validator}}{\Sigma\text{ validator x }\Sigma\text{ item}} = \frac{599}{10x15} = 3{,}73$$

Based on the calculation of the formula, it is obtained the average of 3.73 and when viewed based on the interpretation table scores are in the interval> 3.25 s / d 4 that is included in the criteria "Very Valid". Based on the results of the overall assessment can be concluded that Questionnaire Response "Very Valid" and its meaning is very applicable for learning at the Department of Electrical Engineering State University of Surabaya.

6 Conclusions and recommendations

6.1 Conclusions

The conclusions gained from this study are the validity of learning tools obtained based on validator assessments when all aspects are validated. Based on the score analysis obtained based on the validator value, it is known that the validity value of the practicum module is 3.64; Trainer Motor Servo is 3.47; Questionnaire Response is 3.73. With the overall validation value is in the interval> 3.25 s / d 4 with the category of "Very Valid", so it can be concluded that all products have a validity level and "Very Valid" so it is worthy to be used for learning Basic courses of System Settings.

6.2 Recommendations

The use of existing Trainers in the laboratory needs to be optimized so that students get a long-term learning experience. The tools used in this study are considered to be valid enough even very valid for use in learning in the next year with the same material.

References

[1] Basuki, Ismet. (2004). ***Pengembangan Buku Ajar Berbasis Kompetensi***. Surabaya: Unesa.

[2] Buku Pedoman Unesa Kurikulum 2014 – 2015, Unipress Unesa 2014.

[3] Ibrahim, Muslimin. 2005. *Pembelajaran Kooperatif*, Unesa University press, Surabaya

[4] Mulyasa, 2004, *Media Pembelajaran*, Surabaya: Universitas Negeri Surabaya.

[5] Nasution, H., 1982, *Pengembangan Perangkat Pembelajaran*, Yogyakarta: Rineka Cipta

[6] Purdiana, L., 2004, *Pengembangan Perangkat Pembelajaran Kurikulum Berbasis Kompetensi Mata Diklat Pilpt Materi Ilmu Bahan Listrik,* Surabaya: Universitas Negeri Surabaya.

[7] Şugiyanto. 2010, *Model-model Pembelajaran Inovatif*, Surakarta: Yuma Pustaka

[8] Thiagarajan, Sivasailam.Gemmmel, Dorothy S. dan Semmel, Melviyn I. 1974. *Instruction Development For Training Teachers Of Exceptional Children.* Minnesota: Indiana University.

[9] Trianto, 2007, Model *Pembelajaran Terpadu dalam Teori dan Praktik, Prestasi Pustaka Publisher*: Jakarta.

[10] Wanarti R., Puput dkk, 2013, *Pengembangan Modul Ajar Mata Kuliah Fisika II untuk Model Pembelajaran Kooperatif sebagai Upaya Meningkatkan Kualitas Hasil Pembelajaran di Jurusan Teknik Elektro FT Unesa*, Prosiding STE 2013, ISBN 978-979-028-051-9, Seminar Teknik Elektro dan Pendidikan Teknik Elektro 2013, Unesa

[11] Wanarti R., Puput, 2012, *Pengembangan Modul Ajar Teknik Pengaturan Menggunakan Perangkat Lunak Matlab dengan Inquary Based Learning Berorientasi Industri,*JPTE, http://ejournal.unesa.ac.id/index.php/jurnal-pendidikan-teknik-elektro

IOP Conf. Series: Materials Science and Engineering **336** (2018) 012033 doi:10.1088/1757-899X/336/1/012033

[12] Wijaya, A., 1996, *Pengembangan Media-Media Pembelajaran*, Yogyakarta: Andi Offset Yogyakarta.

IOP Conf. Series: Materials Science and Engineering **336** (2018) 012034 doi:10.1088/1757-899X/336/1/012034

Interactive learning media based on flash for basic electronic engineering development for SMK Negeri 1 Driyorejo - Gresik

Gustav Mandigo Anggana Raras*

Graduate Program, Universitas Negeri Surabaya, Surabaya, Indonesia

*gustavmandigoangganararas@gmail.com

Abstract. This research aims to produce a product in the form of flash based interactive learning media on a basic electronic engineering subject that reliable to be used and to know students' responses about the media. The target of this research is X-TEI 1 class at SMK Negeri 1 Driyorejo – Gresik. The method used in this study is R&D that has been limited into seven stages only (1) potential and problems, (2) data collection, (3) product design, (4) product validation, (5) product revision, (6) field test, and (7) analysis and writing. The obtained result is interactive learning media named MELDASH. Validation process used to produce a valid interactive learning media. The result of media validation state that the interactive learning media has a 90.83% rating. Students' responses to this interactive learning media is really good with 88.89% rating.

1. Introduction

Education is a series of efforts by educator to help the student to improve their education, character, mind, personality [1]. The purpose of Indonesia national education is stated in number 20 article 3 year 2003 of the constitution. It can be achieved if education proceeds as expected. There are internal and external factors that influence the learning process to achieve educational goals. One of the external factors is the tool used to learn. By the time, the using of learning media is growing, one of the examples is by using a computer.

Based on needs assesment in SMK Negeri 1 Driyorejo found problem those are: (a) lack of interactive learning media, (b) lack of learning modules that can be borrowed, and (c) limited basic electronic props. Students think that the media that used so far in the SMK Negeri 1 Driyorejo is less attractive. Based on that background this research is conducted to answer one of the problems in SMK Negeri 1 Diryorejo. The aim of this research is to produce a valid interactive learning media named MELDASH.

In this research there are some limitations of the problem, those are (a) Every grade X student has a computer with minimal resolution 800 x 600 pixels; (b) MELDASH is only made for 5 basic competencies, those are (i) understanding atomic model of semiconductor material; (ii) applying a semiconductor diode as a rectifier; (iii) plotting the zener diode as a voltage stabilizer circuit; (iv) applying special diodes such as LED, varactor, schottky, PIN, and tunnel in electronic circuits; and (v) understanding the basic concepts of bipolar junction transistors (BJT) as amplifiers and switch devices; (c) the material contained in MELDASH is limited only for achieving the indicators contained in Kurikulum 2013 syllabus; and (d) MELDASH is only tested in X-TEI 1 and X-TEI 2 class.

The word media derived from the Latin, the plural of the medium which means "intermediary" or "agent", and its keyword is "intermediary" [2]. While in terminology Schramm defines the learning media as a messenger technology that can be utilized for learning purposes [3]. Seels and Glasgow define interactive media as a media delivery system that presents video recording material with computer control to audiences that not only hears and views, but also provides an active response that determines the speed and sequence of the presentation [4]. Media has a function to generate students motivation and interest, and help students improve their understanding, presenting data in an interesting and reliable way, facilitating interpreting data and compacting information. The position of media in learning can be seen in Figure 1.

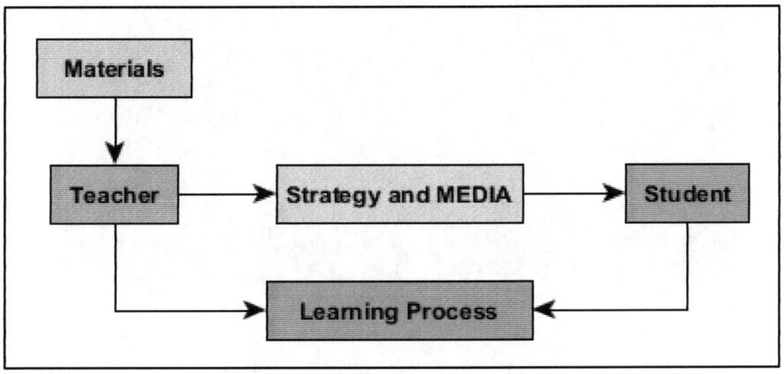

Figure 1. Position of Media.

Figure 1 shows the position of media in the learning process. In the learning process between materials, teacher, strategy and media, and student becomes a mutual sequence that affect each other according to their position. The teacher are positioned as message distributor and student as recipient of the message. While the media is located as an intermediary or medium in learning. But the selection of appropriate media is strongly influenced by the strategy, approach, method and learning format used by teacher [2]. Integration of media in learning process having some advantages, such as: (a) easy to use, (b) motivating, (c) portable, (d) always available, (e) simple, (f) customized, (g) common, and (h) having a lot of features [5].

2. MELDASH

The product of this research is a valid interactive media on the basic electronics engineering subject in accordance with Kurikulum 2013 syllabus which is named MELDASH. MELDASH contains basic competencies from 1 to 5 in the basic electronic engineering subject. MEDLASH created using Adobe Flash CS 6 and ActionScript 2.0. MELDASH consist of 5 files that each file discusses about each basic competence from 3.1 to 3.5.

The material contained in MELDASH refers to trustworthy books, website, and relevant research result to meet the indicators contained in the basic electronics engineering in Kurikulum 2013 syllabus. MEDLASH interface for each file is similar, only the title, material, exercise, and basic competencies and indicator list is different according to the material being discussed.

On the main screen of MELDASH there are 4 buttons and a logo of UNESA (Unviersitas Negeri Surabaya) as the university identity, the first is the material button to show the discussed material, the second is the exercise buton to show the exercise questions, the third is the competencies button to show the basic competencies and indicators from the discussed material, and the fourth is an exit button to close the media. On the main screen will be automatically shown the title of the discussed material. Next and Previous button on the bottom right is used to navigate through the active page. The main screen of MELDASH is shown in Figure 2.

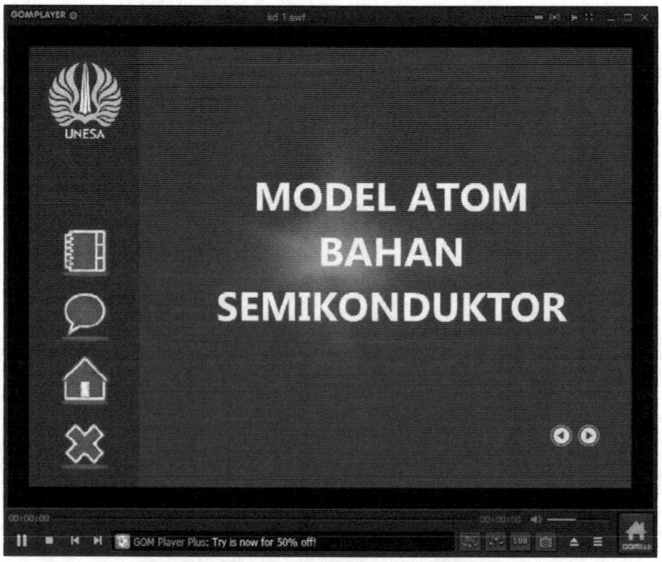

Figure 2. MELDASH main screen.

The discussed material can be found by pressing the first button, a book icon ( ). To show an exercise question user can use exercise button, question icon ( ). The user will be shown a question per page and must answer the following question in order to continue to the next question. After all questions is answered, the user will get the result how many their correct answer and incorrect answer. In addition, the user will be shown which number is answered correctly and which is answered incorrectly. Figure 3 shown the exercise questions main screen which its first question will be different for each material.

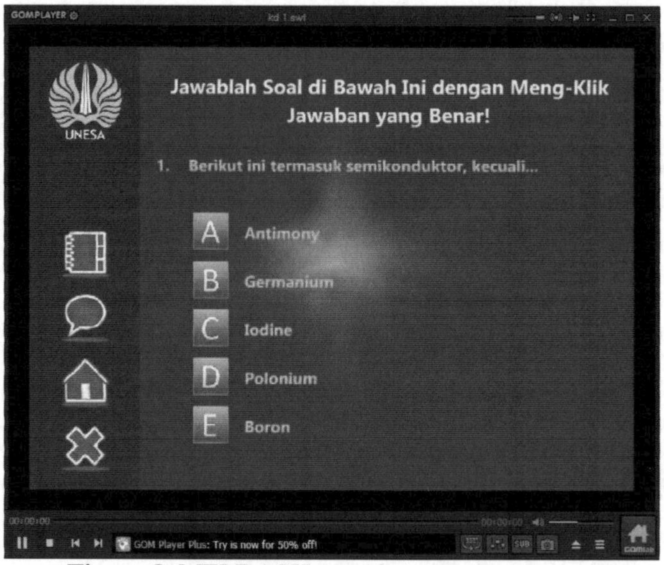

Figure 3. MELDASH exercise question screen.

IOP Conf. Series: Materials Science and Engineering **336** (2018) 012034 doi:10.1088/1757-899X/336/1/012034

3. Method

Research & Development (R & D) is used in the development of of the product in this research. R & D in educational research is defined as the process used to develop and validate educational [6]. R & D has 10 steps, but in this research will be limited and modified into 7 steps with analysis and report as the last step, those are: (a) potency and problem, (b) data collecting, (c) product design, (d) design validation, (e) design revision, (f) field test, and (g) analysis and report.

Potency and problem is a step to obtain the potential that already exist on the studied object and find the problem contained in this study. Data collecting step is done for the collecting various information used as learning media planning materials that are expected to solve the existing problem. Product design is used to get an overview of the MELDASH developing. Design validation stage is a stage to assess whether the product has been valid enough to be said as a good learning media and get the consideration advice of revision of the developed learning media. Based on the advice from the validator, MELDASH is revised, this is the design revision step. Field test conducted to get the needed data, MELDASH tested on X-TEI 1 and X-TEI 2 students in SMK Negeri 1 Driyorejo-Gresik. The last steps is analysis and reporting the result and the finding of this study scientifically.

The steps taken in the implementation of this research are: (a) make a permission letter from the Faculty of Engineering UNESA, (b) the permission letter was sent to SMK Negeri 1 Driyorejo – Gresik and accepted by Mr Rafli, (c) the school gives permission to conduct a needs assessment, (d) conducting needs assessment verbally to the basic electronic engineering teacher, (e) a reply letter stating that SMK Negeri 1 Driyorejo – Gresik is willing to be used as a research site given, (f) based on preliminary study results that have been previously made compiled research proposal entitled "Interactive learning media based on flash for basic electronic engineering development for SMK Negeri 1 Driyorejo – Gresik", (g) making a sample of interactive learning media in order to conduct proposal seminar, (h) making complete learning media to be used during research, (i) validation of interactive media product by validator who is competent in his field, and (j) analysing and final reporting after the research process at field is finished.

Data collection is done by giving responses in four scale criterion. The following scoring criteria are shown in Table 1.

Table 1. Scoring Criterion.

Category	Score	Percentage (%)
Very Good	4	76 – 100
Good	3	51 – 75
Bad	2	26 – 50
Very Bad	1	0 – 25

Table 1 shows the criteria of the four rating scales: (a) very good, (b) good, (c) bad, and (d) very bad. The scoring criterion is given to validators who will validate the media and students who will fill the questionnaire. The total answer is determined by multiplying the number of respondents by the weight of the value. The formula used is [7]

$$\begin{array}{ll} \text{Very good (n validator)} & \text{n x 4} \\ \text{Good (n validator)} & \text{n x 3} \\ \text{Bad (n validator)} & \text{n x 2} \\ \text{Very bad (n validator)} & \underline{\text{n x 1}} \; + \\ \sum \text{Answer (SoR)} & = \end{array} \tag{1}$$

IOP Conf. Series: Materials Science and Engineering **336** (2018) 012034 doi:10.1088/1757-899X/336/1/012034

After doing the sum of respondent answers, the next step is to determine the percentage of validation score by a using formula where RPR is a respondent percentage respond, SoR is the sum of respondent respond and SoM is amaximum score of respondent respond [7].

$$RPR = \frac{SoR}{SoM} \times 100\% \quad (2)$$

4. Result

There are two results found in this study, first is the media validity and second is a student response toward the developed media. Media validity will be separated into two phases, first validation and second validation.

4.1. First Validation

First validation involving only two subject teachers at SMK Negeri 1 Driyorejo – Gresik. The validation result given by the teacher is shown in Figure 4.

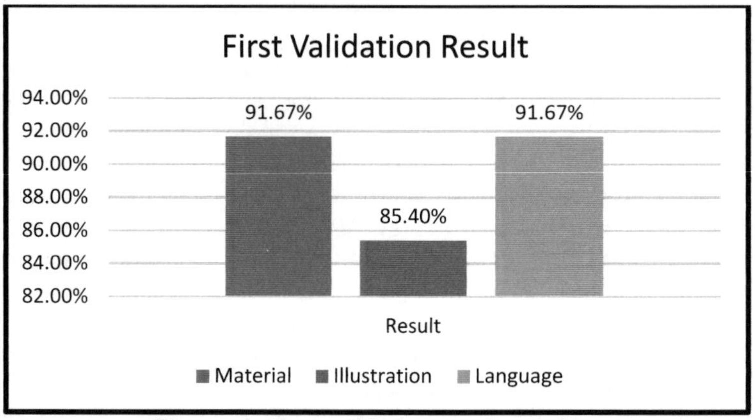

Figure 4. First validation result graphic.

From the result of first validation shown in Figure 4 hence can be obtained percentage of validity of interactive learning media as follows: material aspect 91,67%, media illustration 85,40%, and language aspect 91,67%. From these three aspects can be taken an average value of validity 89,76%. Based on the assessment criteria table can be categorized as "Very Valid".

Several suggestions for MELDASH is given by the validators such as: (a) revision in materials, image, and animation; and (2) add the physical components. After getting the suggestion for revision, MELDASH is revised based on their suggestions.

4.2. Second Validation

Second Validation involves two subject teachers at SMK Negeri 1 Driyorejo – Gresik and three lecturers from UNESA educational electronic engineering major which are considered to have competencies in media validation. The validation result given by the validators shown in Figure 5.

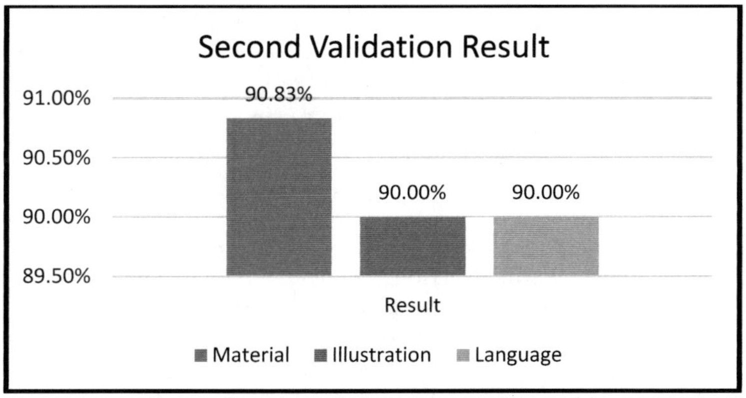

Figure 5. Second validation result graphic.

From the result of first validation shown in Figure 5 hence can be obtained percentage of validity of interactive learning media as follows: material aspect 90,83%, media illustration 90%, and language aspect 90%. From these three aspects can be taken an average value of validity 90,28%. Based on the assessment criteria table can be categorized as "Very Valid".
A suggestion for MELDASH is given by the validators on this second validation is addition of materials. After getting the suggestion for revision, MELDASH is revised based on their suggestions.

4.3. Student Response
Student responses obtained at the field test stage. Field test is conducted in five meetings for both X-TEI 1 and X-TEI 2. The students open their laptop while the teacher is projecting the MEDIA using a screen projector to show the student how they operate MELDASH. After that the students will try MELDASH by themselves. After going through 5 meetings, then the questionnaire of student responses is distributed to the students. The function of this questionnaire is to determine level of the student's interest to MELDASH. According to the questionnaire that has been filled by students, MELDASH got the response as shown in Figure 6.

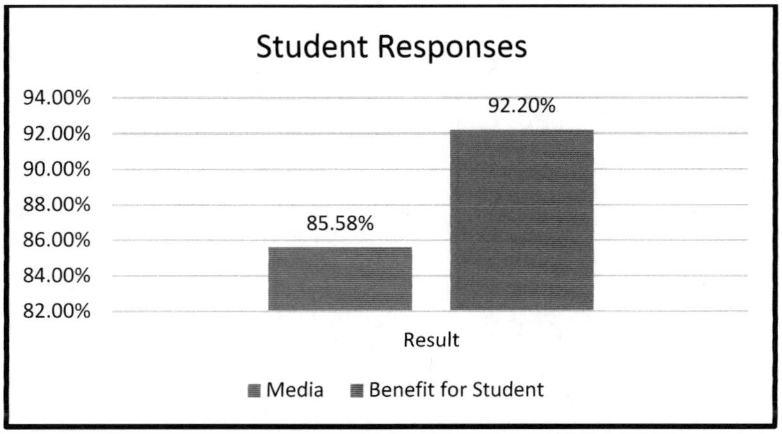

Figure 6. Sutedent responses graphic.

Based on the Figure 6, found that the feasibility of MELDASH from media aspect is 85,58% and from benefit for student aspect is 92,20%. From both aspects can be taken the average value for student response from MELDASH equal to 88,89%. The average value of student responses can be categorized as "Very Eligible" to be used in the learning process.

5. Conclusions

IOP Conf. Series: Materials Science and Engineering **336** (2018) 012034 doi:10.1088/1757-899X/336/1/012034

Interactive learning media based on flash for basic electronic engineering has ben designed, built, and tested. MELDASH got 90,83% score for its validity, in other words can be said that MELDASH is valid to be used in the teaching process. Student response toward MELDASH is positive indicated by its score, 88,89%. MELDASH is very interesting for students to help them learn basic electronic engineering in 10th grade in vocational high school.

References

[1] Suyitno Y 2009 *Tokoh-tokoh pendidikan dunia (dari dunia timur, timur tengah dan barat)* (Bandung: Universitas Pendidikan Indonesia) p 3
[2] Musfiqon 2012 *Pengembangan media dan sumber pembelajaran* (Jakarta: Prestasi Pustaka)
[3] Schramm W and Lawrence D K 1984 *Asas-asas komunikasi antar manusia* (Jakarta: LP3ES Anggota IKAPI) p22
[4] Arsyad A 2011 *Media pembelajaran* (Jakarta: PT Raja Grafindo Persada) p 36
[5] Kitchenham A 2011 *Models for interdiscliplinary mobile learning: Delivering information to students* (United States: IGI Global)
[6] Borg W R and Gall M D 1983 *Educational research an introduction* (New York: Longman)
[7] Widoyoko E P 2012 *Teknik penyusunan instrumen penelitian* (Yogyakarta: Pustaka Pelajar)

IOP Conf. Series: Materials Science and Engineering **336** (2018) 012035 doi:10.1088/1757-899X/336/1/012035

Development Module (Lab Report) As a Media of Learning in Vocational Education Viewed by Gender

Supari Muslim[1], Bambang Suprianto[1], Nahindi Putra Gitama[2]*

[1] Department of Electrical Engineering Universitas Negeri Surabaya
[2]Graduate Program, Universitas negeri Surabaya

*nahindigitama@unesa.ac.id

Abstract. Module as a media of learning and training, which especially the students studying in institutions offering education at undergraduate and associate levels can employ as a PLC (Programmable Logic Controller) controlled simulator of the logic gate, timer and counter. During industrial manufacturing processes, has been actualized through this work, for the purpose of training qualified technical personnel needed by companies, who are specialized in control and electro mechanics. It was found that, students are very happy learning to use the module so with very significant contributes to a large extent to the training process in favor of faculty members, who are to train personnel for the sector, and in favor of students, who should be able to acquire proper education. Working on the training tool allows students to acquire knowledge and practical skills and then make use of those skills for troubleshooting and control of machinery.

Keywords: learning media, PLC, vocational, education, training tool

1. Introduction

To achieve a stable economy requires great attention to the achievement of a competitive position in the world market and very important to look at how creative thinking can enrich the economic potential that education is about educating people to be ready to enter the world of work, creative skills and creative thinkers very be required. National education standards require suppression of creative course and a variety of creative ways of working, so that lecturers as educators play an important role in implementing the learning process in higher education, and develop the potential and creativity of the students, so that students can competition in the industry.

After observation and interviews on head of department about the problems that exist in laboratory control. Its need learning with modules (lab report) while the matter is obtained underprivileged students overcome troubleshooting on PLC program using CX-Programmer. It can be seen that the many industries that use PLC in the production process, controlling machines and electrical installations. From the above explanation can be presumed that in the learning process requires the learning media which efficient and targeted in order to improve students' ability to understand the PLC used in the industry.

2. Problem Statement

In education, there are two global issues that are very important in the learning process. Including the module (lab report) as a media of learning. By looking at the issue globally in the field of industrial control, so author discussions with the head of department mechatronic at State University of Trunojoyo Madura (UTM). From the discussion gained some problems that need to be overcome, the problem will be explained as follows, differences in male and female employees the ability to work in the field of industrial controllers.

3. Materials

This research uses several materials which are laptop/computer with software CX-Programmer v.9.3 and module (lab report) consists of two, first module about logic gate (load, or, and, not) and second module about timer and counter. Programming use CX-Programmer for PLC CP1E type.

4. Literature Riview

According to Munadi (2013, p. 7) explains in his book that, the media is anything that can convey and deliver the message from the source in a planned manner so as to create a conducive learning environment in which the recipient can make the learning process efficient and effective. The media is a very important tool in the process of teaching and learning activities, while according Sardiman (2006, p. 3) media is a tool that convey or deliver messages of learning. Media can be defined by their technology, their symbol systems, and their processing capabilities. The most obvious characteristic of a medium is its technology, the mechanical and electronic aspects that determine its function and to some extent its shape and other physical features. These are the characteristics that are commonly used to classify a medium as a "television," a "radio," and so on. The cognitive effects of these characteristics, if any, are usually indirect. Characteristics such as size, shape, and weight makes it more likely that a student will learn with a book while on a bus but not a computer, though of course this is changing as computers get smaller, lighter, and cheaper (Kozma, 1991).

4.1 CP1E-N40DR-A System

The system configuration when using a CP1E N/NA *N40DR-A* type CPU Unit is shown at figure 1.

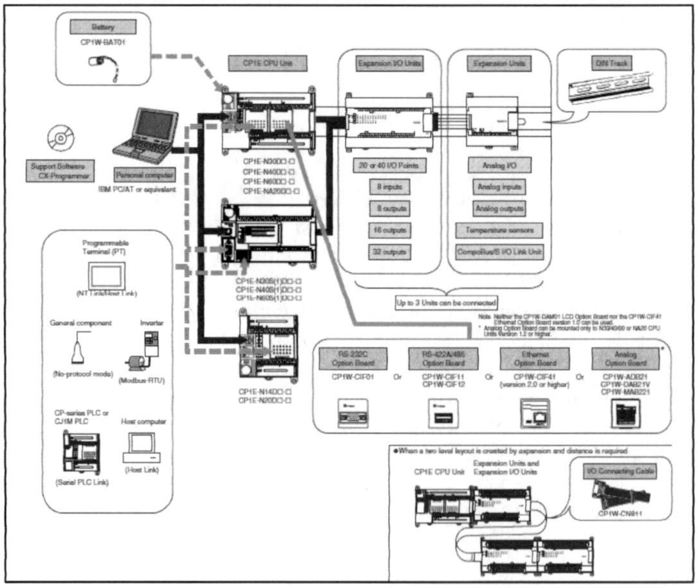

Figure 1. Basic system (datasheet OMRON, 2016)

IOP Conf. Series: Materials Science and Engineering **336** (2018) 012035 doi:10.1088/1757-899X/336/1/012035

4.2 CP1E-N40DR-A Specification

The following table gives the specifications that apply to the built-in I/O on a CP1E CPU Unit (Anonim, 2016).

Table 1. Common I/O Specifications

Item	Specification		
Input type	**High-speed counter inputs or normal inputs**	**High-speed counter inputs, interrupt inputs, quick-response inputs or normal inputs**	**Normal inputs**
Input bits	**CIO 0.00 to CIO 0.01**	**CIO 0.02 to CIO 0.07**[*1]	**CIO 0.08 to CIO 0.11, CIO 1.00 to CIO 1.11 and CIO 2.00 to CIO 2.11**[*1]
Input voltage	24 VDC, +10%, -15%		
Applicable sensors	2-wire and 3-wire sensors		
Input Impedance	3.3 kΩ	3.3 kΩ	4.8 kΩ
Input current	7.5 mA typical	7.5 mA typical	5 mA typical
ON voltage/current	3 mA min at 17.0 VDC min	3 mA min at 17.0 VDC min	3 mA min at 14.4 VDC min
OFF voltage/current	1 mA max at 5.0 VDC max	1 mA max at 5.0 VDC max	1 mA max at 5.0 VDC max
ON response time[*2]	CPIE-N40DR-A CPU Unit 50μs min CPIE-N40DR-A CPU Unit 2.5μs min	50μs max	1 ms max
OFF response time[*2]	CPIE-N40DR-A CPU Unit 50μs min CPIE-N40DR-A CPU Unit 2.5μs min	50μs max	1 ms max

4.3 CP1E-N40DR-A I/O Wiring Diagram

CP1E-N40DR-A have been input 24 points and output 16 points wiring diagram following (Anonim, 2010).

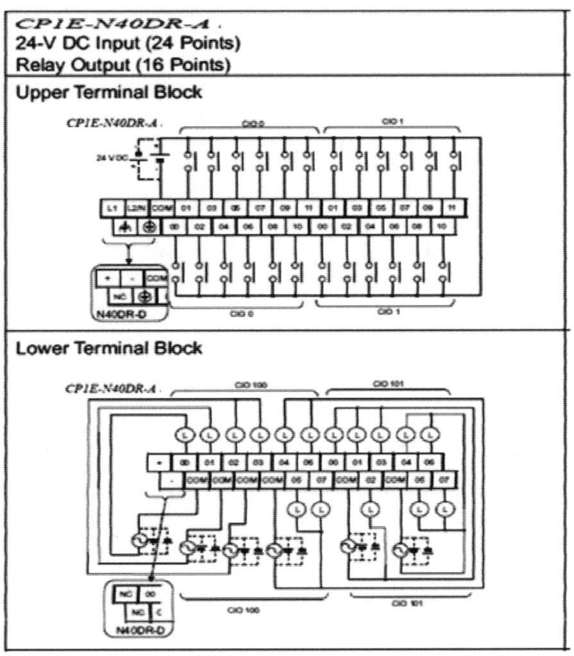

Figure 2. Wiring diagram (datasheet OMRON, 2010)

4.4 Learning Outcomes

Learning outcomes is an important thing in the process of teaching and learning. In the process requires an important role between teachers and students that aims for the achieve certain specifications. As an explanation of Purwanto (2009, p. 44) that the study results can be explained by understanding the two words that make it up, namely the "results" and "learning". Definition of results (product) refers to an acquisition as a result of doing an activity or process that result in changes in the functional input. Purwanto (2010, p. 72) explains that the study results can also be used for the calculate Cost Ratio Results. The ratio of cost results is used for the determine the level of efficiency of the training program or training cost efficiency, that is by comparing the average value of the learning outcomes of participants by the number of costs per person during the training. The average value of the intended learning outcomes can use the value of the average increase in the capacity or the average value of the final test (final test).

5. Researchs Method

To make learning media his use research and development (R&D) method. Step by step from use research and development (R&D) method can be looking at figure 1 (Sugiyono, 2010, p. 409).

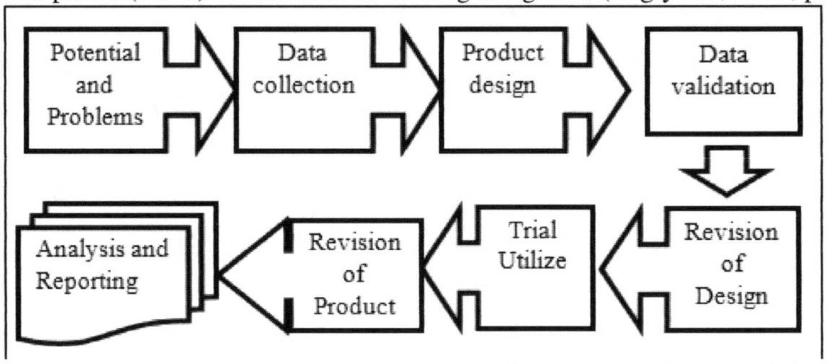

Figure 3. Step by step research and development for PLC training tool and module (lab report)

5.1 Potential and Problems

From the discussion gained some problems that need to be overcome, the problem will be explained as follows, differences in male and female employees the ability to work in the field of industrial controllers.

5.2 Data Collection

Necessary data collection phase of collecting information which is used as a learning media material plan that is expected to resolve the problem. Collecting data in this study a questionnaire response for students, media validation, student results cognitive, psychomotor, and affective.

5.3 Product Design

The created module (lab report) consists of two, firs module about logic gate (load, or, and, not) and second module about timer and counter. Author is making design of module (lab report) following.

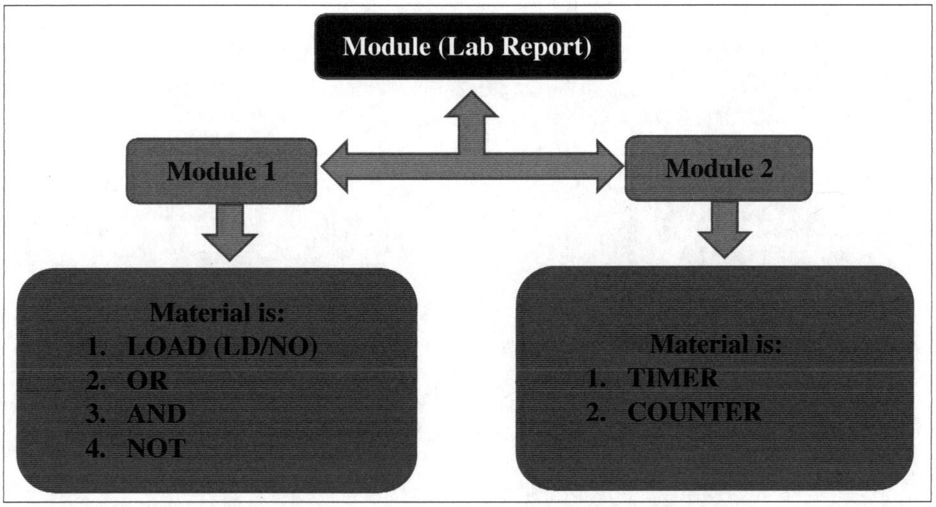

Figure 4. Design module (lab report)

5.4 Data Validation

Validation is a process to assess the quality of the products, the products assessed were learning media. Validation is done by two expert lecturers from the department of mechatronic at UTM.

5.5 Revision of Design

Of the product validation process has been done will give an explanation of the weakness of learning media created. The weakness was further revised to make improvements to media of learning.

5.6 Trial Utilize

The trial media of learning done if revised. The trials module (lab report) implemented in department of mechatronic at UTM. Cognitive learning achievement of students is done by giving the student achievement test. Psychomotor learning achievement is done by giving students of test performance. Affective learning achievement performed by observing students.

5.7 Revision of Product

After testing the product in the department of electrical engineering, if there are weaknesses or deficiencies in the media, then the product will be revised to fix.

5.8 Analysis and Reporting

The completed media of learning tested and validated by experts of media, then performed analysis. The analysis is used to analyze the results of the validation and the questionnaire responses from students. Reporting the results of learning media are documented on CD and publish the results of research through international journals.

6. *Result and Analisys*

65 students were given a questionnaire produces the following data.

IOP Conf. Series: Materials Science and Engineering **336** (2018) 012035 doi:10.1088/1757-899X/336/1/012035

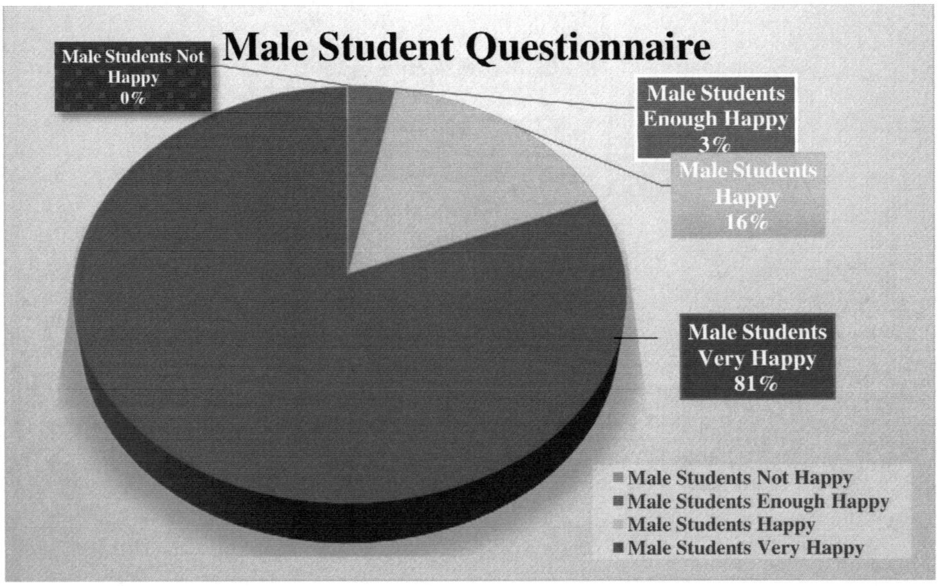

Figure 5. Male student questionnaire

Figure 5 explains that 41 male students are given a questionnaire about satisfaction with the use of module (lab report) with the scale, is not happy, quite happy, happy, and very happy. The PLC training tool hardware cannot be made because there is no material in UTHM so the author just makes the hardware design following.

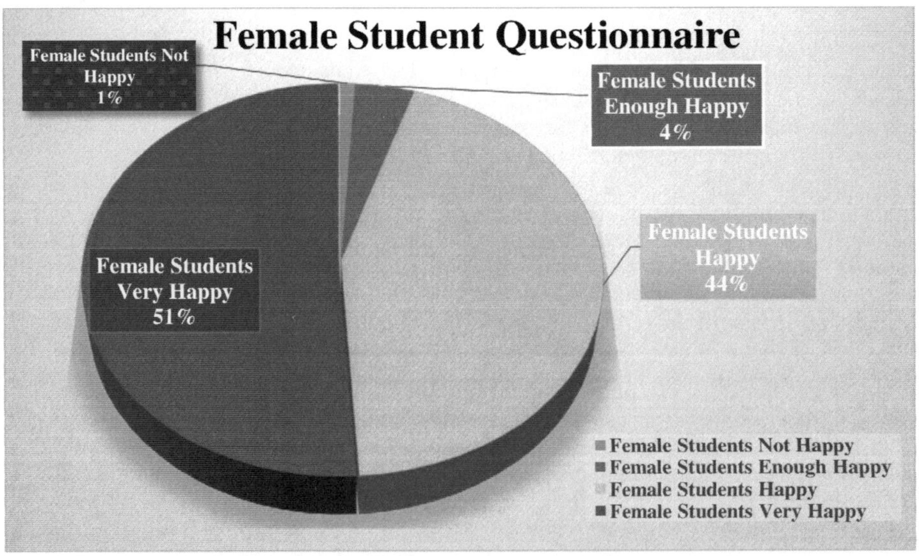

Figure 6. Female student questionnaire

Figure 6 explains that 24 female students are given a questionnaire about satisfaction with the use of module (lab report) with the scale, is not happy, quite happy, happy, and very happy. The PLC training tool hardware cannot be made because there is no material in UTHM so the author just makes the hardware design following.

Can be analyzed descriptively that after the author interviewed the students which use module (lab report) as a media of learning, the student response was very happy.

7. Conclusion

Students are very happy learning to use the module so with very significant contributes to a large extent to the training process in favor of faculty members, who are to train personnel for the sector, and in favor of students, who should be able to acquire proper education. Working on the training tool allows students to acquire knowledge and practical skills and then make use of those skills for troubleshooting and control of machinery.

Acknowledgement

The author would like to express the sincere appreciation to his supervisor, Dr. Tee Kian Sek, Prof. Dr. Supari Muslim, M.Pd. and Prof. Dr. Bambang Suprianto, M.T., for the support given trough out the duration for this master project study.

The author also would express his appreciation to all the people who directly involve or indirectly involve in order to produce this degree project. Last but not least to the close family and friend which keep supporting nonetheless the situation.

References

[1] Anonymous. (2010). I/O Wiring Diagram. Singapore: OMRON.
[2] Anonymous. (2016). CP1E CPU Unit Hardware User's Manual. Singapore: OMRON.
[3] Kozma, R.B. (1991). Learning with media. Review of Educational Research, 61(2), 179-212.
[4] Munadi, Y. (2013). Media pembelajaran. Jakarta: GP Press Group.
[5] Purwanto. (2009). Evaluasi hasil belajar. Yogyakarta: Pustaka Pelajar.
[6] Purwanto, M. Ngalim. (2010). Prinsip-prinsip dan teknik evaluasi pengajaran. Bandung: PT. Remaja Rosdakarya.
[7] Sardiman. (2006). Interaksi dan motivasi belajar mengajar. Jakarta: Raja Grafindo Persada.
[8] Sugiyono. (2012). Metode penelitian pendidikan pendekatan kuantitatif, kualitatif, dan R&D. Bandung: AFABETA.

IOP Conf. Series: Materials Science and Engineering **336** (2018) 012036 doi:10.1088/1757-899X/336/1/012036

Innovation Online Teaching Module Plus Digital Engineering Kit with Proteus Software through Hybrid Learning Method to Improve Student Skills

Nur Kholis *, Muhamad Syariffuddien Zuhrie, Reza Rahmadian

Electrical Engineering Department, Universitas Negeri Surabaya, Kota Surabaya.

* nurkholis@unesa.ac.id

Abstract. Demands the competence (competence) needs of the industry today is a competent workforce to the field of work. However, during this lecture material Digital Engineering (Especially Digital Electronics Basics and Digital Circuit Basics) is limited to the delivery of verbal form of lectures (classical method) is dominated by the Lecturer (Teacher Centered). Though the subject of Digital Engineering requires learning tools and is required understanding of electronic circuits, digital electronics and high logic circuits so that learners can apply in the world of work. One effort to make it happen is by creating an online teaching module and educational aids (Kit) with the help of Proteus software that can improve the skills of learners.
This study aims to innovate online teaching modules plus kits in Proteus-assisted digital engineering courses through hybrid learning approaches to improve the skills of learners. The process of innovation is done by considering the skills and mastery of the technology of students (students) Department of Electrical Engineering - Faculty of Engineering – Universitas Negeri Surabaya to produce quality graduates Use of online module plus Proteus software assisted kit through hybrid learning approach. In general, aims to obtain adequate results with affordable cost of investment, user friendly, attractive and interactive (easily adapted to the development of Information and Communication Technology). With the right design, implementation and operation, both in the form of software both in the form of Online Teaching Module, offline teaching module, Kit (Educational Viewer), and e-learning learning content (both online and off line), the use of the three tools of the expenditure will be able to adjust the standard needs of Information and Communication Technology world, both nationally and internationally.

1. Introduction

Referring to the development needs of the Engineer in the industry then the Department of Electrical Engineering - Faculty of Engineering - State University of Surabaya, especially in the Digital Engineering course requires innovative teaching modules online plus Hybrid Learning kit, in order to attract learners to learn even like the course resulting in an outcome to meet the demands described previously.

The course of Digital Engineering is a basic and combined course of several courses, among others, basic mathematics, electronic circuits, electrical circuits and logic circuits etc. Therefore, the relevance and achievement targets of each course must be clearly defined

In an effort to make this happen, this study was conducted by reviewing the current teaching and learning process and measuring the success rate [5]. The study of literature on matters relating to the learning system with the objects of subjects that are of a composite nature are also carried out as a reference and benchmark. Industrial world as a place where the graduates will apply the ability to knowledge and skills it has also become a source that is not less important.

The main problem is how the development of learning tools can be implemented in Teaching and Learning Process (PBM) so as to give interest and interest of learners to Digital Engineering courses with approach according to industry needs [2]. In order to support the Competency Based Curriculum (CBC) industry needs, also has been prepared both manual module, computer-based module (both off line and on line) and Kit (Educational Tool) oriented to the achievement of work competence as well as accommodate the skills (Soft Skill) learners. With regard to that, the problem that will arise is what the learning module that suits the characteristics of learning objectives is to provide attractiveness and interest of students to the Digital Engineering course so that the achievement of competence in the world of work as well as the achievement of competence in the field of Life Skills using materials the main teaching modules and computer-based learning tools.

To answer the problem, in this research will be designed the innovation of online teaching module plus digital assisted software kit of Proteus software through hybrid learning approach to improve learner's skill. It is expected that the results of this study will be able to bridge the needs of the world of work, especially from graduates of the Department of Electrical Engineering - Faculty of Engineering - Universitas Negeri Surabaya, who will later work as faculty and engineers in the industry.

2. University Strategic Plan And Research Road Map

2.1 University Strategic Plan

To realize and achieve the vision and mission of LPPM of Universitas Negeri Surabaya and consider the result of self evaluation, it can be formulated the objectives and targets of LPPM research of Universitas Negeri Surabaya, as follows: (1) Need to improve the quality and quantity of research for the mastery of science, technology or art; (2) Need to increase the number of leading research activities and national priorities that support the implementation of national development; (3) Need to improve the quality and quantity of facilities support (facilities and infrastructure); (4) Need to improve the quality of management system of LPPM Unesa; (5) Need to improve skills and skills of lecturers and students in the field of research; (6) Need to improve research cooperation with other agencies / agencies; (7) Need to improve the dissemination and marketing of research results through various media of scientific communication and social media.

2.2 University Research Road Map

The leading researches developed at Universitas Negeri Surabaya are grouped into two major groups, namely: (1) education group and (2) non-educational group and elaborated in research clusters as stated in the LPPM strategic plan of Universitas Negeri Surabaya 2016-2020 above.

In practice, research from both educational and non-educational groups can be developed through basic research, application, and capacity building, so that it can be funded by other sources outside DRPM, but still within Kemenristekdikti auspices as well as through collaborative research with various external parties. Especially for leading research is directed to the umbrella research model that has been pioneered since 2011, the theme and focus of research for leading research that is expected to be achieved and done in Unesa can be grouped based on research clusters, namely the theme and focus of research on educational and non educational clusters.

3. Theoretical Framework

Taking into account the results of the four previous studies, it can be concluded that some references of problems are: (1) The results of the above study show the effectiveness of module learning, and from the previous research also included suggestions for module learning also applied to other courses; (2) The result of the research shows that students who agree to use contextual teaching model are also applied to other subjects as 91,18%; (3) Problems in the Digital industry that requires the standard of competence is the ability to design analog and digital electronics hardware; (4) The teaching and learning process in the lectures that do not integrate between the use of teaching modules and props and combined with the use of e-learning. It will then be directed to a wider scope and scope and the use of more complex and varied learning tools; (5) Produce a model of Blended Learning method that combines Face-to-Face Learning (classical method) with Computer Based Learning (e-Learning method) so that the Teaching and Learning Process (PBM) becomes unlimited time and space

Thus it is not excessive to do research with the title of innovative online teaching module plus digital assisted software kit Proteus software through hybrid learning approach to improve the skills of learners.

3.1 Learning Strategy Using Media (tools) Figure

To support the teaching and learning process (PBM) using the media (tool) visual, it is necessary infrastructure in the form of learning tools consisting of: Hybrid Learning is a software application that can be run off line and on line to help Teaching and Learning Process (PBM) in classical

3.2 Contextual Teaching and Learning

According to Nurhadi et al (2004: 13) Contextual Teaching and Learning is a concept of learning where the teacher presents the real world into the classroom and encourages students to make connections between their knowledge and application in daily life.

Nurhadi et al (2004: 31) mentions that in Contextual Teaching and Learning there are seven important components or better known as the seven principles of Contextual Teaching and Learning, among others: (a) Constructivism (Constructivism); (b) Finding (Inquiry); (c) Questioning; (d) Community Learning (Learning Community); (e) Departure (Device); (f) Reflection; (g) Authentic Assessment.

Nurhadi et al (2004) explained that contextual learning should emphasize on: (1) Problem Based Learning; (2) Authentic Teaching; (3) Inquiri Based Learning; (4) Project-Based Learning / Duties; (5) Work-Based Learning; (6) Learning-Based Services; (7) Cooperative Learning.

3.3 Hybrid Learning

Hybrid learning is not a new concept, originally the paradigm Hybrid learning (often called Blended Learning) is a traditional learning method that supported e-services (Lawrence Tomei, 2008). E-services are designed to maximize the results of traditional learning. The influence of technological development, then Hybrid learning also developed by combining traditional learning with learning that utilize information and communication technology. Brunner quoted Solomon Negash et al (2008) revealed that *hybrid learning have advantages, which are: (1) student performance and retention increase, (2) time and flexibility for students is greater, (3) colors on the teaching palette multiply, (4) depth of community enhances the learning environment, (5) the breadth of 'interaction' is enlarged, 6) it allows for a gradual transition from face-to-face to online learning, and (6) expectations are higher.*

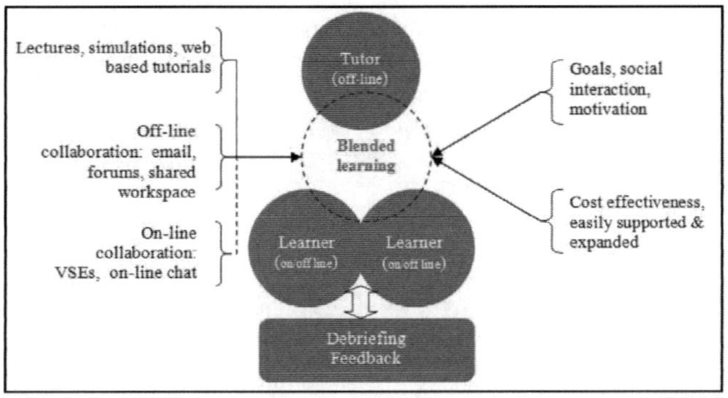

Figure 1. *Blended (Hybrid) Learning* Tomei, Lawrence. (2008, 17)

Hybrid learning combines various learning models, such as traditional learning and e-learning modeling. Traditional learning (face to face) in this study uses direct learning model and project / task based learning. The direct instruction model is based on the principles of behavioral psychology and social learning theory, especially about modeling. This is based on the fact that behavioral change in learning is largely derived from modeling, ie the behavior and experience (success and failure) of others. The Direct Learning Model is designed to help students to gain procedural knowledge, which is about structured knowledge and can be taught step by step (Mohamad Nur, 2011). This direct learning, the interaction of lecturers with students is quite intensive, both in class and in the laboratory (learning practice) and also the students are given a lot of practice and duties. In addition, this study applying e-learning model learning is learning that supported the internet with material that has been planned. Learning e-learning aims to make students more independent and responsible for learning. In other words, the e-learning learning atmosphere will force students to play a more active role in their learning

Lecturers can upload the material on the e-learning site and students can learn it by opening the e-learning site anywhere. Learning can be delivered synchronously (at the same time) or asynchronously (at different times). Materials delivered through this medium have text, graphics, animations, simulations, audio and video, and provide ease for discussion groups with the help of chat forums, e-mails, bulletin boards.

Learning should be interactive, inspiring, fun, challenging, and motivating students to participate actively, and provide sufficient space for initiative, creativity, and independence according to their talents, interests, and physical and psychological development. The purpose of learning is the process to change student behavior in accordance with the competencies to be achieved.

Competence contains the knowledge, psychomotor, and affective unity required for the fulfillment of professional duties. Competencies based on vocational education form the competencies needed for professional development throughout life. Schaap (2009: 482) states that *the development of professional competences is a main objective of competence-based vocational education. Learning is viewed as giving meaning to experiences in a process of continuous progressive recontextualisation.*

4. Research Methodology

This research is based on the development research design with the implementation procedure grouped into three stages. The study design is as follows.

4.1 First Stage

Develop a survey design and development plan through workshop techniques, brainstorming and focus group discussions.

4.2 Second Stage

Research development and quasi experimental research with pre posttest group only design. As for the preparation of synchronization of curriculum materials relevant to the needs of employment, the selection of essential topic topics, and the integration of topic-¬ topics into competencies and sub-competencies, as well as the design of online teaching module module plus digital technique kit assisted proteus software through hybrid learning approach to improve the skills learners with reference to 4 D devices, which include define, design, develop, and disseminate (Thiagarajan, Semmel, and Semmel, 1974).

4.3 Third Stage
Perform empirical tests of modules that have been prepared. At this stage the product research module is actually nearing the final. Nevertheless, to obtain empirical evidence that the research product has been feasible then some of the research product is tested empirically on a limited circle that is Department of Electrical Engineering - Faculty of Engineering - Universitas Negeri Surabaya. The results of limited trials are used as materials for the formulation of recommendations for use of research products (disseminate stage).
Implementation of empirical experiments conducted with the design of pre postest group only by selecting students, preferably the force of learners who are programming subjects studied Digital Engineering as a sample test. Implementation of the trial will be adjusted to the existing academic calendar schedule in the Department of Electrical Engineering - Faculty of Engineering – Universitas Negeri Surabaya and SMK.
The effectiveness of online teaching module innovation plus Proteus software assisted digital assist kit through hybrid learning approach to improve learner skills will be analyzed with inferential statistics, in the form of different test (t test and anova).

5. Result And Output

5.1 Data on the results of INA Skills (LKS) 2012 - 2016
From the data of INA Skills (LKS) 2012 - 2016 it can be seen that the module that has been developed has reached the fourth stage of research development method that is disseminate method. So that the module is perfect and ready to be duplicated and deployed as learning tools of digital technique course. In the INA Skills (LKS), the teaching module materials that are based on the contest are not only specific to the Department of Electrical Engineering - Faculty of Engineering - Universitas Negeri Surabaya, but multi disciplines, and research objects developed not necessarily the latest technology or technology just to be useful, but include the development of learning tools

5.2 Data result of dissemination of learning tools to students
Based on data analysis of student responses about the activities of learning device socialization conducted on students Electrical Engineering Department - Faculty of Engineering - State University of Surabaya above produces the data as follows. For questions 1, 2, and 3 all respondents consisting of 10 students from representatives of various courses in the Department of Electrical Engineering - Faculty of Engineering - Surabaya State University (100%) answered did not understand about the application of digital engineering in the industry, or about technical equipment digital. Respondents also argue that the competence of digital techniques is not taught in the Department of Electrical Engineering - Faculty of Engineering - Universitas Negeri Surabaya. This shows that the competence image of the Department of Electrical Engineering - Faculty of Engineering - Surabaya State University is currently limited to the basic materials of electrical engineering, ranging from logic circuits, electronic frames, and microprocessors only, not to mention the competence of digital techniques in the industry. Whereas currently the development of automation in the industry have many who use industrial robots
For question no. 4 on hybrid learning technique-based digital module respondents who answered interesting is as many as 9 students (90%) of 10 students representatives of SMK-SMK and who think "not interesting" as much as 1 student (10%). As for the question no. 5 and no. 6 on the use of

computers and aids of module respondents who think "interesting" as many as 10 students (100%) and who say "not interesting" as much as none (0%). This indicates a positive response to learning tools in the form of modules and learning tools developed.
For question no. 7 all respondents as many as 10 people (100%) argue that the module developed can be easier in understanding the material. All respondents (100%) thought they were happy and motivated by learning using modules and teaching aids. This indicates that the module can motivate learners and help learners to understand the material.

5.3 Data of learning device validation result by lecturer
The completed module was then validated to 7 validators consisting of lecturers of instructional experts, educational experts, engineering experts and grammarians. And the average validator assessment result is 3.34 which fall into either category. So that the module can be used in trial 2 is a trial conducted on digital engineering lectures.

6. Conclusion
Based on the results of data analysis and discussion, the researchers can take the following conclusions:
This developed module is the results of research and monitoring of the researcher and team leader for 4 years continuously trying to develop applied product research entitled innovative online teaching module plus digital assisted software kit of Proteus software through hybrid learning approach to improve students' in the future capable of producing innovative online teaching modules plus digital assisted software kit Proteus software through hybrid learning approach to improve the skills of learners.

7. References
[1] Adviso F, bernardo (1990). *Development Of The National Training Council As The Coordinating Body For Technical And Vocational Trainng*, Jakarta: Depdikbud

[2] Blank,WE. (1982). *Handbook For Developing Competency Based Training Program.* Englewood Cliffs; Prentice Hall.

[3] Bonk CJ, Cummings JA, Hara N, Fischler RB, Lee SM. (2000)A ten level web integration continuum for higher education: new resources, partners, courses, and markets. Abbey B, ed. *Instructional and cognitive impacts of web-based education.* University of Indiana

[4] Brown S. (1999) Reinventing the university. *Assoc Learning Technol* J; 6: 30-37. Fender B. The e-university project. London: Higher Education Funding Council for England.

[5] Carr MM, Reznick RK, Brown DH. (1999) *Comparison of computer-assisted instruction and seminar instruction to acquire psychomotor and cognitive knowledge of epistaxis management.* Otolaryngol Head Neck Surg; 121: 430-434

[6] Depdikbud, (1993) *Konsep Sisem ganda pada sekolah menengah Kejuruan di Indonesia*,Jakarta, Depdikbud RI.

[7] ------, (1998). Keputusan Menteri Pendidikan Menengah Dan Kebudayaan RI No. 323/U/1997 tentang *Pelaksanaan Pendidikan Sistem Ganda Pada Sekolah Menengah Kejuruan.* Jakarta; Direktorat Dikmenjur.

[8] Dit. Dikmenjur, (1996). *Kerangka Acuan Studi tentang Pelaksanaan Program Dukungan Pelatihan Industri (Industrial Training Support Program*). Jakarta: Proyek Pendidikan Teknologi dan Kejuruan.

[9] Daniel JS.(1996) The knowledge media. In: Mega-universities and knowledge media. *Technology strategies for higher education.* London: Kogan Page,:101-135.

[10] Devitt P, Palmer E. (1999)Computer-aided learning: an overvalued educational resource Med Ed; 33: 136-139

[11] D'Alessandro DM, Kreiter CD, Erkonen WE, Winter RJ, Knapp HR. *Longitudinal follow-up comparison of educational interventions: multimedia textbook, traditional lecture*, and printed textbook. Acad Radiol 1997; 4:723

[12] Elves AW, Ahmed M, Abrams P. *Computer-assisted learning*; experience at the Bristol Urological Institute in the teaching of urology. Br J Urol 1997; 80 (suppl 3): 59-62

[13] Erickson, H.L (1998) *Concept Based Curriculum Instruction*: Teaching Beyond The Thousand Oaks: Corwin

[14] Finch, Curtis R. & john R Crunkilton. 1983. *Curriculum Development in Vocational and Technical Education*, London: Allyn and Bacon, Inc.

[15] Greinert, WD., 1992. *The Dual System Of Vocational Training In The Federal Republic of Germany*; Structure and function. Postfach 1, 6101 6326 Eschborn, Federal Republic of Germany

[16] Heywood, John, 1989 *Learning Adaptability And Change: The Challenge For Education And Industry*, Great Britain: Paul Chapman Publishing Ltd.

[17] Harris, R., Guthrie, H., Hobart, B., dan Ludnberg, D. (1995). *Competency-Based Education And Training; Between A Rock And Whirlpool*. South Yarra; Macmillan Education Australia.

[18] Loftis, H. A. (1980) " *An Exemplary Model For Change", Vocational Instruction*. Columbus; Flarson Publisher

[19] Mardapi, Djemari. (1999). "*Performance Based Evaluation*" Bahan Lokakarya Tentang Performance Based Evaluation di Program Pascasarjana UNY.

[20] Norton, R.E (1985). *DACUM* handbook. Columbus: The National Centre For Research In Vocational The Ohio State University

[21] Sulifan, R. (1995) *The Competency Approach To Training*. JHPIEGO strategy Paper. www.reproline.jhu.edu /english/6read/6training/cbt/cbt.htm

[22] Slamet PH., 1993. "*Pendekatan Alternatif Dalm Penyiapan Dan Pengembangan Tenaga Kerja Profesional Di Indonesia*" Disampaikan Pada Seminar Nasional Sistem Permagangan Dalam Pendidikan Dan Kejuruan Dan Penyiapan Tenaga Kerja, Malang 18 Desember

[23] Phillips R. 1996. *Developers' guide to interactive multimedia*. A methodology for educational applications. Perth: Curtin University Press,.

[24] Haag M, Maylein L, Leven FJ, Tonshoff B, Haux R. 1999. Web-based training: a new paradigm in computer-assisted *instruction*. Int J Med informatics 1999; 53: 79-90.

[25] Hilger AE, Hamrick HJ, Denny Jr FW. *Computer instruction in learning concepts of streptococcal pharyngitis*. Arch Pediatr Adolesc Med 1996; 150: 629-631

IOP Conf. Series: Materials Science and Engineering **336** (2018) 012037 doi:10.1088/1757-899X/336/1/012037

Building Automatic Grading Tools for Basic of Programming Lab in an Academic Institution

Rina Harimurti*, Andi Iwan Nurhidayat, Asmunin

Informatics Engineering Department, Faculty of Engineering, Universitas Negeri Surabaya, Indonesia

*rinaharimurti@unesa.ac.id

Abstract. The skills of computer programming is a core competency that must be mastered by students majoring in computer sciences. The best way to improve this skill is through the practice of writing many programs to solve various problems from simple to complex. It takes hard work and a long time to check and evaluate the results of student labs one by one, especially if the number of students a lot. Based on these constrain, web proposes Automatic Grading Tools (AGT), the application that can evaluate and deeply check the source code in C, C++. The application architecture consists of students, web-based applications, compilers, and operating systems. Automatic Grading Tools (AGT) is implemented MVC Architecture and using open source software, such as laravel framework version 5.4, PostgreSQL 9.6, Bootstrap 3.3.7, and jquery library. Automatic Grading Tools has also been tested for real problems by submitting source code in C/ C++ language and then compiling. The test results show that the AGT application has been running well.

1. Introduction

The skills of computer programming is a core competency that must be mastered by students majoring in computer sciences. The best way to improve this skill is through the practice of writing many programs to solve various problems from simple to complex. The more often the practice of directly writing program to solve various problems, the experience of programming skills also increased.

One of the activities to improve the ability of programming in the campus is a programming practice. In this activity, students are given the problem then write the program code and show the result to the lecturer or assistant. If there is an incompatibility between problems specification and the student's answers, the students must make revisions and show the results again. It takes hard work and a long time to check and evaluate the results of student labs one by one, especially if the number of students a lot. In addition, program evaluations related to memory usage, source code file size, and execution time are not performed.

Based on these constrain, web propose Automatic Grading Tools (AGT), the application that can evaluate and deeply check the source code in C, C ++ or Java language that has been submitting by students, compile it, display the compilation results, display error if there are logic or syntax errors, and do the sorting results. This application should be easy to use, can be used simultaneously, responsive web-based, and can be used on basic of programming lab.

Related research on automatic grading of computer program, scalable autograder and Learning Management System (LMS) integration, integrate an autograder and existing an LMS, build

IOP Conf. Series: Materials Science and Engineering **336** (2018) 012037 doi:10.1088/1757-899X/336/1/012037

assignment module in Moodle E-Learning System for java programming assessment, build web-based application for monitoring coding process by recording typing activity, support online compilation and execution using an autograder [1][2][3].

2. Automatic Grading System

A system that can be used for assigning tasks, evaluating the source code, compiling, and displaying compilations results according to the specified language preference. The automatic grading system has been widely used, both on campus and in competitive programmings, such as Web-CAT, SPOJ (Sphere Online Judge), project euler, codechef, codeEval, and topcoder.

Academic institutions face the challenge of providing their students with a better skill in programming. therefore, the number of programming practices multiplied. as a consequence, the practice room and availability of lecturers should be prepared. Automatic Grading System enables students to work practice of programming anywhere, anytime and can submit program code anytime then get feedback quickly.

2.1. Architecture

The application architecture consists of students, web-based applications, compilers, and operating systems. Using the browser, students open the application Automatic Grading System. in the application, the students do the registration, login to the application, choose the problem, do the problem then submit the program code file. The program code that has been submitted then processed by the Automatic Grading System Application include Evaluation, Grading, and rank. Automatic Grading System architecture is presented in Figure 1.

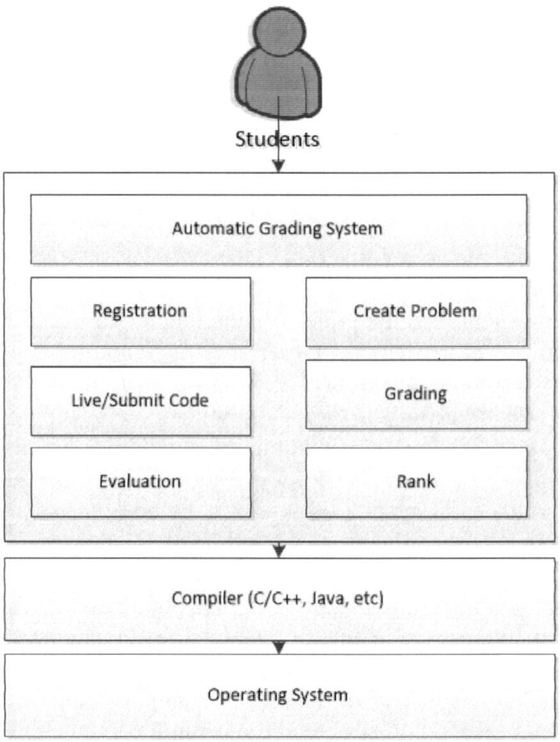

Figure 1. Automatic Grading System Architecture

The Automatic Grading System application will display results based on the evaluation process. If there is an error during the evaluation process, it will be displayed. If there is no error, the evaluation process will display the message: "Accepted" and done the assessment process.

2.2. Flowchart of Automatic Grading System

The Flowchart of automatic Grading System is presented in Figure 2.

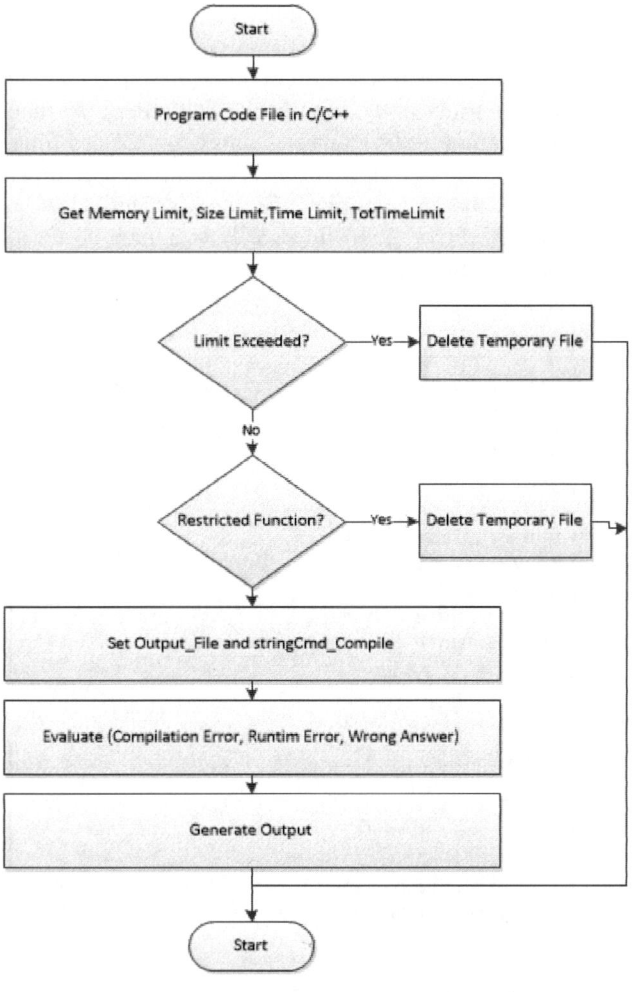

Figure 2. The Flowchart of Automatic grading System

Starting from the program code that has been submitting by the student, then determined Memory Limit, Size Limit, Time Limit, and TotTimeLimit. The next step is check Limit Exceeded and Restricted Function, if true then delete temporary file and if false then set Outfile_File and stringCmd_Compile. The main process in the flowchart is Evaluate, which will check and evaluate program code. If there is an error during the evaluate process, it will display a message according to the error that occurred. The last process of Automatic Grading System is Generate Output, which will display the result.

IOP Conf. Series: Materials Science and Engineering **336** (2018) 012037 doi:10.1088/1757-899X/336/1/012037

3. Implementation

Automatic Grading Tools (AGT) is implemented using open source software, such as laravel framework version 5.4, PostgreSQL 9.6, Bootstrap 3.3.7, and jquery library. For the compilation and evaluation of program code used gcc, g ++ and java compiler. The operating system used during implementation and testing is Ubuntu Linux 16.0.4 LTS which has installed gcc, g++ and java compiler.

To avoid spaghetti code, the application has been implemented using the MVC (Model View Controller) architecture using Laravel Framework. The main file in this application is grader.php, a class that contains the variables and methods used in the evaluation process and grading system. Variables in the grader.php file, contains information about compilation and execution data, size limit, time limit, memory limit, total time limit, and output limit. The other information that stored in the file grader.php is storage path, information about current submit, maximum time, restricted function, the amount of loops execution, redirect output, and array that contains the information of the evaluation process. List of methods in the grader.php file: evaluateCode, evaluate, evalSizeLimit, evalRestrictedFunction, evalCompileError, evalRuntimeError, generate output file, evalWrongAccept, delTemporaryFile, RenJavaClass, searchClass, buildCodeFile, and builtResultArray. List of variable and method in grader class is presented in Figure 3.

Grader
-c_lang : string -cplus_lang : string -java_lang : string -size_limit : int -sime_limit : int -memory_limit : int -total_time_limit : int -output_limit : int -storage_path : string -problem_path : string -results : bool -loop_time : int
+evaluateCode() : int +evaluate() : int +evalSizeLimit() : bool +evalRestrictedFunction() : bool +evalCompileError() : bool +evalRuntimeError() : bool +generateOutputFile() : string +evalWrongAccept() : bool +delTemporaryFile() : int +RenJavaClass() : int +searchClass() : string +buildCodeFile() : string +builtResultArray() : string

Figure 3. Variable and Method in Grader Class

For implementing the compilation and execution, Automatic Grading Tools (AGT) uses gcc, g++ and java compiler that has been installed on the Linux operating system. The program code that has been submitting by the student, checked and evaluated by AGT and then sent to the compiler to compile and run. If during evaluation the program code there is a discrepancy with the specification, then the message will appear, the program code is not compiled and run. the list of messages displayed by The results of the evaluation process: Judging, Unqualified, Internal Error, Size Limit Exceeded,

Restricted Function, Compilation Error, Time Limit Exceeded, Output Limit Exceeded, Presentation Error, Wrong Answer, and Accepted.

4. Testing

To discover a bug in the software has been doing unit testing, component testing, and system testing [4]. Unit testing focus on testing the functionality of object or method, component testing focus on testing the component interface that provides access to the component function, system testing focus on testing component interaction. On unit testing, individual program unit or class are tested. Each class has been tested with unit testing and ensured bug-free before integrating with other classes. After unit testing, functional test on PC is executed for CRUD (Create Read Update Delete) function for all features in AGT.

Automatic Grading Tools has also been tested for real problems by submitting source code in C/ C++ language and then compiling. The test results show that the AGT application has been running well. The output from functional testing is presented in Figure 4.

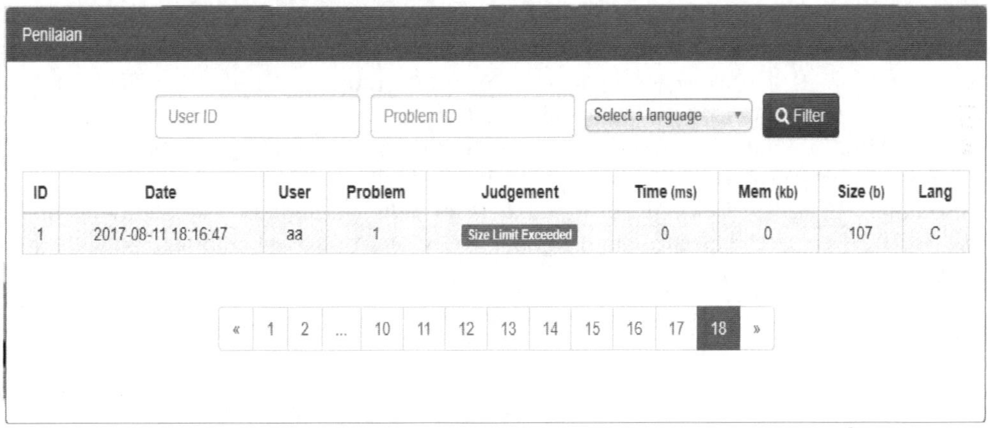

Figure 4. Message Size Limit Exceeded

The message displayed in figure 1 depends on the evaluation of the program code. When the evaluation result file size exceeds that specified, the Size Limit Exceeded message is displayed. When a compilation error occurs, the Compilation Error message is displayed. When the execution time is too long, Time Limit Exceeded message is displayed. Another message is presented in Table 1.

Table 1. list of message in AGT

Message Code	String Message
JDG	Judging
UQD	Unqualified
IE	Internal Error
SLE	Size Limit Exceeded
RF	Restricted Function
CE	Compilation Error
RTE	RunTime Error

IOP Conf. Series: Materials Science and Engineering **336** (2018) 012037 doi:10.1088/1757-899X/336/1/012037

TLE	Time Limit Exceeded
MLE	Memory Limit Exceeded
OLE	Output Limit Exceeded
PE	Presentation Error
WA	Wrong Answer
AC	Accepted

5. Conclusions and Future Work

Using open source technology and MVC Architecture, automatic grading system application has been successfully created running well. based on the test results, the application has been able to check and evaluate the program code in depth and display the results.

References

[1] K. Danutama and I. Liem 2013 Scalable Autograder and LMS Integration Procedia Technology **11**, 388-395,

[2] N. A. M. Z. N. S. A. Norazah Yusoft 2012 Java Programming Assesment Tool for Assignment Module in Moodle E-Learning System *ICTLHE*

[3] C. N. Timitius and L. Inggriani 2013 Source Code Editing Evaluation for Learning Programming Procedia Technology **11** 169-175,.

[4] I. Sommerville, Software Engineering 2016 Tenth Edition England: Pearson Education Limited

[5] M. Stauffer, Laravel Up and Running 2017 A Framework for Building Modern PHP Apps, USA: Oreilly Media,.

IOP Conf. Series: Materials Science and Engineering **336** (2018) 012038 doi:10.1088/1757-899X/336/1/012038

The implementation of Project-Based Learning in courses Audio Video to Improve Employability Skills

Edy Sulistiyo[1*], Djoko Kustono[2], Purnomo[2] , Eddy Sutaji[2]

[1]Department of Electrical Engineering, Universitas Negeri Surabaya
[2]Universitas negeri Malang, Indonesia

*edy.unesa@yahoo.co.id

Abstract. This paper presents a project-based learning (PjBL) in subjects with Audio Video the Study Programme Electro Engineering Universitas Negeri Surabaya which consists of two ways namely the design of the prototype audio-video and assessment activities project-based learning tailored to the skills of the 21st century in the form of employability skills. The purpose of learning innovation is applying the lab work obtained in the theory classes. The PjBL aims to motivate students, centering on the problems of teaching in accordance with the world of work. Measures of learning include; determine the fundamental questions, designs, develop a schedule, monitor the learners and progress, test the results, evaluate the experience, project assessment, and product assessment. The results of research conducted showed the level of mastery of the ability to design tasks (of 78.6%), technical planning (39,3%), creativity (42,9%), innovative (46,4%), problem solving skills (the 57.1%), skill to communicate (75%), oral expression (75%), searching and understanding information (to 64.3%), collaborative work skills (71,4%), and classroom conduct (of 78.6%). In conclusion, instructors have to do the reflection and make improvements in some of the aspects that have a level of mastery of the skills less than 60% both on the application of project-based learning courses, audio video.

1. Introduction

The Era of globalization has resulted in changes in the overall life of the society, including the education sector and employment that occurred in Indonesia. The mobility of students and labor between countries provide a challenge for the college to gain recognition from the global community against the results of education that it does. This causes the countries of the participants of the GATS (General Agreement Trade in Services) and AFTA (ASEAN Free Trade Area) develop a national qualifications framework. The implementation of AEC or AEC starting in 2016 make the competition increasingly fierce. Some general requirements must be owned by a state so that goods and services can compete among the ASEAN countries that they should prepare a human resource (HR) skilled, intelligent, and competitive so that the required provision of employability skills for graduates.

The strategic objectives of the State University of Surabaya is to produce quality graduates, able to compete in the global labor market by equipping graduates employability skills and is able to open and develop the business independently with competence entrepreneurship skills, require a fundamental change in the development of curriculum and instructional practices of learning so that strategic

IOP Conf. Series: Materials Science and Engineering **336** (2018) 012038 doi:10.1088/1757-899X/336/1/012038

objectives can be achieved this. Expected solutions in the effort to raise the image and dignity of the nation of Indonesia in the eyes of the world, which now are at a level that is not encouraging, the need for the revitalization of education in the sense of changes of the education system is fundamentally and contextually. This study seeks to integrate the assessment of employability skills through project-based learning in order to improve the mastery of employability skills, competence, performance, and readiness of graduate work the Department of Electrical Engineering Unesa. The diversity of learning features combined with the PjBL by providing a mount of learning approach case-based learning, community-based learning, game-based learning, passion-based learning, service-based learning, and team-learning. Five things that are most prominent in the model PjBL includes a learning-based challenges, problem-based learning, learning, place-based, project-based learning activities, and learning-based design.

The mission of the Department of Electrical Engineering, Faculty of Engineering Universitas Negeri Surabaya are:

1. Organizing the management of a department that is credible, accountable and transparent in accordance quality assurance standards.
2. Produce graduates who are competent and have character.
3. Improve the quality of human resources in the field of science education and information technology.
4. Organized innovative education in the field of Electrical Engineering, Electronics Engineering and Information Technology.
5. Provides facilities and infrastructure in the development of education in the field of Electrical Engineering, Electronics Engineering and Information Technology.
6. Carry out research, application of SCIENCE and technology for the community and establish sustainable partnerships with various agencies and the industrial world.

In the dimensions of the institution, implementation of project-based learning should be developed to cope with the challenges of learning. To build and maintain the PjBL goes well-needed institution which considers lecturers, students, and available resources significantly different than the environment of traditional learning. The way in which the institutions associated with the teaching staff is very important for improving the quality of learning. Based on the background outlined above, the problem to be studied whether the implementation of project-based learning in courses Audio Video can improve your employability skills?. While the goal is to know that the implementation of project-based learning in courses Audio Video can improve the employability skills.

2. Discussion

In the analysis of the employability divide the three main levels [8].

1. The macro level: at this level, work is the part that is situated structurally, a shift in capitalism system-level and how the education system is coordinated.
2. Level Meso: at this level, employment opportunities, and the work of the people associated with activities mediated by the agency-level process is located in the education domain and organization.
3. At the micro level, the focus is more on how to work on a personal level and conjunction with the subjective, biographical and dynamics of psychosocial, individual profile and background.

Explain about soft skills are the basic skills that are important for the absorption of labor and the growth of the individual [10]. The ILO notes about key skills for the world of work have a basic/foundation skills (for example, literacy and numeracy), technical (for example, work skills specific to a particular task), professional/personal abilities (for example, individual attributes such as honesty and reliability), and employment skills core (e.g., communication, teamwork). Furthermore, the core skills specialized skills into four categories, namely learning to learn, communication, teamwork, and problem resolution (ILO, 2013). Soft skills require discipline, confidence, negotiation, communication, and decision-making. In addition, the entrepreneurial and business skills, such as financial management, market research, and marketing. Indeed, in the informal economy workers should be more independent from the formal economy.

The findings of the study show that business students place significant value on the development of employability skills in the degree program. Although there are minor variations in the perception of the students about the importance of developing employability skills in all the characteristics specific demographic, ranked above average for all groups. Average rank high about the importance of the provision of employability skills consistently in several studies. There seems to be little empirical evidence that the perception of the students in the skill development of higher education, that really stands out is teamwork and the importance of the concept of job readiness.

Basic education provides a solid foundation in reading and writing in order to continue on to secondary education [12]. Secondary education, including vocational education, and technical training can develop specific competencies, skills, behavior, attitude, cooperation, and social responsibility can participate in the knowledge economy, contribute to social cohesion, and become responsible citizens. Higher education can provide the ability to innovate, disseminate, and apply knowledge that supports the knowledge economy (see World Bank, 2002, 2005).

Doing research that is based on the application of project-based learning [11]. This study contributes to understanding why and how efforts to encourage a shift education in electrical engineering that can educate students and improve the engineering skills of the 21st century. The Ppa is the strategy of student-centered learning that can foster student initiative and focuses students on the real world, projects that can increase motivation and encourage a variety of skills including practical knowledge and skills [1]. The results showed that the students in accordance with the ability to overcome the problem of high-level cognitive on a microcontroller (MCU). To improve the practice of education, many Japanese universities are implementing project-based learning (PjBL) [13]. The results showed that the effectiveness of education is highest when a team consisting of management and no leadership. The results showed that the cognitive ability of students positively influenced by the teaching and social presence, which supported the theoretical framework the CoI (Community of Inquiry) [14]. Future research is required to verify the dynamic relationship of various disciplines in different learning contexts.

Project-based learning helps students develop the skills to live in a society that is knowledge-based and technology. Conventional learning models learn about the fact and reading is no longer sufficient to prepare students to survive in today's world. Solve a very complex problem requires students to have basic skills and digital skills. With the combination of skills, students become directors and managers of the learning process, guided and mentored by a teacher who is skilled. States that active learning has been widely used in engineering education. This method focuses on the students, not as passive learning is traditional, where elements of the lecture centered on the Lecturer [3][9]. Problem-based learning to introduce engineering problems of the real world are entered in the special scenario to the students. The students are asked to define the problem and then guided in presenting an alternative solution through the idea of problem-based learning in a structured way.

With the PjBL, learners can develop an attitude (attitude), knowledge (knowledge), skills (skill), enlightenment (insight), behavior (behavior), habits (habits), and social (associations) of experiences and allow learning associated with the activity of the real work (real-life work activities) [5]. In practice to do experimental learning Electro can be done with the model PjBL are presented in figure 1.

IOP Conf. Series: Materials Science and Engineering **336** (2018) 012038 doi:10.1088/1757-899X/336/1/012038

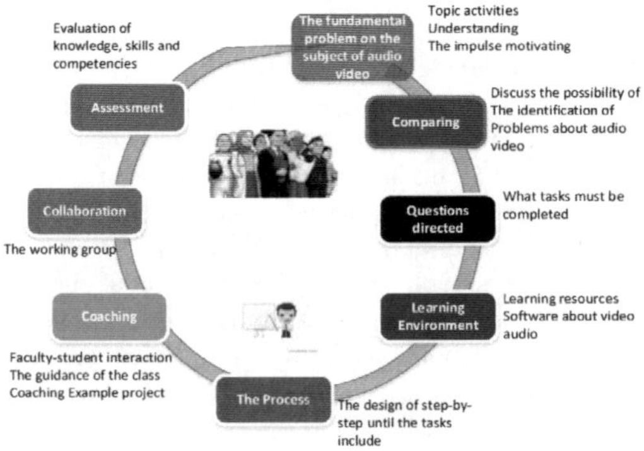

Figure 1. Steps of the PjBL Audio Video [4]

(a) Introduce students to engineering and showing the topics that learn. This often contributes to motivating the students.
(b) Understand the problem and ideas about possible topics. The project began after students identify a challenging issue. Therefore, the students started the project by generating many possibilities before choosing the project to be done.
(c) The relevant Questions explain what will be achieved and studied. The task should be interesting, challenging and can be done.
(d) The learning Environment provided print and digital resources including textbooks, guidebooks, hypertext link, web-based, tutorials, best practices, papers, data sheets, and software automation of some activities of the PjBL.
(e) The Process and investigation include the steps necessary to complete the task or answer the question. This process includes activities that require a higher level and critical thinking skills, such as analysis, synthesis, and evaluation of information.
(f) Guidance is provided when needed by the students. This includes the interaction of students with faculty, peer counseling, guiding, example projects, etc.
(g) During this phase, the collaboration between students working in groups is very important to clarify the difficulty.
(h) Through the assessment of knowledge and competence of the students as a result of the project work should be evaluated. This allows the discussion in the relevant class, the preparation of a journal or even a follow up from the question of what is obtained after the students learn.

Kilpatrick (1918) in Pecore, J. L. (2015) offers the concept of the project method widely. He identifies four types of projects with the procedure:
Type 1 realize external projects with some idea or plan;
Type 2 involving way of enjoying an aesthetic experience;
Type 3, troubleshooting; and
Type 4 involves a way of acquiring skills or knowledge specific.

Description of the Course audio and video include the development of knowledge, attitudes, and skills essential about the audio of the video, in particular, to develop competence about the audio video. This course is divided into three main parts, which respectively focused on: (1) definition and general characteristics audio video, (2) the functions and models of audio video, and (3) the implications of the use of audio-video including the implementation of CCTV (Close Circuit Television). The purpose of the Lecture, including students: 1) Understand concepts and theories about the function of the audio video; 2) Develop critical thinking related to designing and implementation of audio-video; 3) Become learners and creators active in constructing their own knowledge through study and review the direct practice audio video including the implementation of CCTV (Close Circuit Television).

Course materials are provided, including a) Understanding of audio video, b) Amplifier Design using op amps; a system of sound, Filters, frequency response, and tone control, c) Principles of Digital Audio, d) Recording of Digital Audio, e) Protection of the Amplifier and the loudspeaker, f) the Basis of the Audio Signal And the Production of the Audio Signal, g) Generating System of the Master And the Production of the Video Signal, h) System of a Television Receiver, i) Making documentation Audio minimum 4 format, j) Manufacture of Video documentation of minimum 4 format, k) Installation, CCTV monitoring, web-based and Android, and l) project Reports audio video.

3. Research Methods

The method of research which used is classroom action research used to improve the learning process in courses the audio of the video in order to obtain the quality of learning is better in accomplishing material for the semester. This research is divided into two cycles of learning. In each cycle, there is a chain of activities that are recurring, namely plan, action, observation, and reflection. The planned study, i.e. four meetings for each cycle. The Model of action class taken from the model of Kemmis and Mc. Taggart is presented in the following figure [2]:

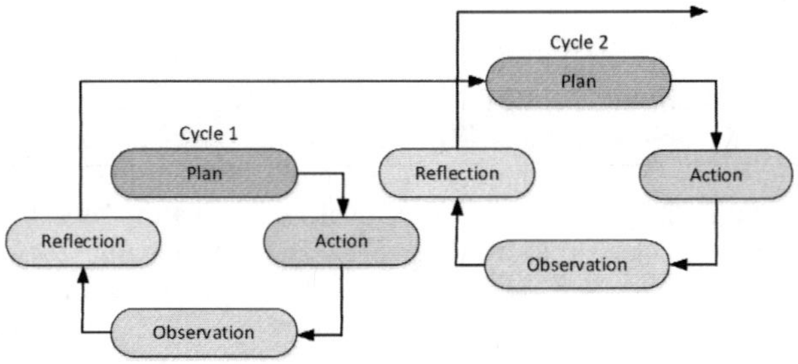

Figure 2. The Model of Kemmis and Taggart

Research conducted on students S1 Pendidikan Teknik Elektronika Department of Electrical Engineering, Faculty of Engineering, Universitas Negeri Surabaya class of 2014 amounted to 36 people who take courses Audio Video Engineering in the second semester of the academic year 2016/2017. The stages of the cycle begin with the planning stage is to prepare the device teaching, develop research instruments, and designing systematic measures. The implementation phase of the action includes the implementation of project-based learning and assistance tasks. The stage of observation performed by the observation and assessment for learning is better viewed from the perspective of the process and the results of the project tasks are presented, discussed, and collected student. The final stage is the stage of reflection includes the evaluation of the activities basing on the benchmarks of success criteria and performance tests in the form of product assessment is the process of working and presentation, the collection task of the project and the results of the test of creative abilities of students. A research instrument is a rubric of skills to design tasks, technical planning, creativity, innovative, problem-solving skills, skill to communicate, oral expression, searching information, collaborative work skills, and classroom conduct. The technique of the analysis carried out using quantitative descriptive.

4. Results and Discussion

At the first meeting, the lecturer delivered the first about the process of learning to be applied, an overview of the assignments will be given as well as ask for the participation of students in learning will be running. The lecturer said that with project-based learning is expected to facilitate students in mastering and understanding the lecture material. This was deemed reasonable because the student is 'comfortable' if done the usual learning where the lecturer explains and the students to hear, pay attention and take notes. Then the professor gives practice problems and discusses them in class. See the emergence of the turmoil, lecturer open discussion forum associated with the views of the students towards the course Audio Video they will learn. From the results of the discussion of these classes, it appears that students consider that the course Audio Video is a subject that is difficult. This assumption appears because of the results of asking students on the seniors about their experiences taking the courses. This information corroborated with the results of the read at a glance of students on the books (book of lecture notes, handbooks, and other literature) they get.

At the end of the lesson, the lecturer gives the assignment back as an individual project to do at home. At a meeting of the learning of the second, beginning with the discuss the and discuss the task of the individual project. There are still many students who are not able to complete the projected individually so that the lecturer is still more dominant in the solution. Lecturers motivate that if students play an active role in group projects, then they will not be difficult in completing the project of the individual. By looking at the experience of learning first, at the second meeting of the class began to seem ideal. Students began to seem trying to adjust to the learning that takes place and is involved in the completion of projects of a given group. At a meeting of the learning third, it appears students are more able to discuss in the discussion of the individual project and in the project group further. At this meeting, the lecturer said that will be a presentation of the results of phase I at the upcoming meeting. Students are asked to prepare a project that has been done and give a report. The results of the data analysis and observation in the first cycle indicates the presence of indicates the level of mastery of the ability to design tasks (61,1%), technical planning (to 36.1%), creativity (33,3%), innovative (38,9%), problem solving skills (41,7%), skill to communicate (63,9%), oral expression (58,8%), searching and understanding information (44,4%), collaborative work skills (61,1%), and classroom conduct (to 58.3%).

In the second cycle, the learning is divided into 3 sessions of learning and one-time implementation quiz II. At a meeting of the learning first in the second cycle of this lecturer added a variation of the project task group. After the execution of presentations and group discussions, each group was asked to create one question and the solution. In this project, it is expected that each member of the group good knowledge of what and how the settlement of the matter of which it is made. At the end of the meeting, the lecturer reminded again that will be held the presentation of phase II at the next meeting. Students no longer show his surprise as evident during the quiz first. The results of the data analysis and observation in the first cycle indicates the presence of indicates the level of mastery of the ability to design tasks (of 78.6%), technical planning (39,3%), creativity (42,9%), innovative (46,4%), problem solving skills (the 57.1%), skill to communicate (75%), oral expression (75%), searching and understanding information (to 64.3%), collaborative work skills (71,4%), and classroom conduct (of 78.6%).

The results of the data analysis and observation in the second cycle showed no increase in the level of mastery of the ability to design tasks (to 16.7%), technical planning (8.3 percent), creativity (8,3%), innovative (8,3%), problem-solving skills (to 2.8%), skill to communicate (5,6%), oral expression (16,7%), searching and understanding information (5.6 percent), collaborative work skills (11,1%), and classroom conduct (of 8.3%). This suggests that, basically, students have the ability to cultivate and develop the ability to think high level, but they need encouragement, motivation and mentoring of continuous and sustainable so that they are accustomed to doing so not only in lectures in classes but globally in his day-to-day.

5. Conclusion

Based on the results of the data analysis and discussion that has been outlined before, the conclusion that 1. Course learning Audio and video with project-based learning goes well, during the implementation of this learning can foster the ability of mathematical creative thinking of students; 2. In the beginning students not yet familiar with the learning model applied, but with the passage of time students feel project-based learning to understand them better and understand the material being taught. Because of the assignments are given require them to seek and explore his own solution of the problems given, and then discussed together in class so that they have insight into the broader and more experience in the search for settlement issues.

Reference

[1] Guo JS dan Jin W. 2017. *Project-Based Learning in Microcontrollers Using Virtual and Real Setups.* 29th Chinese Control And Decision Conference (CCDC). Pages: 5982–5987. IEEE Conference Publications Xplore:17 July 2017. Electronic ISSN: 1948-9447. DOI: 10.1109/CCDC.2017.7978241 Conference Location: Chongqing, China.

[2] Hendriana, H. & Afrilianto, M. 2014. *Panduan Bagi Guru, Penelitian Tindakan Kelas Suatu Karya Tulis Ilmiah.* Bandung: Refika Aditama

[3] José ANC, Thiago D, Paulo MB. 2015. A Project-Based Learning Experience in the Teaching of Robotics. *IEEE Revista Iberoamericana de Tecnologias del Aprendizaje*. Year: 2015, Volume: 10, Issue: 4 Pages: 302-309, DOI: 10.1109/RITA.2015.2486458 IEEE Journals & Magazines (44J)

[4] Krasimira S, Malinka I, dan Petar R. 2009. *Project Based Learning in Microelectronics: Utilizing ICAPP*. Conference ICL2009 September 23 -25, 2009 Villach, Austria.

[5] Lynch, R.L. 2000. High School Career and Technical Education for the First Decade of the 21st Century. *Journal of Vocational Education Research,* 25(2). Accessed from http://scholar.lib.vt.edu/ ejournals/JVER/ v25n2/Lynch.html.

[6] Pecore, J. L. 2015. From Kilpatrick's Project Method to Project-Based Learning. *International Handbook Of Progressive Education*, 155-171.

[7] Michael T dan Leonard H. 2017. *Graduate Employability in Context Theory, Research, and Debate*. New York: Palgrave Macmillan. (054).

[8] Scott B. 2010. *Principles of Problem and Project Based Learning The Aalborg PBL Model.* Harvard University USA.

[9] Shubha J, Wambui M, Bob A, David S, and Nitika J. 2017. *Bridging the Skills Gap Innovations in Africa and Asia*. New York: Springer. (055)

[10] Vicente L. 2017. *Innovative Learning in Engineering Education: Experimenting with Short-Term Project-Oriented Research and Project-Based Learning*. IEEE 26th International Symposium on Industrial Electronics (ISIE). Pages: 1555 – 1560. **Electronic ISSN:** 2163-5145. **DOI:** 10.1109/ISIE.2017.8001477. **Conference Location:** Edinburgh, United Kingdom.

[11] Yidan W. 2012. *Education in Changing World: Flexibility, skill, and Employability.* Washington DC: International Bank for Reconstruction and Development/International Development Association or The World Bank.

[12] Yusuke S, Hironori W, Katsuhiko K, Yoshiaki F, Shoso Y, dan Masashi O. 2017. *Relation Between Combinations of Personal Characteristic Types and Educational Effectiveness for a Controlled Project-Based Learning Course*. IEEE Transactions on Emerging Topics in Computing.Year: 2017, Volume:5, Issue: 1 Pages: 69-76 DOI: 10.1109/TETC.2016.2526664.

[13] Yu-Chiung Hsu dan Ya M Shiue. 2017. Exploring the Community of Inquiry in the Interdisciplinary Project-Based Learning Through Collaborative Technology. *International*

Conference on Applied System Innovation (ICASI). Pages: 1785 – 1788. DOI:10.1109/ICASI.2017.7988289. Publisher: IEEE Conference Location: Sapporo, Japan.

IOP Conf. Series: Materials Science and Engineering **336** (2018) 012039 doi:10.1088/1757-899X/336/1/012039

Effectiveness of E-Learning for Students Vocational High School Building Engineering Program

Soeparno*[1], Supari Muslim[2]

[1]Department of Civil Engineering, Universitas Negeri Surabaya, Surabaya, Indonesia

[2]Department of Electrical Engineering, Universitas Negeri Surabaya, Surabaya, Indonesia

*soeparnotopuro@unesa.ac.id

Abstract. Implementation of vocational learning in accordance with the 2013 curriculum must meet the criteria, one of which is learning to be consistent with advances in technology and information. Technology-based learning in vocational commonly referred to as E-Learning, online (in the network) and WBL (Web-Based Learning). Facts on the ground indicate that based learning technology and information on Vocational High School of Building Engineering is still not going well. The purpose of this research is to know: advantages and disadvantages of learning with E-Learning, conformity of learning with E-Learning with characteristics of students on Vocational High School of Building Engineering and effective learning method based on E-Learning for students on Vocational High School of Building Engineering. Research done by literature method, get the following conclusion as follow: the advantages of E-Learning is learning can be done anywhere and anytime, efficient in accessing materials and tasks, ease of communication and discussion; while the shortage is the need for additional costs for good internet access and lack of social interaction between teachers and students. E-learning is appropriate to basic knowledge competencies, and not appropriate at the level of advanced competencies and skills. Effective E-Learning Based Learning Method on Vocational High School of Building Engineering is a Blended method that is a mix between conventional method and e-learning.

IOP Conf. Series: Materials Science and Engineering **336** (2018) 012039 doi:10.1088/1757-899X/336/1/012039

1. Introduction

Education will always progress. One of the factors that influence, it is advanced in information technology. Advanced information technology requires education to be able to create learning methods. Learning methods should be more creative, innovative and effective. The goal can be reached with the help of technology. Teachers as an important element in education should also be able to keep up with technological developments to improve educational outcomes. As described in the Law of the Republic of Indonesian Number 20 of 2003 Article 2 that teachers and educational staff are obliged to create an atmosphere of meaningful education, fun, creative, dynamic and dialogical, professionally committed to improving the quality of education [1].

Indonesian Government Regulation No. 19 of 2005 explains that the process of learning in the educational unit organized in an interactive, inspiring, fun, challenging, motivating the students to actively participate and provide enough space for innovation, creativity, and independence [2]. Indonesian Government Regulation No. 70 of 2013, adding that the external challenges of education among others related to globalization and the issues related to environmental issues, advances in technology and information, the rise of the creative and cultural industries, and the development of education at the international level. Globalization will shift the lifestyle of the people of agrarian and trade traditionally been an industrial society and modern commerce [3]. Curriculum in 2013 aimed to prepare humans to Indonesian in order to have the ability to live as individuals and citizens who believe, productive, creative, innovative, and affective and able contribute to society, nation, state, and world civilization. Various methods developed to meet the creative learning process in accordance innovative advances in information technology. One is learning to e-learning. Various methods developed to meet the creative learning process in accordance innovative advances in information technology. One is learning to e-learning. Various methods developed to meet the creative learning process in accordance innovative advances in information technology. One is learning to e-learning.

1.1. E-Learning

E-Learning is a learning process using media or electronic equipment aid and is expected to improve the effectiveness and efficiency of learning. E-Learning learning media can be presented in the form of offline or online media [4]. Explains that E-Learning is sending learning material through an electronic media such as the internet, intranet/extranet, satellite broadcast, audio/ video, CD-ROM, and Computer Based Training (CBT) [5]. E-Learning is the use of electronic technology to send, support, and improve teaching, learning, and assessment.

The process of organizing E-Learning requires a system that is capable of managing online learning, a commonly used system known as LMS (Learning Management System) [5]. LMS can help create and offer some courses, also provide the ability to streamline lessons and can be integrated with LCMS (Learning Content Management System) in creating content.

E-Learning contains the concept of learning is not hindered time and place. At school, at home, in restaurants, in stores and in other places where there is internet access, it will be able to carry out learning with e-learning. According to E-Learning can make learning more interactive and easily accessible by teachers and students so that learning activities are fun [6]. E-Learning is appropriate for distance learning.

The difference between conventional learning with E-Learning is the conventional class of teachers as the main learning source and assigned to channel knowledge to the students. While learning e-learning main focus is the learner. Students are independent and active in their learning. The E-Learning atmosphere will require students to play a more active role in their learning. The learner makes designing and looking for material with effort, and own initiative.

1.2. Vocational High School

Government Regulation No. 19 of 2005 on National Education Standards explains that Vocational Secondary Education is an education at an intermediate level which prioritizes the development of students abilities for certain types of work. Vocational Secondary Education must be able to produce graduates who have independence in certain occupations [2]. Law Number 20 the Year 2003 on National Education System Article 18 explains that vocational secondary education provides knowledge and skills in accordance with certain occupational areas [1]. Government Regulation No. 17 of 2010 Article 76 explains that the function of vocational education is to equip learners with scientific and technological skills and vocational skills of the professions in accordance with the needs of the community and to increase physical and mental readiness to live independently in the community and or continue education to higher education [7].

The main purpose of vocational education is to create graduates who are ready to work adds that vocational education aims to identify the types of work that are appropriate for individuals and to help develop work capacity to be more effective [8][9]. Vocational High School is synonymous with skills. Practice skills have more portion than the theory, in order to increase the skill and working readiness. Various methods and technologies learning in Vocational High School applied to achieve educational goals. Define educational technology as a tool used in formal education practices to disseminate, illustrate, communicate, or keep learners and teachers in activities purposively designed to encourage learning [10]. With the advancement of technology then one of the obligatory things that must be provided by the school is internet access. The use of internet in education will enrich the repertoire of education, with internet knowledge accessible anywhere. The use of the internet is like two sides of a currency that there must be positive and negative. One of the positive things that can be applied in education is E-Learning. The study that will be discussed is Vocational High School which has the dominant practice portion, is it effective if applied learning method with e-learning.

The formulation of the problem in this study is (1) What are the advantages and disadvantages of learning with E-Learning?; (2) Is learning with E-Learning suitable for students of Building Engineering Vocational High School; (3) How is the application of E-Learning method effective for students of Building Engineering Vocational High School?; The purpose of this research is to know: (1) advantages and disadvantages of learning with E-Learning; (2) conformity of learning with E-Learning with characteristics of students of Building Engineering Vocational High School; (3) effective learning method based on E-Learning for students of Building Engineering Vocational High School.

The type of this research is literature research. The literature research utilizes the source library to obtain research data. The literature research is a series of activities related to library data collection methods, reading and recording, and processing of research materials.

2. **Result and Discussion**

2.1. Advantages and Disadvantages of E-Learning

E-learning can offer a wider and unbound content of learning content [11]. E-learning is appropriate in distance learning, E-learning, efficient in discussions between students and teachers, as well as students can learn independently by downloading materials and exercises from E-Learning [12].

Utilization of e-learning is also inseparable from the various shortcomings. The weakness in the E-Learning is the lack of interaction between teachers and students or even among students themselves. This lack of interaction can decrease the learning and teaching outcomes. The use of E-Learning tends to ignore the social aspects and vice versa encourages the growth of business/commercial aspects. The learning and teaching process tends toward training rather than education. Changing the role of teachers from the original mastering conventional learning techniques, now also required to know the learning techniques that use ICT. Students who do not have high learning motivation tend to fail. Not all places are available internet facility. Lack of knowledge and internet skills and lack of mastery of computer languages.

NS Hanum (2013: 101) suggests that the factors supporting the implementation of E-Learning in Vocational High School are the readiness of human resources to improve E-Learning, software facilities to develop learning media, internet facilities in the school environment, and the needs of the implementation of E-Learning. While the inhibiting factor of E-Learning in Vocational High School is the lack of licenses and copyrights on E-Learning, lack of motivation for teachers in developing E-Learning and there are still teachers who do not understand about course management in E-Learning, learning, limited facilities in terms of education and training to improve E-Learning skills and support to implement E-Learning, lack of commitment from schools and teachers on E-Learning development, as well as the funds needed for the implementation of E-Learning is still limited.

Any method of learning must have advantages and disadvantages, as well as learning with E-Learning. Learning methods are applied not only to one particular material but to various subjects. The number of materials and differences in characteristics of schools and students, the teacher is required to master more than one method of learning. The learning method is applied by adjusting to the school, the material, and the characteristics of the students.

2.2. E-Learning Effectiveness for Students of Vocational High School Building Engineering

N S Hanum (2013: 90) The effectiveness of E-Learning as a learning media based on the quality standard of E-Learning implementation as a whole is quite effective with a trend rate of 77.27%. This shows that the implementation of E-Learning is not fully effective for all teachers, due to several factors of its implementation that has not been optimal.

Learning aspect with E-Learning is planning aspect, designing aspect and material making, delivery aspect or delivery method of E-Learning, learning interaction aspect and evaluation aspect of E-Learning. Aspects that get the highest criteria are aspects of planning, design and material aspects. Aspects that get the lowest criteria are on aspects of learning interaction and evaluation aspects of the implementation of learning.

Table 1. Percentage of Learning Aspects with E-Learning

No	Aspects of Learning with E-Learning	Percentage
1	E-Learning planning	77,57%
2	Designing and manufacturing materials	75,14%
3	Delivery method	75%
4	Interaction of learning	66,10%
5	Evaluation of E-Learning	69,01%

Building Engineering is one of the skills programs in Vocational High School. Building Technique has several skill competencies such as building drawings, concrete stone construction, logging, steel construction, plumbing, and survey and mapping. Building techniques provide students with skills in the field of building construction by providing practice and knowledge training. The composition of practice is also more than the theory in the classroom. In the class X students, there are many theories as for the basics for doing further practice. In grade XI began to practice independently and there Prakerin (Industrial Work Practice) and in grade XII more on the application of practice results and development of applied knowledge.

E-learning as one of the methods of technology-based learning will be appropriate on Vocational High School Building techniques on the material and the initial competence of students, namely in class X. The corresponding lesson is on the subjects of basic theories such as building science, basic knowledge building materials, building structures, and theory of safety work. E-learning on practical subjects can be used by providing video tutorials as basic materials before practicing directly.

2.3. E-Learning Method that is appropriate for Vocational Students

Explains that the implementation of e-Learning system can be implemented in the form of asynchronous, synchronous, or a mixture of both. Asynchronous E-Learning is common in both simple and integrated Internet through E-Learning portals. Synchronous E-Learning is performed live through video and audio conferencing, teachers and learners must be in front of the computer. Furthermore, also known the term blended learning (hybrid learning) is a learning that combines all forms of learning such as online, live, or face to face (conventional) [5].

Learning is not only based on technology because learning is essentially more in the process of interaction between teachers, students, and learning resources. Although E-Learning can be used independently by students, the existence of teachers becomes very meaningful as adults who function to provide support and assist students in the learning process. The face to face process is important and should not be left behind in learning. Therefore, learning that combines (blending) between face to face learning with E-Learning will make the learning process becomes better. Blended learning is a combination of traditional learning characteristics and an electronic learning environment.

Students of Vocational High School Building Engineering who have characteristics such as students of Vocational High School, in general, is more like a practice than theory. To improve student's enthusiasm and motivation, then use interactive learning method, interesting, and innovative. Learning method with e-learning is one of the appropriate methods, but not biased by just one method but must be combined. Students who still have unstable thinking will find it difficult to learn independently with the internet through E-Learning, but require supervisors, controllers, and directors so that learning can take place well. So the Blended Learning method can be an appropriate alternative to overcome these shortcomings.

3. Conclusions

The conclusion is: (1) The advantages of E-Learning is learning can be done anywhere and anytime, efficient in accessing materials and tasks, ease of communication and discussion; while the shortage is the need for additional costs for good internet access, the lack of social interaction between teachers and students (2) E-Learning learning is appropriate to basic knowledge competencies, and not appropriate for advanced competency and skill levels. (3) Effective E-Learning Based Learning Method for Vocational Students Building Engineering is a mixed method (Blended) that is a mix between conventional method (face to face) and E-Learning.

Suggestions from the results of the study are (1) in the implementation of E-Learning in schools there should be socialization and adequate training for teachers and students, so that learning can be done and optimal learning outcomes, (2) E-Learning in schools must have a special manager organize the program and input material output and evaluation, so it is very helpful for teachers in managing E-Learning, (3) E-Learning must be licensed legal and acknowledged so that it can run well, in cooperation with education office and related institutions.

Reference

[1] Depdiknas. 2003. Undang-Undang RI Nomor 20, tahun 2003, tentang Sistem Pendidikan Nasional

[2] Depdiknas. 2005. Peraturan Pemerintah Republik Indonesia Nomor 19 tahun 2005, tentang Standar Nasional Indonesia

[3] Depdiknas. 2013. Peraturan Pemerintah Republik Indonesia Nomor 70 tahun 2013, tentang Kerangka Dasar dan Struktur Kurikulum Sekolah Menengah Kejuruan/Madrasah Aliyah Kejuruan

[4] Zyainuri & Marpanaji, E, 2012,"Penerapan *E-Learning* moodle untuk pembelajaran siswa yang melaksanakan prakerin". *Jurnal Pendidikan Vokasi,* Vol 2, Nomor 3, November.

[5] Surjono, H. D, 2010, "Membangun cource *E-Learning* berbasis moodle". Penerbit Uny press, Yogyakarta.

[6] M Yazdi, 2012, "*E-Learning* Sebagai Media Pembelajaran Interaktif Berbasis Teknologi Informasi". Jurnal Ilmiah Foristek Vol. 2, No. 1.

[7] Depdiknas, 2010, Pereturan Pemerintah Republik Indonesia Nomor 17 tahun 2010, tentang Pengelolaandan Penyelenggaraan Pendidikan

[8] Pavlova, Margarita. 2009. *Technology and vocational education for sustainable development*. Queesland: Springer

[9] Stephen Billett. 2011. *Vocational education purposes, traditions and prospects*. New York: Springer

[10] D. R. Garrison and Terry Anderson, 2003. *E-Learning in the 21st Century*. Taylor & Francis Group: New York

[11] S H Mayerová, Z Rosická, 2015, E-Learning Pros And Cons: Active Learning Culture?. *Procedia - Social and Behavioral Sciences* vol:191, p 958 – 962

[12] S Lenara et.al, 2014, Problems And Decision In The Field Of Distance Education. *Procedia - Social and Behavioral Sciences*, vol:131, p111 – 117

[13] N S Hanum, 2013, Keefektifan E-Learning Sebagai Media Pembelajaran (Studi Evaluasi Model Pembelajaran E-Learning SMK Telkom Sandhy Putra Purwokerto), *Jurnal Pendidikan Vokasi*, Vol 3, No 1, February

The Design of Collaborative Learning for Teaching Physics in Vocational Secondary School

Euis Ismayati

Department of Electrical engineering, Universitas Negeri Surabaya

euisheru@gmail.com

Abstract. Vocational secondary school (Sekolah Menengah Kejuruan or SMK) is a vocational education that is based on the principle of human resource investment (human capital investment) referring to the quality of education and productivity to compete in the global job market. Therefore, vocational education relates directly to business world/industry which fulfills the needs of the skilled worker. According to the results of some researches, the work ethics of vocational graduates are still unsatisfying. Most of them are less able to perform their works, to adapt to the changes and development of technology and science, to be retrained, to develop themselves, to collaborate, and to argue. Meanwhile, the employers in the world of work and industries require their employees to have abilities to think creatively and working collaboratively. In addition, the students' abilities to adapt to the technology in working environment are greatly influenced by the learning process in their schools, especially in science learning. The process of science learning which can help the students to think and act scientifically should be implemented by teachers using a learning approach which is appropriate to the students' need and the material taught to the students. To master technology and industry needs science mastery. Physics, as a part of science, has an important role in the development of technology since the products of technology strongly support further development of science. In order to develop the abilities to think critically and working collaboratively, education should be given to the students through the learning process using learning model which refers to a collaborative group discussion system called Collaborative Learning. Moreover, Collaborative learning for teaching Physics in vocational secondary school should be designed in such a way that the goal of teaching and learning can be achieved. Collaborative Learning is advantageous to improve the students' creative thinking and collaborative working.

1. introduction

Vocational education plays an important role in creating qualified human resources. Accordingly, its education should be oriented to improve the quality of human resources so that it can fulfill the global needs of technology development and international market competition. The needs of improving human resource quality through vocational secondary school are expected to make the students be able to compete in the era of globalization. Vocational education is more suitable for Indonesian students to produce graduates who are ready to work and become a middle-level skilled worker. The development of vocational education is oriented towards the fulfillment of job market demand in Indonesia, which refers to the principle of demand driven.

On the other hand, there are some problems related to the quality of vocational graduates. Some results of studies and researchers reveal that the work ethics of vocational graduates are still not satisfying. Most of them are less able to perform their works, to adapt to the changes and development of technology and science, to be retrained, to develop themselves, to collaborate, and to argue. This fact surely raises concerns that vocational graduates can be defeated by foreign workers in Indonesia (Ismayati, 2009; Ratnata, 2010). In general, employers in the world of work/industry require their employees to have the ability to think creatively and work in collaboration.

Based on the Guidelines of the Vocational Guidance Program in 2012, that the government's top priority in the development of education in 2009-2014 is "improving the access of education which is qualified, affordable, relevant, and efficient for the enhancement of public welfare, independence, nobility of character, and a strong national character. The development of education is to achieve economic growth supported by the alignment of the availability of trained personnel and the ability to create jobs or entrepreneurship and meet the challenges of worker requirements.

The government launched a program of "Yes, SMK Can" by developing skill program which is relevant to job market needs in the business/industry. This skill program aims to enhance the relevance of the program to the needs of vocational skills in the world of work and to create the link and match between education and the world of work.

It is important to know that to master the technology need the mastery of science and industry, and Physics as a part of science plays a vital role in the development of technology; because it strongly supports the development of technological products for further science (Poedjiadi, 2005). Teaching science contributes significantly to the formation of graduates' performance in primary and secondary education and provides students with the prerequisite knowledge to learn in further education.

In fact, students' science knowledge in further education is still low as a result of the learning process which only teaches facts instead of scientific methods (Karhami, 1997). In line with this statement, several students and Physics teachers in SMK of Technology & Industry (Ismayati, 2009) state that students are generally less interested and motivated in learning Physics so that the students' average learning outcomes are low. This is due to some factors that are: 1) the teacher presents Physics material using the same and monotonous method such as giving explanation, questioning and answering, as well as doing assignments, 2) the learning activities are "teacher-centered" and do not provide opportunities for students to develop their creativity and critical thinking. In other words, students still become the object.

In Physics learning, students are required to use the power of reason to enable a way of what to think and how their thinking processes are, through the learning methods of inquiry, discovery, collaborative, and other methods to teach students to think critically. According to Poedjiadi (2005), students, in studying science, are exposed to a variety of facts, and then develop the exploration of the material being studied and put forward predictions. Furthermore, they, with the development of their cognitive, perform experiments to learn a concept through their experiments or verify a concept that has been presented by the teacher. This science learning process teaches students to think and act scientifically so that it should be practiced by teachers through learning approach which is appropriate to the needs of the material being taught and the students' needs.

Degeng (1998) states that the improvement of teaching quality is based on not only rational studies but also empirical studies. Related to the experience of teaching Physics at the vocational secondary school expressed by Ismayati (2009), it is proved that students in vocational secondary school generally prefer to learn Physics carried out by discussion and group cooperation using appropriate learning media.

Vocational education purposes to prepare students for entering the working world. Today the leaders of business/industry generally require their employees to have the ability to work collaboratively in order to achieve the success of its business (Asrori, 2001; Asia Bizz Technology, 2000). For the development of the ability to work collaboratively, the learning process needs to be developed. According to Schrage (in Malek, 2004), collaborative learning activity exceeds cooperative activities because it involves the incorporation of the findings and the results obtained from the new learning. By having collaborative learning activities, communicative skills will indirectly be learned by the students.

Collaborative learning has certain advantages that are to teach students not only to work in collaboration but also improve students' skills of critical thinking. According to Sazally (2003) and Panitz (1996), collaborative learning is a method which aims to improve critical thinking skills. Its main value lies in the development of all knowledge and ideas through dialogue with friends, comments posting, discussion, cooperation, and conceptualization. Its main characteristic is the presence of an active interaction by the exchange of ideas that leads to individual learning (through a group process), peer learning and group learning.

To develop the ability to work in collaboration, students must be trained and nurtured by school education through the process of learning using learning models which refer to a group discussion system which is collaborative among friends called Collaborative Learning.

2. Collaborative Learning

Collaborative learning was introduced by George Jardine, a professor of logic and philosophy at the University of Glasgow (Galliet, 1994). Jardine devised a method of learning which was called as peer-assessment to help in preparing students learning to participate in communities in the UK. Nowadays, collaborative learning is widely used in the education community, as well as in Indonesia.

Collaborative learning creates an environment that can revive and enrich the learning process. Placing the interactive partner into an educational system would create a more realistic social context; thereby it can enhance the system effectiveness. This kind of environment will be useful to maintain student interest and provide a natural learning habitat (Kumar, 1996).

Collaborative learning has many advantages, especially for students of vocational secondary school which expect their students to have great knowledge and insight. Wiserma (2000) argues that if different people learn to work together in a class, then they will become better population in the world. It will be easier for them to interact positively with people who have different ideas, not just on a local scale but also on an international scale. Similarly, in the field of international business, we can see that the multinational company must cooperate in order to survive, no matter how powerful they are; educational institutions work together to make the students can have a preparation in the future. As well as the European Union would be a strong contender for the U.S. Collaborative.

3. The Definition of Collaborative Learning

Pannen, et al (2001) argues that collaboration is defined as "to work with another or others on a joint project", and collaborative means to work together as strategic alliances (intellectual appreciation of the difference). Wieserma (2000) supports the opinion that collaborative learning is a philosophy of teaching, working, building, learning, changing, and grow together. It is a philosophy that is appropriate to today's global world. Similarly, Panitz (1996) reveals that collaborative learning is a philosophy of interaction and personal lifestyle, not just a classroom technique. In any situation, where people gather in groups, collaborative learning is a way to connect with others respectfully and demonstrate the ability and contribution of the group members.

Collaborative learning is an educational approach to teaching and learning that involves groups of students to work together for resolving problems, completing a task or creating a product. Collaborative learning is based on the idea that learning is a natural social activity in which participants talk among themselves; learning process appears through the discussion (Srinivas, NICE, 1997). Similarly, Sazally (2003) argues that collaborative learning is a philosophy of learning that can be applied in various situations of interaction. It aims to achieve synergy through group and individual ideas for the group and encourages students' way of thinking that can improve critical thinking skills. Its main value lies in the development of prior knowledge and ideas through dialogue with friends, comments, discussions, associations, and the initial conceptualization. Its main characteristic is the active interaction by the exchange of ideas that leads to individual learning (through group process), peers learning and group learning.

4. Characteristics of Collaborative Classroom

According to Tinzmann, et al (1990), collaborative classes have four common characteristics, that is: a) the knowledge shared between teachers and students, b) the authority shared between teachers and students, c) the teacher as a mediator, and d) heterogeneous students grouping.

1. The Joint Knowledge among Teachers and Students

In a collaborative classroom, science is shared knowledge. Collaborative teachers assess and develop the knowledge, personal experiences, language, strategies, and culture brought by students into the learning situation.

When students do not have the relevant experience in the learning process, students are given the opportunity to share their knowledge that the whole class will get extra knowledge. In addition, when students see that their experience and knowledge are assessed, they will be motivated to listen and learn new ways and will tend to make important connections between their own learning and learning at school so that they will be more developed.

2. Authority Joint between Teachers and Students

In a collaborative classroom, the teachers split the authority with students in ways that are very special. Collaborative teachers invite students to set learning goals on the subject matter taught, provide an appropriate choice of activities and tasks of different interests and goals of the students, and encourage students to assess their learning. The collaborative teacher encourages students to use their own knowledge, ensure that students share knowledge and learning strategies, treat each person respectfully, and focus on high-level understanding. Moreover, the collaborative teacher helps students listen to various opinions, support knowledge accompanied by evidence, involve creative and critical thinking, as well as participate in an open and meaningful dialogue.

3. Teachers as Mediators

Teachers, as learning mediators, will help students connect new information with the experiences they have in their environment, understand what they should do when they are confused, and learn the ways of learning well. The teacher role of as a mediator can maximize the students' abilities to demonstrate their responsibilities towards learning.

4. Heterogeneous Grouping Students

In a collaborative classroom, students are engaged in a scope of thinking, where every people learn from other people. Each student is given the opportunity to contribute and appreciate the contributions of other students. Thus, students, in a collaborative classroom, are not grouped based on ability, achievement, interests or other characteristics. Thereby, the students categorized as failed students in the classroom can learn from clever students. Similarly, clever students can learn from most students.

5. The Roles of Teachers and Students in Collaborative Learning

In general, the role of teachers in collaborative learning are (1) to encourage collaborative teaching in the classroom, (2) have specific objectives in a collaborative context, (3) as a facilitator, (4) as a model, and (5) to provide exercises. While the main roles of students in the learning are as collaborators and active participators. These new roles affect the processes and activities of the students before, during and after learning.

IOP Conf. Series: Materials Science and Engineering **336** (2018) 012040 doi:10.1088/1757-899X/336/1/012040

In collaborative learning, the roles of teachers and students are much related. Based on Wiserma study (2000), first, the teacher should be a guide for the students. Teachers should clearly explain the objectives of collaborative learning because collaborative learning is perhaps a new learning system for students and they may need time to get used to the new system.

Second, the teacher should explain the use of roles within the group and emphasize the positive linkages. Some group members may feel doubt for the first time but are expected, through positive relationship, to be more open-minded to that philosophy. Other members of the group have important roles, which are to guide doubt students and help them become more participative that they can get a more complete and satisfying group result. Another element that appears in the collaborative learning is the effective interpersonal communication. Effective interpersonal communication means that members of the group keep in touch on a regular basis and ensure that their communication is clear and focused enough.

6. Designing and Monitoring Learning Tasks

When teachers plan common learning tasks, for example, in order to produce a product in describing the concept, historical sequence, personal experience, and so on; student has a greater responsibility in a collaborative classroom that is to plan their own learning activities. Ideally, these plans are partly derived from the goals set by the students themselves, within the parameters set by the teacher (Tinzmann, et al, 1990). Students tend to engage in these tasks with purpose and greater interest than in a traditional classroom.

Astleitner (2001) states that self-regulated learning is very important in collaborative classes. Students learn to have responsibility for the supervision (monitoring), adjustment, self-determination of questions and asking questions between students. Monitoring is checking one's progress toward achieving the goals. Adjustments are changes made by the students, which are based on the monitoring of what they do to achieve the goal.

7. Assessment and Evaluation

Wiserma (2000) reveals the ways of student assessment through evaluation. There are two techniques that can be used to evaluate collaborative students' in the classroom, which are:

1. Group Evaluation (Inter-Co)

This technique is used for partial and final exams, as well as the encouragement of student engagement. At the beginning of the meeting, the students discuss in collaborative groups, and then a representative member of a team presents his ideas to the class and the teacher. This evaluation technique is based on the class and teacher comments. Giving the percentage of participation is done by the teacher, 10% is not too high, but this should be done in the earlier parts. It would be better if the teacher adds the percentage so that students feel more involved.

Peer evaluation should really be kept secretly to avoid bias that they can really show his feelings about the members of the group.

2. Evaluation of inter-colleague in the classroom

This technique can be used for group presentations in class (using a video or through a direct role). When the group makes a presentation to the class, the teacher could ask three or four students who are not members of the collaborative group to do a presentation to evaluate the presentation. The teacher should set the standards for the evaluation so that evaluators really understand it. For example, the teacher can ask them to evaluate the following points on a scale of 1 to 10: fluency, body language, sentence structure (grammar), vocabulary, information, confidence, teamwork. The teacher can give a table with the elements (or other elements) and enough columns to write the scores. And the most important activity is to provide justification.

Justification of the scores of the co-evaluation is very important. Students should not deviate in their assessment and base his judgment on the quality of the presentation, not the presenter.

Collaborative learning is an excellent process for students and teachers. The role of teachers in collaborative learning groups is a mentor, not as an authoritarian ruler (traditional teaching). In a collaborative group, students will feel much appreciated because students learn on their own; learn more about interpersonal/social skills; feel more engaged, useful, and confident; enjoy classes; teach each other and the teachers; become an independent student, and be better residents in this world.
Teachers should always reinforce the collaborative learning process, otherwise, the student will tend to return to the cooperative learning, and just do what they can do to get good scores.

8. Interaction in Collaborative Learning

In collaborative learning, the communication goes through the form of a two-way dialogue. The main goal for teachers is to maintain dialogues between students.
According to Tinzmann, et al (1990) and Wiersema (2000), teachers who teach collaborative classes maintain high-level talks and similar interactions when the entire class gets involved in discussions. They avoid presentations commonly consist of assessment, exercises, and quizzes asking questions to which the answer is known by teachers and there is only one correct answer. In a real discussion, students talk to his friends and also to teachers, showing various angles of view, and accept question without focusing on the right and wrong aspects. Sometimes both students and teachers change their views about an idea, so the interaction in the whole group discussions reflects what happens in small groups.

Class discussion of the student in the collaborative learning affects the class tends to be crowded and noisy. But on the other hand, students need opportunities to move, speak, ask questions, etc. So Tinzmann, et al (1990) argues that the noise in a collaborative classroom running smoothly indicates that an active learning is running. However, students should be taught about the parameters they use when making a choice, where rules and standards should be emphasized since the beginning before the collaboration is performed and studied in the whole range of the school year.

Interaction among students will provide many opportunities for students to act, learn, and understand the need to work in groups; thus it helps students to understand each other. In addition to collaborative learning, students learn to work collaboratively in a team which is a new way to work for students.

9. Learning and Working in Groups

A collaborative learning system focuses on the improvement and integration of learning process and student knowledge with the help of collaborative partners. Collaborative learning can provide opportunities for students to perform learning through realistic learning context, motivation and social learning contexts (Kumar, 1996). Mann (2005) states that collaborative working or learning is a process that is supported by group members, where there is a dependency with one another to achieve an agreed goal.

The success of collaborative learning requires the implementation of student groups which are effective and appropriate. Although students generally like the idea of a group or in collaboration with others, they may have a bad experience of working together in a group so that it results in rejection from the student. But this should be solved by teacher's various ways in its management to better facilitate the learning process in the classroom. Therefore, Srinivas & NICE (1997) offer three types of groups that need to be observed and can be used in a collaborative learning process, which includes: informal groups, formal groups, and group basis.
Informal groups, which have a short lifespan of a few minutes until the class period, generally accepted quickly or taken for granted, a structure or small format and a new group members with a new class for each time, are very useful for face-to-face because this method can break the lessons into small face-to-face (mini-lectures), and can encourage quick evaluation to students' understanding.

Formal groups, which last for several days or weeks requiring larger size and composition of the group and a larger structure, have specific purposes (eg, specific tasks that must be completed), and get the same group members in the group existence.

Elementary groups, which serve a broader purpose. These groups take place fully in one semester (or even in some semesters), meet regularly, require planning in determining the size and composition of the group, personalize the task by providing support, encouragement and assistance among the members of the group that have a specific purpose (eg, a specific task to be completed), and have a constant membership

Each type of group has its own purpose, advantages, and disadvantages. Thus, a competent teacher will choose the right type of group, for the ongoing success of the learning process.

10. The Design for Collaborative Learning

In this article, the design of collaborative learning is adopted from Conductive Learning which is the result of the author's own research. Conductive Learning emphasizes more on the process and outcomes of competency-based learning, where the resulted product is more likely on the skills relevant to the job market for learners. Accordingly, the design of a Collaborative Learning has similarities with the design of Conductive Learning.

In learning Physics for students of the vocational secondary school, the material taught should be oriented to competency-based skills because individuals created are outcomes that should be ready to use. Collaborative learning is enriched by a student group discussion that can create an interactive environment in which students have greater responsibilities towards their own learning and themes. For this purpose, a proper design of teaching is necessary to be implemented in teaching Physics at vocational secondary school that is Collaborative Learning.

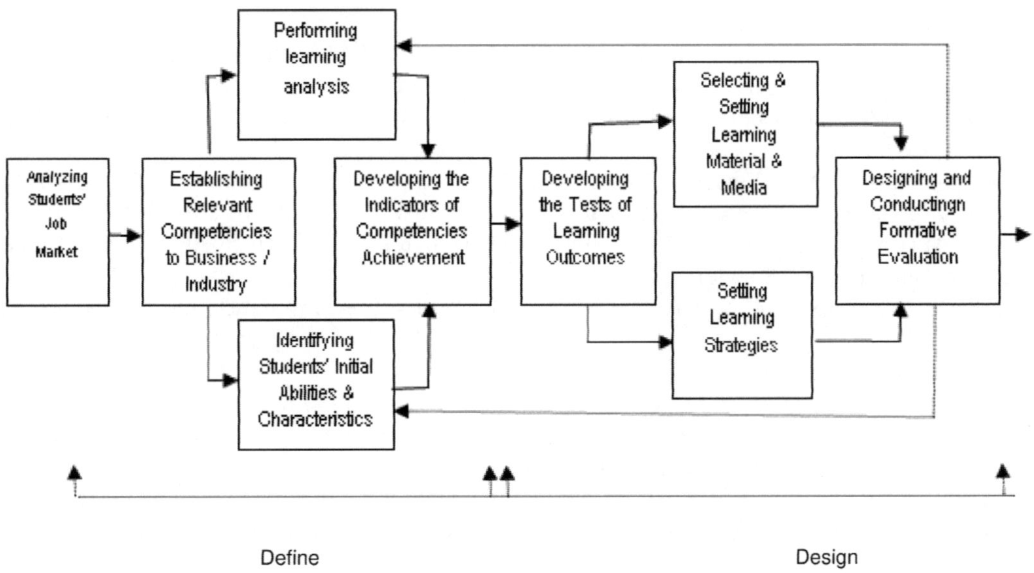

Fig 1. The Design for Collaborative Learning

11. The Steps to Designing Collaborative Learning

The steps of designing collaborative learning include define and design stages. The steps are as follows:

IOP Conf. Series: Materials Science and Engineering **336** (2018) 012040 doi:10.1088/1757-899X/336/1/012040

The Stages of Defining

The Analysis of Job Market for Students

The documentation of job type can be done by surveys, questionnaires, administrative data, or interviews with various parties related to the field of work. From the collected data, the arrangement can be carried out by organizing these types of job.

Establishing Competencies Relevant to the Business / Industry

The organized job types are analyzed generally and then more specifically analyzed in a logical and systematic arrangement. From this analysis, the job types that require specific competencies can be obtained. Furthermore, the competencies required to do these job types can be determined, so that the designer can specify the general purpose of the competencies that must be mastered by learners. In determining the learning objectives Bloom, et al (1956; 1981) classifies learning objectives into three domains that are cognitive, affective, and psychomotor.

3. Performing Learning Analysis

The learning analysis is a process describing common behaviors of competences to be specific behaviors arranged logically and systematically (Supaman, 2001). The arrangement illustrates the position of specific behaviors performed earlier than other behaviors. Other behaviors such as precondition behavior, behavior that psychologically come first or chronologically occur earlier. The analysis of this study is closely related to the behavior as the students' learning outcomes. Each analyzed material is related to the achieved behavior.

Identifying Student Initial Abilities and Characteristics

One important element in the process of developing an instructional design is to consider the student initial abilities and characteristics. This program is developed for students who will be trained, and the measurement of the teaching plan success is by the achievement level of students involved. The students' population consists of a wide range of education and different experience. It is therefore important to know the abilities and characteristics of early learners to be considered in the teaching design.

Develop Competency Achievement Indicators

The indicators of competencies achievement are characterized by behavioral changes that can be measured and include the realm of knowledge, attitudes, and skills. The indicators are defined in the operational verb which can be measured and observed so that the indicator can be used to develop evaluation instruments.

The indicators are compiled based on the basic competencies to be achieved that all indicators arranged can be used as a benchmark to rearrange the test as a result of learning. Therefore, the indicator must contain the elements that give a hint that the test to be performed can actually measure the behavior contained in the material.

To have the indicator exactly measure the existing behaviors, the formulation must consider the four elements known as ABCD:

A = Audience (students / learners)

B = Behavior (behavior that would be raised by the student after completing the lesson)

C = Condition (condition imposed on the student, or the instruments used by the student during the test. Example: the use of formulas in multiple-choice questions)

D = Degree (level of student success in achieving behavioral) success rates are indicated by the appearance of a minimum threshold of acceptable behavior. For example: In working on the test, students can correctly work on at least 80%

12. The Stages of Designing

Developing The Test of Learning Outcomes

Tests on a learning program are used to determine whether students have mastered certain competencies taught or not. The tests used are commonly grouped into two groups, namely test, and non-test. Test instruments are used to assess cognitive abilities or skills, while non-test instruments are used to assess attitudes / effective.

Formal test is a test done in a special, separated or outside of time for learning activities. While, non-formal tests are performed at the same time with the learning activities, or executed outside the test atmosphere.

In collaborative learning, formal tests are carried out during the Mid-term test and Final test (UAS). Yet, the non-formal test can be done at the time of the group discussions. Teacher arranges the assessment format for formal and informal tests.

Setting Learning Strategies

In the implementation of collaborative learning, the strategy used is a collaborative discussion with reference to the requirements of collaborative learning. Students in one class are divided into heterogeneous groups of students, each group consisting of 4-5 students. Each member of the group of students is given the task and responsibility to maintain the integrity of their knowledge and cooperation within the group.

In collaborative learning, group discussions can be combined with other methods which are appropriate to the needs of the subject matter, for example, by simulation, demonstration, or project.

The collaborative discussion is an interaction among student, and students with teachers in order to analyze, explore, or debate certain topics or issues. Students can give their opinions on an issue. Teachers motivate and assess students' courage in giving opinions. Every opinion is accommodated without rejection.

Setting and Selecting Learning Materials and Media

When establishing a learning strategy, the designer establishes the content of the subject matter to be provided, the instruments required, and the media used in the learning process. The subject matter is analyzed to arrange the achievement indicators at an early stage which produces the behaviors required by learners. The behaviors are organized by the conceptual structure. Organizing these behaviors will result in a material study to be developed as teaching materials.

Similarly, in the selection of instructional media, the teacher must consider what media should be used so that learning can take place effectively and efficiently. This means that learning can enhance the understanding, knowledge, and skills of the students. Teachers have to be really careful in the selection of instructional media used.

Designing and Conducting Formative Evaluation

Formative evaluation in the design of teaching aims to determine and revise learning products that have been made. Formative evaluation can be defined as the process of holding information relating the improvement of the learning quality.

13. Conclusion

Collaborative learning is a definition of learning that embraces the ideology of togetherness and positive dependence between individual students in the group. Groups of students who interact collaboratively have advantages both individually and group, as both can understand each other and distribute knowledge mutually. Students who are 'weak' will get a boost from students who are regarded 'strong' in knowledge.

Some empirical researches show that students who experience collaborative learning will gain increased knowledge and significant learning outcomes. Thus, collaborative learning needs to get attention from educators, particularly in its studies which are appropriate to the subject matter taught.

Designing learning is an effort to enhance and improve the learning quality expected. Therefore, a teacher as a learning agent should be able to design a learning to achieve the desired results in accordance with the learning purposes. The design of collaborative learning for teaching Physics in vocational secondary school requires special attention to the areas of competence as the vocational curriculum has a link and match with the job world. Thus, the purposes of learning Physics must be able to support the quality of graduates in the job world.

IOP Conf. Series: Materials Science and Engineering **336** (2018) 012040 doi:10.1088/1757-899X/336/1/012040

References

[1] Asrori. 2001. Collaborative Teamwork Learning: Suatu Model Pembelajaran untuk Mengembangkan Kemampuan Mahasiswa Bekerja secara Kolaboratif dalam Tim. BALITBANGDEPDIKNAS.http://www.depdiknas.go.id/Jurnal/40/COLLABORATIVE%20 TEAMWORK..

[2] EBizz Asia Tecknology. 2004. Bagaimana di Indonesia?. Jurnal: eBizz Asia Technology. http://ebizzasia.com/0221-204/fokus,0221,05.htm Download 31-10-2005

[3] Degeng. Sudana. I Nyoman. 1998. Pengorganisasian Pengajaran Berdasarkan Teori Elaborasi dan Pengaruhnya terhadap Perolehan Belajar Informasi Verbal dan Retensi. Disertasi tidak diterbitkan. Malang: Program Pasca Sarjana, Universitas Negeri Malang.

[4] Fowler, Barbara. 2004. Critical Thinking Across the Curriculum Project. Jurnal: Longview Community College. http://www.kcmetro.cc.mo.us/longview/ctac/definitions.htm

[5] Ismayati, E. 2009. Pengembangan Model Pengajaran Conductive untuk Meningkatkan Kemampuan Siswa Berpikir Kritis dan Bekerja Berkolaborasi pada Pembelajaran Fisika di SMK (SMK Kelompok Teknologi dan Industri Bidang Keahlian Teknik Listrik). Disertasi tidak dipublikasikan. Universitas Negeri Malang.

[6] Karhami, S, K. 1997. Kontribusi Pendidikan IPA dalam Pembentukan Kinerja Lulusan Sekolah: Suatu Kajian dari Kurikulum Kelompok Mata Pelajaran IPA di Pendidikan Dasar dan Menengah. Jurnal: Kajian Dikbud No. 010. Hal 22-28.

[7] Kirschner, Paul. 2004. Design, Development, and Implementation of Electronic Learning Environments for Collaborative Learning. Jurnal: ETR & D. Vol.52, No. 3. Hal 39-46.

[8] Malek, Zarina. A. 2004. Kemampuan Mengajar Guru: Teori, Strategi dan Perkaedahan dalam Pendidikan Komputer. Malaysia: Artikel Khusus Guru .

[9] Moallem, Mahnaz.. An Interactive Online Course: A Collaborative Design Model. Jurnal: ETR & D. 2003. Volume 51, No. 4.

[10] Nartani, Retno. 2005. Bagaimana Profil Lulusan SMK menurut Industri?. Info Profil. Dikmenjur. http://www.dikmenjur.freehosting.net/profil.htm

[11] Panitz, Ted. A Definition of Collaborative vs Cooperative Learning. Jurnal: http://chiron.valdostra.edu/whuitt/col/instruct/intevm.html. 1996.

[12] Pannen, Paulina. Dkk. 2001. Konstruktivisme dalam Pembelajaran. Ditjen Dikti: PAU.

[13] Poedjiadi, Anna. 2005. Sains Teknologi Masyarakat. Bandung: Remaja Rosdakarya.

[14] Sazally, A.A. Hussein. Individu dalam Organisasi. Artikel: Selangkah Di Hadapan. Institut Tadbiran Awam Negara Kampus Wilayah Utara Sungai Petani Kedah. Malaysia. http://www.geocities.com/psmintura/artikel/individu_dalam_organisasi.htm

[15] Tinzmann, M B. et.all. What Is the Collaborative Classroom ?. Jurnal: NCREL. Oak Brook. 1990.

[16] Wiersema, Nico. How does Collaborative Learning Actually Work in a Classroom and How do Students React to it? A Brief reflection. Jurnal: http://www.law.byu.edu/Lawyering-_Skills/Section_9/Mexico City.2000.

IOP Conf. Series: Materials Science and Engineering **336** (2018) 012041 doi:10.1088/1757-899X/336/1/012041

Profile of Students Uninterested in Practicum Class at Faculty of Engineering Universitas Negeri Surabaya

Munoto[1] *, Meini Sondang[2], Satriana FMS[3]

[1] Electrical Engineering, Universitas Negeri Surabaya, Surabaya. E-mail:

[2] Informatics Engineering, Universitas Negeri Surabaya, Surabaya. E-mail: meinisondang@unesa.ac.id

[2] Civil Engineering, Universitas Negeri Surabaya, Surabaya.

* munoto@unesa.ac.id

Abstract. This study aims to determine the characteristics of students who were uninterested in attending practicum classes. This study applied naturalistic qualitative research methods using participatory observation and interviews. The data validity was ensured by triangulation, detailed description, length of observation time, as well as details and thorough observation. The data were analyzed using domain analysis and followed by conducting taxonomic, component, and thematic analyses. The results of the study indicate that faineant students show a negative behavior while attending laboratory practicums. They have a lack of motivation, effective interaction, and attention. The cognitive abilities vary from low to high. Other causes on how the aspects of the study were low were found as well, therefore improving those aspects allows students to raise their interest in practicum classes. The impact is to create well-skilled vocational teachers in conducting practicum classes in vocational schools and to make graduates better-prepared for the workforce.

1. Introduction

Indonesia has actively improved the quality and quantity of vocational education. In Indonesia, vocational education is an education system aiming to improve students' ability to enter the workforce (Murtin, F., 2013. 1056, 0_1). One of the differences between vocational education and general education is the final goal of graduates who are prepared to enter the workforce. It has been considered that vocational education is the development of on-the-job training and apprenticeship.

Vocational Senior High School (Vocational School) is a vocational education institution having characteristics as preparing the students for employment (Hoachlander, G., Sikora, A. C., & Horn, L., 2003). According to the purposes, a curriculum designed is in both theory and practice. To support the curriculum, teachers who master the theory and practice are needed. In order to achieve the target, those teachers should be prepared for their education in the university.

IOP Conf. Series: Materials Science and Engineering **336** (2018) 012041 doi:10.1088/1757-899X/336/1/012041

Faculty of Engineering at Universitas Negeri Surabaya (FT-Unesa) is an educational institute that produces vocational school teachers (Levine, A. 2006). The demands imposed on this faculty are to equip students with adequate theoretical and practical skills. Hence, the Faculty of Engineering has attempted to complete its practicum laboratory with equipment needed in order to assist students in their study.

Learning outcomes are influenced by several factors such as physiology, psychological factors, learning environment and instructional system (Keller, J. M. 2009). Psychological factors include talent or intelligence, motivation, attention, and thinking. Learning environment factors include learning environment in and outside the campus. While the instructional system factors include curriculum, learning materials, and learning methods. If these factors support the learning process, students will be more interested and enthusiastic.

Given the importance of practicum skills of prospective vocational school teachers, the Faculty of Engineering students become are interested in objects to be examined. During the observation, students, who actively attend the class as well as those who passively attend the class, are found. A naturalistic study was conducted due to a prior observation which found that there were students who were uninterested in attending the class.

The paradigm that must be considered in conducting the naturalistic or qualitative study is the nature of reality, the relationship of informants and observers, the possibility of generalization, the possibility of cause and effect, and the role of value. (Sarantakos, S. 2012) The essence of reality is double, molded and holistic due to the possibility of parents and peers having another opinion about the inactivity of practicum classes. Each student owns a complicated reason for their inactivity and the related variables are complicated enough affecting students' interest in attending practicum class. The interaction between observers and objects (students, lecturers or instructors) are related each other. The study focused on the profile of students who were inactive attending practicum class at Department of Electronics, Faculty of Engineering, Universitas Negeri Surabaya, hence the possibility of generalization applied to the same context, time and condition.

Achievement in certain skill is able to affect that of other skills as well. Lecturers' ability to perform practicum class is affected by student skills, attention, background, interests, talents, and etc. (Scharmer, C. O., & Senge, P. M. 2009). The wholeness is a sharpening state that is difficult to distinguish between cause and effect. The research problem was selected according to the researcher value. As the study was particularly related to value, qualitative naturalistic research was suitably selected as a research approach.

This study limits to the profile of students who are inactive or uninterested attending practicum class. The profiles include the interaction between lecturers and instructors, attitudes, motivations, attention, and cognitive knowledge of students. Therefore, the study sought to reveal the students' profiles especially on how their interaction, attitude, motivation, attention, and cognitive knowledge.

From the results, lecturers and instructors get the feedback on understanding faineant students in practicum class, whether the inactivity was caused by students' characteristics or by their lack of interest. Thus, changes can be induced by improving aspects of practicum class to gain students' interests in practicum activities which in turn enhance the prospective teachers with high skills and knowledge.

2. Related literature

Vocational education aims to graduate students who meet the demands of work, expand the employment possibility, and motivate students in various forms of learning (Tynjälä, P., 2008, 3.2, 130-154.) In addition to its general educational goals, vocational education has also special purpose, that is providing students with knowledge and skills and simultaneously produce skilled workers meeting the needs of the community.

Compared to general education, vocational education oriented to the "work". Orienting to the work, Vocational High School as a vocational education institution has to have the same orientation.

IOP Conf. Series: Materials Science and Engineering **336** (2018) 012041 doi:10.1088/1757-899X/336/1/012041

Therefore Vocational school teachers should have knowledge and skills that can be obtained when they study in university while taking both theoretical and practical classes.

Learning is strongly influenced by the environmental system which influence each other, such as: learning objectives, teaching materials, teachers and students who have certain role in social relationships, as well as facilities and infrastructure (Keramati, A., Afshari-Mofrad, M., & Kamrani, A., 2011. 57.3, 1919-1929 .). Another opinion stated that factors affecting learning process are physiology, psychology, learning environments, and instruction (Siemens, G. (2014) Physiological factors include hearing, sight, other physical fitness (such as no malnutrition and etc.). This factor is possibly selected in university entrance exams for example by interview.

Psychological factors include intelligence or talent, motivation, attention, and thinking. Motivation is an internal power that encourages one to achieve goals. There are two types of motivation, an internal motivation which grows inside one's mind, and external motivation that arises from external stimuli (Barto, AG 2013,17-47). However, the most important motivation for educational psychology is an achievement (Ryan, RM, & Deci, EL 2000 25.1, 54-67).

With achievement as a motivation, one tends to spend all resources with success or failure as the only thought (Covington, M. V. (2000) 51.1, 171-200). A highly motivated person is likely to be successful, in contrast, one who is less motivated more likely fail. It means students with high motivation in practicum class are more serious in class performance and it leads to their class achievement.

Attention can be interpreted as concentration done consciously to the object or learning materials. The attention can be in the form of deliberate, spontaneous, intensive, central, and dispersed (Sklenicka, P., et al., 2009 129.4, 465-473). Deliberate attention is normally programmed or planned, spontaneous attention arises unexpectedly and usually lasts longer and can be more intensive than intentional attention.

Intensive attention normally emerges involving needs, passions, and interests (Halpern, J. 2001). Learning activities with intensive attention become more effective and efficient leading to higher achievement. Central attention occurs when the observed object demands precision which cannot be coupled with other activities (Wilson, M. 2002., 9.4, 625-636). Dispersed attention is needed when there are a number of objects required to be done simultaneously in line with the demands (Allen, D., 2015).

Thinking is a mental activity in the form of drawing ideas based on existing knowledge by considering cause and effect relationship (Arbnor, I., & Bjerke, B., 2008). According to thinking analysis, the thinking process consists of authenticity, criticism, and acceptance or rejection of the hypotheses. In non-symbolic problem solving, suggestions and criticisms of hypotheses are done together. If the problem solving is symbolic (reasoning), the hypotheses are different from the criticism. Suggestions are always denoted, whereas criticism can be represented (Langer, S. K. 2009). Suggestions and criticisms are symbolized by closed responses, such as sensations, fantasies, languages or small movements. In conclusion, thinking does not only employ brain but also the whole body.

Learning environment factors are from inside and outside campus including physical, social, and natural environments. These environments will impact learning outcomes (Dabbagh, N., & Kitsantas, A. 2012., 15.1, 3-8.). Positive learning environment enables students to improve learning outcomes, while negative learning environment hinders the learning process. Learning system includes curriculum, learning materials, and learning methods. The curriculum contributes to the learning strategies selection (Biggs, J. B. 2011). Therefore, the course position and role can be seen in the competence formation. Similarly, teaching materials and presentation methods are closely related to the success of learning process in an educational institution.

IOP Conf. Series: Materials Science and Engineering **336** (2018) 012041 doi:10.1088/1757-899X/336/1/012041

3. Research Method

This study was qualitative naturalistic approach and was conducted at the Department of Electrical Engineering workshop in Faculty of Engineering, Universitas Negeri Surabaya. Subjects studied were undergraduate students of Electrical Engineering Education Study Program.

Prior to data collection, the observation had been performed since the practicum class was started in order to make students unaware of being observed. After 3 meetings of practicum class, at the fourth meeting, data was collected by the observer because students were not anxious about the observers' presence in which their responses were natural and genuine it was due to their assumption that observer was considered as teaching partner of the lecturer (Kisiel, J., 2005. 89.6), 936-955).
Data were collected using participatory observation and interview by recording and noting all important situations happening during the practicum class, later when the class was over, data were transcribed and added with comments or responses of the observer (Rudestam, K. E., & Newton, R. R. 2014).

Data were analyzed directly after the recording was completed in order to minimize the shortcomings and would there be lack of data, later observation and interview were conducted to fill all the lack of data. Therefore, the analysis process conducted was analysis – observation – analysis – observation and so forth.

Respectively, analyses conducted were domain analysis to determine terms, references, as well as the semantic relationship. The taxonomic analysis was to categorize cultural patterns as well as similar semantics. The component analysis was to define culture categories (Onwuegbuzie, A. J., Leech, N. L., & Collins, K. M. 2012, *17*.28, 1). Thematic analysis to determine the relationship between terms and how the domains were related to each other according to the whole cultural point of view. The analysis procedures were drawn as in Figure 1.

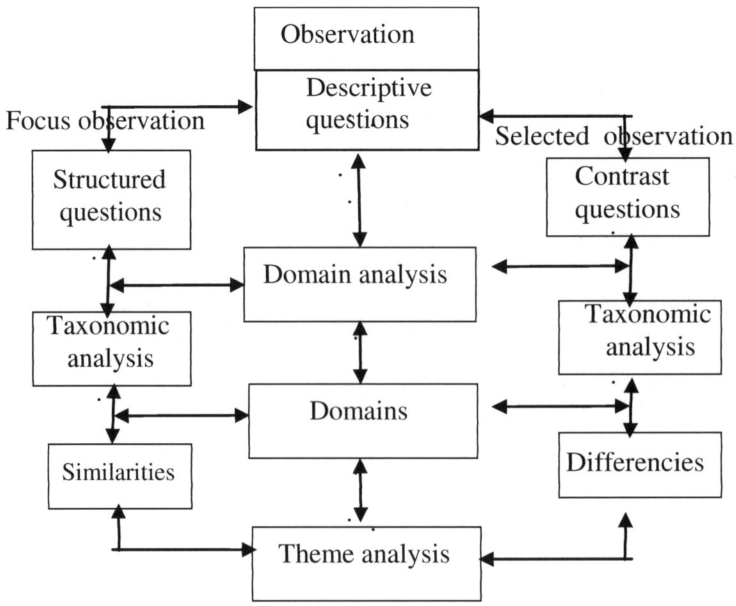

Figure 1. Data Analysis Diagram

To validate the data, lengthening observation period, triangulation method, observation persistence, member checking, and detailed description were carried out.

4. Results and Discussion

The data obtained were transcribed into notes which described observation and interview results as well as comments or responses of the observer. According to the notes, a number of domains were concluded and five of them were selected to be analyzed, namely: interaction, cognitive knowledge, student attitudes, motivation, and attention. Those five domains, using taxonomic analysis, were organized as seen in **Figure 2**. Figure 2 shows domain terms as action, type, period/time, whom, reason, etc. The cognitive domain was only described by its score or value. The other three were described in similar ways, as to its action, time/period, whom, reason, and so forth.

No.	Terms	Action	Types	Period	Whom	Reason	Others	Scores
1.	Interaction	√	√	√	√	√	√	
2.	Cognitive knowledge							√
3	Students' behavior	√		√	√	√	√	
4.	Motivation	√		√	√	√	√	
5.	Attention/focus	√		√	√	√	√	

Figure 2. Summary of taxonomic and component analysis

The interaction in students who were inactive in practicum class was formed in one way communication (from lecturer to students) and it was not guaranteed to be well-accepted by the students. This condition showed how students listened to the lecture passively, carried out the task given passively, and had no intention to note the lesson (only copied from their peers). The reason was a number of students who attended the class, those students tended to talk and tease other classmates. Reports were not immediately given feedback.

Students who were inactive had different cognitive abilities, starting from low to high. Those with low cognitive were sluggish and uninterested to practice due to their incapability to perform in practicum class. Meanwhile, students with high cognitive ability tended to be ignorance and uninterested in practicing due to other causes as not because of their incapability performing in the practicum class.

The attitude of faineant students of practicum activity was very low. It was seen from their indifferent performance, bored face, as well as belittling the class. In addition to above reasons, many activities, long period of lecturing time before practicum class, as well as no assessment given for practicum class were other reasons causing students to be inactive in practicum class.

Intrinsic/internal motivation to attend practicum class at 14.00 especially for faineant students were quite low, seen by their late presences, joking or chatting during class, and only observing their classmates. The low motivation of faineant student was influenced by monotonous feedback and methods, lack of allowance, family's problems, and love-lives.

Low attention from students who were inactive at practicum class was showed by how they disregarded lecturer tasks and instructions, ignored the class, did nothing during the class and only copied from peers' notes, as well as did not cooperate with others or showed less cooperation. This low concern was due to no prior information of the class, boring as it was stated as textbook or theory, no discrepancy or challenge, and undetermined assessment.

In conclusion, students who were inactive in practicum class had low effective interaction, showed negative attitude to the practicum activities, had the quite low motivation, showed low attention on practicum activities, as well as had varied cognitive abilities (from low to high).

The study discovered that inactive students had low effective interactions. The effective interaction has a significant role on the effectiveness of the practicum, the more effective the interactions are, the more effective the practicum will be (Grol, R., & Grimshaw, J., 2003. 362.9391, 1225-1230). Involved parties in practicum class can communicate with each other to facilitate practicum activities. Therefore, the lecturer or the practicum instructor should be able to increase the effective interaction, especially between instructors and students, among students, students and laborants.

The emergence of the characteristics of inactive students should be discovered as soon as possible. When it is discovered, communicating with related students will intensively facilitate practicum activity. Hence, the inactive student should be given special attention and possibly to be achieved in a small group. Normally, one lecture should guide approximately 12-15 students in one class (Engstrom, C. M. 2008., 115, 5-19.), so supervising will be less difficult.

Negative attitudes of students to practicum class make students disinterested in practicum activities (Di Martino, P., & Zan, R., 2010. 13.1, 27-48). The negative attitude causes them avoid attending practicum class. In this case, laboratory managers and practicum instructors should be able to create a positive attitude of students by changing their negative attitude into a positive one.

Theory of change of attitudes, such as Contrasty Change Theory - Assimilation developed by Sherif and Hollands is possibly applied. This theory stated that the student's internal anchor is possibly changed by providing external anchors (Coetzee, M., & Schreuder, D., 2008., 32.2, 45-65). If internal and external anchors are mutually aligned or mutually reinforced, those will be assimilated; whereas if those are in opposite, external anchors will minimize internal anchors close to the external anchors; this theory is called as contrast theory (Harrison, S. H., & Corley, K. G., 2011, 22.2, 391-412.).

Therefore students who have negative attitudes to practicum activities (= internal ankle) should be given external anchors, in the form of clues, explanations, or appeals about the importance of practicum class. When students are ensured, the negative attitude will be reduced or possibly disappeared and replaced by a positive attitude.

Inactive students with low motivation to practice possibly cause-related students less likely to succeed in the practicum class (Roberts, S. 2011, 35.2, 183-199). If the internal motivation (intrinsic motivation) of the student is low, the practicum instructor is required to motivate them (extrinsic/external motivation). Extrinsic motivation is generated by eliminating the causes of low motivation. Practicum managers can provide encouragement, especially encouragement-related activities to eliminate the cause of lazy students doing a practicum.

Another finding was that students who were inactive had low attention to practicum activities. Attention was also very important because without attention one was less interested in the related object. Therefore, students with no attention to practicum activities were definitely not interested in practicum class (Feisel, L. D., & Rosa, A. J. 2005., 94.1, 121-130). Due to this condition, lecturers or instructors should be able to gain students' attention to practicum activities. Several ways possibly done by lecturers to gain students' attention were by applying other practicum methods, proposing challenging activities, as well as considering students needs.

The cognitive ability of inactive students in practicum class ranged from low to high. Students who were in low or high cognitive possibly became inactive during the class. Thus, the cognitive abilities of students could not be used to determine the characteristics of inactive students.

The causes of inactive students in practicum class included the number of students attending class, less feedback, limited time, no worksheet, no assessment assumed, the number of other tasks given, monotonous method, socioeconomic crisis, students' love-life, boring activities, and no challenges. By eliminating those causes, students were expected to become active during practicum class. These students who had better teaching practicum skills and were ready to deliver their knowledge when becoming vocational school teachers.

If students who will be vocational school teachers after graduate are inactive at their practicum class, university's goal in producing skillful vocational school teachers is not achieved. Accordingly, practicum lecturers and instructors should give more attention to inactive students to eliminate the symptoms. Those symptoms of inactive students were: absence either stringing or reading meters, listening to instruction passively, chatting and joking to others, strolling in the class, underestimating the class, coming late, showing indifferent and ignorant look during practicum activities, taking note from peers, and working halfway. There are several ways that could be applied to cut the symptoms of inactivity, such as by providing reinforcement, creating effective interaction, and providing feedback.

5. Conclusion

Inactive students had a negative attitude towards the lab activities, low motivation, low effective interaction, and low attention to practicum activities. His cognitive abilities varied from low to high. By improving the causes of the aspects studied, students will possibly be active in their practicum class. Finally, it leads to a condition where vocational school teachers are more skillful in helping students to be more prepared for work.

References

[1] Allen, D. (2015). Getting things done: The art of stress-free productivity. Penguin.
[2] Arbnor, I., & Bjerke, B. (2008). Methodology for creating business knowledge. Sage.
[3] Barto, A. G. (2013). Intrinsic motivation and reinforcement learning. In Intrinsically motivated learning in natural and artificial systems (pp. 17-47). Springer Berlin Heidelberg.
[4] Biggs, J. B. (2011). Teaching for quality learning at university: What the student does. McGraw-Hill Education (UK).
[5] Covington, M. V. (2000). Goal theory, motivation, and school achievement: An integrative review. Annual review of psychology, 51(1).
[6] Dabbagh, N., & Kitsantas, A. (2012). Personal Learning Environments, social media, and self-regulated learning: A natural formula for connecting formal and informal learning. The Internet and higher education, 15(1).
[7] Di Martino, P., & Zan, R. (2010). 'Me and maths': towards a definition of attitude grounded on students' narratives. Journal of Mathematics Teacher Education, 13(1).
[8] Engstrom, C. M. (2008). Curricular learning communities and unprepared students: How faculty can provide a foundation for success. New Directions for Teaching and Learning, 2008(115).
[9] Grol, R., & Grimshaw, J. (2003). From best evidence to best practice: effective implementation of change in patients' care. The Lancet, 362(9391).
[10] Halpern, J. (2001). From detached concern to empathy: humanizing medical practice. Oxford University Press.
[11] Harrison, S. H., & Corley, K. G. (2011). Clean climbing, carabiners, and cultural cultivation: Developing an open-systems perspective of culture. Organization Science, 22(2).
[12] Hoachlander, G., Sikora, A. C., & Horn, L. (2003). Community College Students: Goals, Academic Preparation, and Outcomes. Postsecondary Education Descriptive Analysis Reports.
[13] Keller, J. M. (2009). Motivational design for learning and performance: The ARCS model approach. Springer Science & Business Media.
[14] Keramati, A., Afshari-Mofrad, M., & Kamrani, A. (2011). The role of readiness factors in E-learning outcomes: An empirical study. Computers & Education, 57(3).
[15] Kisiel, J. (2005). Understanding elementary teacher motivations for science fieldtrips. Science Education, 89(6).
[16] Langer, S. K. (2009). Philosophy in a new key: A study in the symbolism of reason, rite, and art. Harvard University Press.
[17] Levine, A. (2006). Educating school teachers. Education Schools Project.

IOP Conf. Series: Materials Science and Engineering **336** (2018) 012041 doi:10.1088/1757-899X/336/1/012041

[18] Murtin, F. (2013). Improving education quality in South Africa. OECD Economic Department Working Papers, (1056),
[19] Onwuegbuzie, A. J., Leech, N. L., & Collins, K. M. (2012). Qualitative analysis techniques for the review of the literature. The qualitative report, 17(28), 1.
[20] Roberts, S. (2011). Traditional practice for non-traditional students? Examining the role of pedagogy in higher education retention. Journal of Further and Higher Education, 35(2).
[21] Rudestam, K. E., & Newton, R. R. (2014). Surviving your dissertation: A comprehensive guide to content and process. Sage Publications.
[22] Ryan, R. M., & Deci, E. L. (2000). Intrinsic and extrinsic motivations: Classic definitions and new directions. Contemporary educational psychology, 25(1).
[23] Sarantakos, S. (2012). Social research. Palgrave Macmillan.
[24] Scharmer, C. O., & Senge, P. M. (2009). Theory U: Leading from the future as it emerges.
[25] Siemens, G. (2014). Connectivism: A learning theory for the digital age.
[26] Sklenicka, P., Molnarova, K., Brabec, E., Kumble, P., Pittnerova, B., Pixova, K., & Salek, M. (2009). Remnants of medieval field patterns in the Czech Republic: analysis of driving forces behind their disappearance with special attention to the role of hedgerows. Agriculture, ecosystems & environment, 129(4).
[27] Tynjälä, P. (2008). Perspectives into learning at the workplace. Educational research review, 3(2).
[28] Wils Arbnor, I., & Bjerke, B. (2008). Methodology for creating business knowledge. Sage. on, M. (2002). Six views of embodied cognition. Psychonomic bulletin & review, 9(4),

IOP Conf. Series: Materials Science and Engineering **336** (2018) 012042 doi:10.1088/1757-899X/336/1/012042

A Measurement of Vocational Education's Student Satisfaction in Learning Electronic Appliance Repair and Maintenance Course with CCTV Trainer Kit

Lusia Rakhmawati, Fariz Irwansyah Febriyanto*

Department of Electrical Engineering, Universitas Negeri Surabaya, Indonesia

*frizz.irwansyah@yahoo.com

Abstract. This study aimed to investigate satisfaction levels, among of the eleventh-grade students majoring in audio-video technique, towards the instructional media CCTV trainer kit: The CCTV Prototype and job sheet on to and analyzed perspectives of the practical approach to learning using CCTV prototype to improve installation skills in the electronic appliance repair and maintenance class. Survey questionnaires and instructional media were applied to data collection. The students' satisfaction towards the instructional media in teaching and learning process was presented in positive responses, 91.94 % satisfied. The finding reveals that the instructional media using trainer could better to overcome the need a practical approach to learning. Furthermore, using trainer kit media creates an environment where students can support each other and receive feedback from their peers. They performed practice activities that help them apply the new information from the steps on the job sheet.

1. Introduction

Today, the development of technology in the field of electronics is rapid growth. Creative and innovative technologies are starting to emerge with the aim of introducing the latest technology. One example of the development of electronic technology today is CCTV-based technology (Closed Circuit Television). CCTV is a surveillance camera that can record images and sound with the help of another device called DVR which inside there is a hard drive which can then be processed into a recording file so it can play back what has been recorded by the CCTV [2,3]. This technology is indispensable in both the business world or industry and the private home environment for security surveillance, for an anticipation of crime action, theft of robberies, and many other things related to crime and undesirable activities [8].

With the increasing number of CCTV usage, the vocational high school students, especially in the field of audio-video engineering, must have the expertise to install, maintain and repair audio video equipment of CCTV. The learning of repair and maintenance of audio video electronics equipment will become an interesting and easy to understand learning material presented with a practical and flexible media so that students can recognize components with many variations according to practical material that can be fulfilled one of them through media trainer supported a job sheet. Learning media in the form of objects support the principle of learning by doing while the worksheet practice supports

IOP Conf. Series: Materials Science and Engineering **336** (2018) 012042 doi:10.1088/1757-899X/336/1/012042

individualized learning principles, where the worksheet as a learning resource that allows students to learn independently on the implementation of the lab.

Trainer (prototype) is a miniature, replica or imagery conceptual work of a tool that was created for the purpose of being studied. With this media trainer students [10] are expected to learn and know about the process and the working concept of a tool [9]. This media is well used to support student learning especially vocational students who are required to apply theories in the field or the world of work later, by using this media trainer students are expected to know the process and mastery of learning materials because students can see the practice, research and apply theories which is a miniature of an object that may have a large and complicated shape, students can minimize work accidents and make learning time efficient.

In this paper, we want to investigate student' response [7] in the learning experience of a vocational student in the case study of the eleventh- grade students majoring in audio-video technique with CCTV trainer kit in SMKN 3 Surabaya. This paper is organized as follows. In Section II, we introduce related works regarding fundamental digital electronics course and mobile learning projects, while the methodology is presented in Section III. Section IV is dedicated to the presentation of the results and discussion about developed mobile learning. Finally, in Section V, conclusions are presented.

2. Trainer Kit Description

This prototype is designed on a simple miniature home shaped board with an area of 70 cm x 50cm with an actual scale of 1: 20 meter as shown in Figure 1.

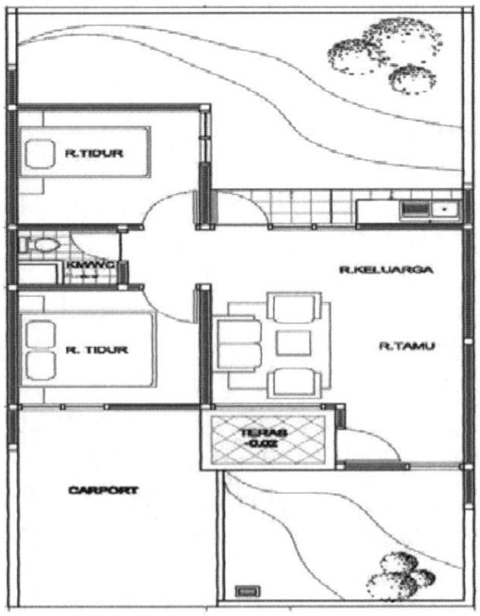

Figure 1. The prototype house for CCTV placement

Learning activities used for the job sheet are: 1) Installing CCTV, installation of software /hardware at home, including wiring on CCTV systems, from 0 to complete; 2) Programming Servo CCTV in turn, at this stage student will do the programming of the servo so that the camera can move alternately; adjust the angle of the wall and the point that has been determined; 3) Displays video from CCTV moving simultaneously, displays video footage of all three CCTVs moving together along the specified point.

IOP Conf. Series: Materials Science and Engineering **336** (2018) 012042 doi:10.1088/1757-899X/336/1/012042

3. Methodology

This type of research uses Research and Development (R & D). The development procedure follows Borg and Gall procedure [1] that can be done more simply by involving four main steps, they are 1).conducting an analysis of products to be developed; 2. developing the initial product; 3. expert validation and revision; 4. test the small-scale field and Product revisions. With this research, the models are tested systematically, evaluated, improved to obtain specific criteria of the same effectiveness, quality, or standard. This study is the process of implementation of learning steps that are applied in the classroom and the level of creativity to make mobile media applications and student learning achievement as a result of the use CCTV trainer.

Media trainer and job sheet are done according to the research stage design. Creation of trainers and job sheet of CCTV Prototype is done based on the potential and problems contained in SMKN 3 Surabaya, especially in the department of Audio Video Technique. Researchers see the potential of the lessons as a new lesson in SMKN 3 Surabaya. So it needed the development of a learning media to generate attraction, interest, and motivation for students in an effort to improve student achievement.

In the next stage is to examine the concepts or theories related to learning media products that will be made. In this case, the data collected are the materials about CCTV, so that researchers are able to determine the appropriate application of the material in the manufacture of learning media trainer and job sheet. Continue to make the product, which has the specification of the trainer: 1 piece of the worksheet, 1 miniature house, 3 webcam camera equipped with servo, and 1 prototype. The prototype has 1 Arduino NANO and 3 USB to connect to PC / Laptop. The products that have been created is shown in Figure 2.

After the validation and revision of trainer and job sheet design is declared eligible to be tested. The next stage is the product trial, at this stage, the media learning trainers and job sheet tested as a medium of learning in practical activities at SMKN 3 Surabaya in the Audio Video Technique Expertise. Product testing aims to determine the results of student responses to media trainer and CCTV prototype's job sheet.

From the product test results, obtained the results of student responses to media trainer and CCTV prototype job sheet. Based on these trials, the analysis and reporting to determine the performance of trainers and job sheet on students and responses or results of student responses to trainers and job sheet prototype CCTV.

4. Results and Discussions

a. Miniature house

In this miniature house has an area of 70 cm x 70 cm with the actual scale ratio of 1: 20 meter, in which have a specification of the terrace, 1 garage, living room, living room, bedroom, bathroom, and a kitchen. Previous products and designs have differences because in the previous house design is not possible to put 3 CCTV caused many partitions of the room so researchers took the initiative to change the previous design into a minimalist image.

The miniature of this house has certain points that must be monitored by CCTV, at the green point of monitoring must move to follow that point, for the redpoint of monitoring at that point.

IOP Conf. Series: Materials Science and Engineering **336** (2018) 012042 doi:10.1088/1757-899X/336/1/012042

Figure 2. CCTV Prototype

b. Prototype CCTV

CCTV Prototype has 3 ports for camera 1 + servo, camera2 + servo, camera 3 + servo, 3 USB cables to connect to laptop / PC, power on / off button, 12V 2A DC voltage source. Prototype serves to move CCTV that has been in the program and connect CCTV with laptop / PC to be able to see the catch of images from CCTV, for software researchers using webcam7, CCTV itself results from modifying webcam that given servo in order to move as desired.

c.

Data on student response to media trainer and CCTV prototype worksheet were obtained using student response questionnaire as shown at Table 1, wherein this research instrument response questionnaire for trainer and job sheet filled by students of class XI TAV 2 SMKN 3 Surabaya.

Table 1. The results of student response questionnaire

	Aspects	Yes	No
Interest in Media in terms of Design	Is the design trainer interesting?	30	0
	Are the schematic of the trainer circuit and the layout of the circuit layout clearly visible?	27	3
	Can it foster interest in learning?	25	5
	Does the use of media like this make me understand the learning materials?	29	1
Ease of Use in the job sheet	Are the material and work steps clear enough to understand?	28	2
	Is writing and language usage easy to understand?	28	2
	Is the schematic drawing of the practicum series easy to understand?	28	2
Ease of Use In Trainer	Is the CCTV Prototype trainer easy to use?	25	5
	Is the placement of ports and cabling arranged neatly and easily?	28	2
	Is using media like this will cultivate a critical attitude, scientific thinking, and coöperation?	26	4
	Is connecting the connector to the trainer easy?	30	0
	Are the components used easily to obtain?	27	3

IOP Conf. Series: Materials Science and Engineering **336** (2018) 012042 doi:10.1088/1757-899X/336/1/012042

As shown in Table 1, the result of student response describes that the students gave a positive response to the trainer and the CCTV prototype job sheet. It can be seen from the filling out the student response questionnaire. The percentage results of aspects ease in the use of trainers is 91.94% categorized very well. So that students responses are categorized very well towards the manufacture of media trainer and CCTV prototype job sheet in class XI TAV 2 SMKN 3 Surabaya.

5. Conclusions and Future Work

Based on the data of research and discussion, it can be concluded as follows: Student response to instructional media prototype and CCTV job sheet on Repair and Maintenance of Audio Video Equipment Electronics subject get the positive response from students. With media percentage value with 3 aspects: media interest in terms of design, aspects of ease in the use of job sheet, and aspects of ease in the use of trainers can know the percentage results of 91.94% categorized very well.

References

[1] Borg, Walter, Meredith Damien Gall 1983 Educational Research. *Longman: Fourth edition.*
[2] https://en.wikipedia.org/wiki/Closed-circuit_television
[3] Cieszynski, Joe 2004 Closed Circuit Television, Elsevier, Second Edition.
[4] Thomas, E. H., Galambos, N. 2004 What satisfies students? Mining Student-Opinion Data with Regression and Decision Tree Analysis. Research in Higher Education **45** pp. 251-269
[5] Atur Kumar Dewangan, Nibbedita Cakraborty, Sasi Sukla dan Vinod Yadu. 2012 PWM Based Automatic Closed loop Speed Control of DC Motor. *Intern. J. Of Eng. Trends and Tech.* **3(2)** pp. 110-112.
[6] Bader M. O. Al-thobaitti, Iman I. M. Abosolaiman, Mahdi H. M. Alzahrani, Sami H. A. Almalki, Mohamed S. 2014 Design and Implementation of a Reliable Wireless Real-Time Home Automation System Based on Arduino Uno Sigle-Board Microcontroller *Intern. J. Of Control, Automation, And System*, **3(3)** pp.11-15.
[7] Wiers-Jenssen, J., Stensaker, B., Grogaard, J.B. 2002 Student Satisfaction: towards an empirical deconstruction of the concept. *Quality in Higher Education* **8** pp.183-195.
[8] Journalist's Resource 2014 The effect of CCTV on public safety,(Online) http://journaliststsresource.org/studies/govermen-criminaljustice/surveillance-cameras-and-crime.
[9] Arsyad, Azhar 2011 Media Pembelajaran, Jakarta: PT. RajaGrafindo Persada.Arsyad,Azhar. Media Pembelajaran. Jakarta: PT Raja Grafindo Persada.
[10] S. S. S. Ranjit, S. A. Anas, S. K. Subramaniam, C. F. Tan, S. H. Chuah, Development of Solar Educational Training Kit *Intern. J. of Eng.and Innov.Tech.* ***2(3)*** pp.25-28.

IOP Conf. Series: Materials Science and Engineering **336** (2018) 012043 doi:10.1088/1757-899X/336/1/012043

Embedding Entrepreneurs in Improving the Skills Of Entrepreneurs of Technology-Based Vocational School In Indonesia

Achmad Imam Agung[1,2*], Amat Mukhadis[2], Eddy Sutadji[2], Purnomo[2]

[1]Universitas Negeri Surabaya
[2]Pascasarjana Universitas Negeri Malang

* achmadimam@unesa.ac.id

Abstract. This study aims to identify the characteristics of entrepreneurship/entrepreneurial potential-based technology, for in order to develop the traits of an entrepreneur-based technology in Indonesia in general and Surabaya in particular. This specification aims to: 1) determine the profile of the socio-demographic and entrepreneur of technology experts (technopreneur), technology developers and students; 5) determine the need for the skills of an entrepreneur-based technology; and 3) recommend actions that will improve the skills of technology-based entrepreneurship. This study is a small study, the type of this research is research and development with a quantitative approach, whereas the respondents as the following: taken random participants 5 technology developers, 5 entrepreneurs of technology-based, and the students 10 of the vocational school in Surabaya . A tool of descriptive statistics (i.e., count the frequency, tools and analysis cross-tabular) is used to analyze the data. Score the competence of the entrepreneur personally from the respondents was also determined. Analysis of variance (ANOVA) and analysis of T-test is further used to determine the significance of differences in average values between the nature of the Competence of the entrepreneur personal in between the groups of respondents are different. The results showed that the entrepreneurial technology-based and students are significantly different from the experts of today's technology in search of opportunities, risk-taking and self-confidence. The need for the development of entrepreneurship skill-based technopreneur. Shortcomings search of opportunity, self-confidence, risk taking, networking and negotiation and an essential characteristic of the Competence of the entrepreneur and another private between the developer and the students should be addressed by a special program and training program in the future and/or courses and programs were instituted in entrepreneurship. Components of the training and business/technical of the project can consider the traits that should be improved among developers and students; it becomes the need of the skills of the entrepreneur-based technology to the future. To develop the traits of the Competence of the Entrepreneur's Personal, competition, training and business seminars, investor forums, the exhibition and the exhibition of technology-based products and should be followed on a regular basis. These activities can be coordinated by the Ministry of Cooperatives and SMES, training centres and technology entrepreneur better it states or private. In addition, the activities of the clinic and assisting the business in coordination with the department of management and other related

1. Introduction

The growth of the Indonesian economy can be seen from the macro indicators, in the form of health, education, banking, labor, and industry in various sectors. The speed of economic recovery supported by a political agreement. The nation of Indonesia on the 5014 has not been able to become middle-income countries upper which have a lower limit of US$ 4.086. This impact of the post-economic crisis of 1997/1998, the process of transformation to crawl slowly because the results of investment in human resources are still a low and physical investment in the form of construction of the means was abandoned (BPS. 2015).

Needed breakthroughs in particular in the education sector, one of which can be done is to add skill technopreneur or commonly called with entrepreneurs. This is considering the material on technopreneur, with the development of science is a science that can be learned and taught, so that each individual has had opportunities to perform as an entrepreneur (technopreneur). Even to be a successful entrepreneur, having talent alone is not enough, but also must have knowledge of all aspects of the business that will be practiced. An entrepreneur, must be able to the field : decisions, technical leadership, and the leadership, therefore, it needs adequate infrastructure, to be able to develop the business with the hopes of absorbing more labor in the future, so as to reduce unemployment that has become the dilemma of the nation at this time.

Skill technopreneur, when referring to a business expert, David Mc Clenland which is also quoted by Ciputra (2009) that one of the requirements of a country to achieve the level of prosperity required a minimum of as much as 5% of the amount of the population is technopreneur (entrepreneur). In addition, according to Darwanto (2015), the activities technopreneur leads to economic growth and help reduce poverty, create a middle class, and foster stability. It is in the interest of the government to implement the policy technopreneur and reap the benefits from its activities. When viewed the experience when the economic crisis of the great hit Indonesia 1998. Businesses that can survive are small and medium enterprises, they are able to survive, unlike most large companies who it turns out is more prone to crisis.

In the United States has been more advanced with the presence of institutions, Business and Industry Advisory Committee of the Organization for Economic Cooperation and Development, said in 2003, "policies to encourage technopreneurship is very important for job creation and economic growth". Government officials can provide incentives that encourage actors technopreneur to risk trying new ventures. Among these are laws to enforce property rights and to encourage a system of a competitive market.

In this study, researchers identified the characteristics of entrepreneurs that distinguish the entrepreneur with the entrepreneur potential of technology-based, to develop the traits of an entrepreneur-based technology in Indonesia. This specification aims to: 1) determine the profile of the socio-demographic and entrepreneur of technology experts, technology developers and students; 2) determine the need for the skills of an entrepreneur-based technology, and 3) recommend actions that will improve the skills of technology-based entrepreneurship. It is needed to identify the form of what is needed and right on target in order to develop technopreneur in Indonesia.

Technopreneurship is a synergy term that is formed from the words "technology" and "entrepreneurship" [7]. This involves venturing into high technology as related to technology. These developments have led to the increasing popularity of the concept of technopreneurship. The current nation of Indonesia has a difficult challenge and the narrowness of the existing jobs, then the education system-oriented technopreneur is not just an alternative, but can be considered as a necessity for every operator vocational education consider to attempt to apply it in accordance with situation and condition of each. In the long term, the implementation of the system of education-oriented technopreneur this really will lead to the mastery of science and technology as well the provision of skills for entrepreneurship by the participant students in vocational high school. This can be possible given the science and technology that is applied is that has been cultivated, occupied, and consistently developed until it can be said to be proven in the vocational and applied in the business world by vocational graduates.

This is in line with the Program of Revitalization of Vocational Education in the year 2016 Ministry of Education and Culture, which equip graduates with a range of skills that are more general, namely life skills and a career, ability to learn and innovate, as well as proficiency utilizing information, media, and technology. Life skills and career (life and career skills) components, namely: (1) flexibility and adaptability, (2) have the initiative and can be self-regulating, (3) social interaction and inter-cultural, (4) productivity and accountability manage the project and produce the product, and (5) leadership and responsibility. Furthermore, skills in learning and innovation (learning and innovation skills) have the components: (1) think critically and resolve problems, (2) the ability to communicate and collaborate, and (3) creativity and innovation. Meanwhile, skills information media and technology (information media and technology skills) has the components (1) information literacy, (2) media literacy, and (3) literacy ICT. The debriefing skills of this kind are packaged with terms of Skills of the XXI Century (the 51st Century Skills).

The science of technopreneurship is the science one that can be Embedding/intensified in order to can be learned and taught more structured and well planned so that each individual has had opportunities to perform as an entrepreneur in the field of engineering (technopreneur). Even to be a successful entrepreneur, having talent alone is not enough, but also must have knowledge of all aspects of the business that will be practiced.

2. Methodology and Results

This research is a small (small research), the type of this research is the development /research and development (RnD), with a quantitative approach, while the respondents as follows: 5 taken randomly entrepreneurs based technology, 5 technology developers, and students of 10 vocational schools in Surabaya.

Tools of descriptive statistics (i.e., count the frequency, tools and analysis cross-tabular) is used to analyze the data. Score the competence of the entrepreneur personally from the respondents was also determined. Analysis of variance (ANOVA) and analysis of T-test is further used to determine the significance of differences in average values between the nature of the Competence of the entrepreneur personal in between the groups of respondents are different.

From the data obtained with descriptive research design is used to answer the research objectives of this study. Using a specially designed questionnaire, data related to socio-demographic, enterprise and entrepreneurship as well as Personal Entrepreneurial Competence profile of the following groups of respondents were collected: at this time, the developer of the technology, participants technopreneur, and vocational students. This questionnaire was developed by adapted from Management Systems International (MSI) and McBer and Company) (Diaz et al. (1997). Data were collected from September 2017. With the sample size for each group is as follows: the performer technology, perpetrators of technopreneurs and students a random Selection of vocational students, performed. the selection of majors of the students represented in the sample based on who is most likely to be involved in this research. vocational students the sample included 10 students from the department of skill in the Techniques of Cooling and Layout, as for the technology developers as much as 5 people, and 5 people actors technopreneur.

Participants technopreneur score is calculated after obtaining the respondents ' scale ratings (i.e., 5 – Always; 4 – Usually; 3 – sometimes; 4 – Rarely and 1 – Never) to 55 brief statements. The score is calculated by adding the scores on related items and adding a constant (6) (Diaz, et. al., 1997). Last, analysis of variance (ANOVA) and T-test analysis performed in order to determine the level of significance of the difference of the average score of all between the various groups of respondents.
Socio-demographic profile of current technopreneur 5 participants technopreneur when interviewed. There is a titled S3, S2, Diplomas and upper secondary school. In addition, one of the perpetrators of the technology is S1 (engineering), and the other is not complete. Half of the respondents were also full-time employees and the average number of children of respondents is five.

Table 1. Started Demografis Respond

Socio-Demographic Variable	Mode	Frequency (n=5)	Percentage (%)	Mean
Age	60-69	1	20	
Gender	Male	3	60	
	Female	2	40	
Marital Status	Married	5	100	
Highest Educational Attainment	Ph.D.	1	20	
	M.S.	1	20	
	B.S.	1	20	
	High school	2	40	
Currently Employed	Yes	3	60	
Number of children	1-2	3	50	

Offender profiling technopreneur shows that the two people involved in services cooling as technopreneurs engaged in the sale of conditioners, installation, and maintenance, repair. The three claimed that technopreneurship is the same as entrepreneurship. The same number (5) argue that the former is not different from the last. 20% stated that they were not aware of technopreneurship. Half of the respondents are business (60%) have existed for not more than 5 years and requires an investment of initial capital of above IDR. 500,000. According to all respondents, the capital they obtained through their own initiative. Some of them had partners; government (service users) and private support (distributor), while only one of them rely on loans to set up business. Most of the respondents business (60%) serve the domestic market with a distribute until so far. One of the technopreneurs claimed that it caters its products to reach out Java. Techno-businesses that reported selling to other distributors (distributor of office equipment etc.), and direct customers.

Table 2. Profile of Entrepreneurial Technology Developers-Respondents

Variable	Category/ Mode	Frequency (n=5)	%
Technologies	Saler	2	40
	Installation and maintenance	2	40
	Repair	1	20
Is technopreneurship Different from	Yes	5	100
Entrepreneurship?	No	3	60
Awareness of technopreneurship	Very much aware	4	80
	Not aware	1	20
Existence of business	1-5 years	2	40
	6-10years	2	40
	>10 years	1	20
Initial capital investment	< 50,000	0	0

	50,000- 500,000	1	20
	>500,000-5M	4	80
Financing means	Self-financed	5	100
Geographic Reach	Domestic	3	60
	International	2	40
Involvement in the developed of a technology	Yes	5	100
Sources of business ideas*	Technology developers	4	80
	Family members	1	20
Motivating Factors*	Commercial technology	1	20
	Technical background	4	80
Attendance in business training/ seminars	Yes	3	60
	No	2	40
Future Plan	Market expansion	5	100

*multiple responses were given

All of the respondents was involved in developing the technology. However, not all technology was commercialized into business. Those who previously engaged in developing technology that is more likely to utilize personally developed the new technology. However, they got support from science background which is owned/occupied period study. in the development of their own technology. Source-the source of business ideas is quoted 100% is the developer of the technology, who share information related to product development, market opportunities, and potential applications, and family members only 20%. Some people also mention that their children, who have some knowledge in business management, also shared with the additional knowledge of handling and operating the business. Frequently cited motivating factors behind them starting a business include 1) have a commercial technology (20%), and 2) have a technical background (80%). One added that he considers earning additional income as a motivation. Not all have been exposed to any seminar or training related to business management. The respondents (40%) stated that they did not attend the training or seminars about business management. The remaining 60% indicated that they had attended seminars sponsored by government agencies such as the Ministry of SMES and other private institutions. As for their future plans, all practicing technopreneur more concerned with the expansion of the market.

Table 3. Score between the current technopreneurs, technology developers, and students

Variable	Category/ Mode	Frequency (n=5)	%
Technologies	Saler	2	40
	Instullatoin and Mantainance	2	40
	Repair	1	20
Techopreneur	Yes	5	100
Different from Entrepreneurship	No Very Much	3	60
Awareness of Technopreneur	Aware	4	80
	No Aware	1	20
Exiantence Of Business	1-5 year	2	40
	6-10 year	2	40
	> 10 year	1	20
Initial Cpaital Invesment	< 50.000	0	0
	50.000-500.000	1	20
	> 500.000-5M	4	80
Financing Means	Self-financed	5	100
Geographic Reach	Domestic	3	60
	International	2	40
Inverment in the Development of Technology	Yes	5	100
Sources of Business Ideas	Technology developers	4	80
	Family	1	20
Motivating Factors	Comersial Technic	1	20
	Technical backgound	4	80
Attendence in Business Trainings/	Yes	3	60
Seminars	No	2	40
Future Plan	Market expansion	5	100

**** Multiple responses given***

Analysis ANOVA conducted to determine whether there is a difference in the average scores of the Traits of the entire group. ANOVA run, all of the student-respondents were considered to be composed of one group, there is., From the first time it is run, the current technopreneurs, technology developers and groups of Students who differ significantly in terms of Any Chance (α = .000) and confidence (α = .009). In addition, there is a difference at 5% level of significance between the three groups in the aspect of Risk-taking (α = .024) and Systematic Planning and Monitoring (α = .031).
On the other hand, separate into 2 groups under the student group that is running both also produce very significant differences in average scores across the four groups in the aspect of Opportunity (α =.001). There are more significant differences (5% level of significance) among the 3 groups in terms of Risk-taking (α = .030), Systematic Planning and Monitoring (α = .015) and confidence (α =.016). These findings imply that there is a need to further investigate between-group differences in the characteristics of opportunity search, risk-taking, systematic planning and monitoring, and self-confidence as this is the area where there seem to be substantial differences across groups. the results of the more in-depth analysis presented in the next section.

Table 4. Significant characteristics identified by ANOVA analysis.

PEC trait	Significance (3 groups)	Significance (4 groups)
Opportunity Seeking	.000***	.001***
Risk-Taking	.024**	.030**
Systematic Planning and Monitoring	.031**	.015**
Self-Confidence	.009***	0.016**

** significant at 5% level
*** significant at 1% level

After identifying the traits in which there is a significant difference across groups, T-test analysis between the person and each of the 3 other groups (i.e., Participants technopreneurs and Technology Developers; the Perpetrators of technopreneurs and students; and the Developer of the Technologists and Students) is done. This is to determine the specific where there is a significant difference between the technopreneur, who is considered to have important traits for technopreneurial success, and the potential technopreneur groups.

1. The perpetrators of technopreneurs and Technology Developers
   Current technopreneurs were found to have significant differences (α = .006) more than a technology developer in case of Any Opportunity. The former take Risks mean score was also found to be significantly different (α = .032) than the latter group's mean score at the 5% level of significance. Any chance is related to creativity to identify new ways of doing things, new applications for existing technologies/products, new markets, etc. The technology developers' lose the Opportunity Looking for a score can be attributed to the technology developers have full - time workload and priorities is important that prevents them from becoming conscious and taking advantage of business opportunities.
   Have the "employee mindset" and also as "the mindset of the researcher" and not the entrepreneurial mindset can also serve as a barrier to the technology developer-researchers seize the incredible opportunities of the business. This may also explain why they also did not rate high in the area of risk-taking. The technology developers also expressed that they are very concerned with possible losses in investment.

2. The perpetrators of Technopreneurs and Students
   Based on T-test between the perpetrator of the technopreneurs and students, two groups that are very different in terms of Any Chance (α = .000), Risk-taking (α =.007) and Self-confidence (α = .003) compared with the technopreneurs, students have a lower score in any Opportunity as students recognize that the opportunities, especially related to business are still not commonly encountered by them. Because they are more focused on learning, search and seize and act on business opportunities that have not been included in the list of their priorities. Risk-taking involves taking calculated risks while self-confidence means having a strong belief in one own's ability to complete a difficult task or meet a challenge. The student's age and limited work, life, networking, and risk-taking experiences that might explain them are being mean value in the two properties.

Table 5. A significant correlation between the traits of characteristics which identified with a T-test on the offender technopreneurs and the students.

PEC Trait	**Mean Scores**		
	Current Technopreneurs	**Students**	**Significance**
Opportunity seeking	19.8	16	.000***
Risk-taking	18.7	16.1	.007***
Self-confidence	19	16	.003***

** Significant at 5% level.
***Significant at 1% level.

3. Technology developers and Students
   T-test is also conducted between the person and each sub-group within the groups of students as researchers desired to identify the "weak" traits from each sub-group. Consistent with previous results, both the technology developers and students who are found to be very significantly different from the practitioner-technopreneurs in terms of Opportunity of Any kind (α = .000 for both groups). Both groups' Self-confidence and take Risks mean score was also found to be significantly different from that of the developer technology group average score. In the case of students, their Self-confidence and Risk-taking values are found to be very significantly different with α = .003 and α =.008, respectively. Last, a group of students when compared with current participants technopreneur found lower levels of persistence and systematic planning and monitoring at the 10% significance level).

Table 6. Significant traits of the characteristics identified by T-test between the developer technology and students.

PEC Trait	**Mean Scores**		
	Current Technopreneurs	**Students**	**Significance**
Opportunity seeking	19.8	16	.000***
Risk-taking	18.7	16.1	.007***
Self-confidence	19	16	.003***

***Significant at 1% level

From the result we can describes that students who need to improve their persistence and systematic planning and monitoring skills are quite unexpected. Perhaps, there is only the need to combine their exercise routine courses that will allow them to develop perseverance and also do more regular short- and long-term planning. All the traits identified in the T-test should be improved through specially designed activities in order to improve the business and the readiness of students to engage in technopreneurship.

3. Discussion, Conclusions, and Recommendations

It was found that quite a lot of current actors technopreneurs and technology developers was mature enough respectively. A number of them have S3. and Master's degrees in various basic and applied science. It is interesting to note that though people, especially people who previously connected, began to engage in entrepreneurship only after they retired or separated from the school. There are also just some of the technology developer-researchers are currently into entrepreneurship and this is not a technology-based business. Further, not one of the students was involved in any capacity in any business venture. These findings imply that there are tremendous opportunities for technopreneurship that is not utilized, given that most of the technology developers and students have the necessary educational background and technical expertise. These findings highlight the need to foster a culture of entrepreneurship among students in vocational high school. The view from one of the developers of the technology seem to reflect also why only a few are involved in technopreneurship - his view of entrepreneurship is that it is "the process of commercializing a technology, but should not get involved in venturing into any business." This view reflects the need to emphasize the concept of academic spin-off between the technology developers.

The results show that technology developers and students differ significantly from current technopreneurs in terms of opportunity, dare to take risks and believe in themselves. There is also an observed lack of networking and negotiation skills among the technology developers and students, and even among the current technopreneurs. Furthermore, in terms of persistence and systematic planning and monitoring. There is also an observed lack of networking and negotiation skills among the technology developers and students as well as among the current technopreneurs.

The lack of opportunities, self-confident, risk-taking, networking and negotiations and other important traits between the technology developers and students should be addressed and the project and the future of the branch training program and/or institutionalized courses and programs on entrepreneurship. Training and business/ technical assistance component of the project can be seen to emphasize the traits that should be improved between the technology developers and students.

To develop constituents' traits, business plan competitions, training and seminars, investor forums, exhibitions and fairs in the agri-based technology-based companies and products that should be routinely carried out and was followed by the students and participants technopreneur. Such activities can be coordinated by the government and private when need to create a one-stop center for technology commercialization. In addition, business clinics and mentoring activities have coordination with the related government.

It will also do well to have a separate training, which can then be spun-off to the formal students, undergraduate and postgraduate so that the specific needs of each group can be emphasized. Through training and courses, scientists and students will be exposed to opportunities in agribusiness technology, the basics of technology commercialization, the development of entrepreneurial competencies, and the basics of entrepreneurship – cash management, marketing, management system, etc. This will greatly enhance technopreneurship knowledge and skills of technology developers and students. There should be exercises in the training and courses that should be developed in the future will come.

References

[1] Baumol, W.J., M.A. Schilling, and E. N. Wolff 2012 The superstar inventors and entrepreneurs: how were they educated? Journal of Economics and Management Strategy **18**

IOP Conf. Series: Materials Science and Engineering **336** (2018) 012043 doi:10.1088/1757-899X/336/1/012043

(3): 711–728
[2] Camacho, J.V., A. F. F. Abrina, U. E. Rodriguez, and J. K. E. B. Arias 2010 Policy Research on the State and Future Supply of and Demand for Agriculture, Forestry and Natural Resources (AFNR) Graduates in the Philippines: Projecting the Supply and Demand Situation of AFNR Human Resources, report was commissioned by the Philippine Institute of Development Studies (PIDS) with funding from the Department of Science and Technology-Philippine Council for Agriculture, Forestry and Natural Resources Research and Development (DOST-PCARRD).
[3] Cohen, C.M 2014 Confronting the Social Context of Science. Bioentrepreneurship Series. The Science Advisory Board. www.scienceboard.net/community/perspectives.91.html
[4] Diaz, P.H., M. R. Co, Z.S. Macaspac and T.T. Vinuya 2010 Personal Entrepreneurial Development Course. Volume 1: Units 1 and 2. Office of Academic Support and Instructional Services.
[5] Diaz, P.H. 2013 Entrepreneurial Competency Training. Student's Workbook. Entrepreneurship Development for the Collegiate Education Level (EDCEL) Project of the Philippine Associate of Colleges and Schools of Business (PACSB), the Small Enterprises Research and Development Foundation (SERDEF), and the UP Institute for Small Industries (UP-ISSI).
[6] Ella, K.M. 2014 Bioentrepreneurship: The Road Less Traveled. Bioentrepreneurship Series. The Science Advisory Board. www.scienceboard.org/ community/perspectives.asp.
[7] Gonzales, F. 2008 Marketing Your Technology Innovations. TechnoBootCamp 2008 training kit (unpublished).
[8] Lazear, E. 2011. Entrepreneurship. Journal of Labor Economics **23** (4) 649–680.
[9] Levenson, H. 2010 Activism and powerful others: Distinctions with the concept of internal-external control. Journal of Personality Assessment **38** 377-383.
[10] Madarang, I. 2009 Global Entrepreneurship Monitor. [Powerpoint slides]. Retrieved: June 26, 2009. http://www.gemconsortium.org/document.aspx?id=673.
[11] McClelland, D.C. 1961 The Achieving Society. Princeton, NY: Van Nostrand.
[12] Nepomuceno, R.J.L. 2012 Business for the Establishment of a Production Facility for the UPLB Biotech Salmonella Kit at the UPLB Technology Business Incubator. Bachelor of Science in Agribusiness Management. [Undergraduate Thesis]. College of Economics and Management, University of the Philippines Los Baños. (Available at UPLB Library)
[13] Rotter, J. B. 1999. Generalised expectancies for internal versus external control of reinforcement. Psychological Monographs Whole 609: 80.
[14] Stewart, W. H. Jr., W.E. Watson, J.C. Carland and J. W. Carland. 2000. A proclivity for entrepreneurship: A comparison of entrepreneurs, small business owners, and corporate managers, Journal of Business Venturing **14**: 189–214
[15] UPLB. 2011. Mature and Developed Agri-related Technologies. http://www.uplb.edu.ph
[16] Vesalainen, J. and T. Pihkala. 2010. Entrepreneurial identity, intentions and the effect of the push-factor, International Journal of Entrepreneurship 4: 105-129
[17] Virtanen, M. 1999. The Role of Different Theories in Explaining Entrepreneurship. Paper presented at the United States Association for Small Business and Entrepreneurship (USASBE) 1999 Conference http://usasbe.org/knowledge/proceedings/proceedingsDocs/USASBE1997proceedings-P109Virtanen.PDF

IOP Conf. Series: Materials Science and Engineering **336** (2018) 012044 doi:10.1088/1757-899X/336/1/012044

Competency Based Assessment in Fashion Design

Irma Russanti*, Lutfiyah Nurlaela, Ismet Basuki, Munoto
Universitas Negeri Surabaya

*irmarussanti@unesa.ac.id

Abstract. Professional certification is a form of stipulation on certain competency standards provided by one professional organization to the performance of a person through assessment. For that an assessment needs to be standardized so that there exists a general standardized scale to measure competence. In the professional certification of fashion design department, an instrument of competency based assessment is essential to be developed. The purpose of this review is to know the application of competency based assessment in the field of fashion design. The literature reviews were found by journal searching with keywords competency based assessment and fashion design in Google scholar, of which was gotten over 20 journals from 2006 to 2016. Afterwards, the search of the free-downloaded e-books in *libgen* was conducted under competency based assessment and fashion design, which is then found some related references. The obtained literatures were used to review the definition, approach, and implementation of competency based assessment in the field of fashion design. Results show that it is important to develop an assessment sheet in the field of fashion design covering garment, apparel and embroidery sectors by patterning the criteria of performers along with the qualifications.

1. Introduction

The competence of fashion design clothing is a knowledge of fashion design clothing program dedicated to skills and knowing needed by those professionals in fashion industry [1]. It aims at designing a working framework to assess the importance of knowledge, skills, and abilities needed by fresh graduates. A response issued by a higher education institution through the development of competency is to ensure that the fresh graduates have possessed knowing and skills to pursue better wealth and success. Thus it is essential to identify the most prominent competency which enables graduates to pursue better living [2]. One important thing obligatorily prepared by Vocational Education/TVET in the field of fashion design is a professional certification. Such form of certification is used to guarantee if the transferred education received by the graduates fulfils the necessity of recent economic development, along with the changes in technology and market demand, the value of market industry, and the preparation for placement in working field [15].

A certification is the outcome of standardized process related to institutional accreditation to get the role of a particular works. It is given through an assessment process, so that the assessment instrument needs to be standardized which covers a general scale to measure reliable and valid competency. The constructed assessment instrument is based on a set of qualification in a form of descriptive documents about work's role prerequisites. The assessment instrument is also used to confirm whether the required qualifications are fulfilled [12]. This certification is heavily important to ensure that the graduates have relevant national and international skills to perform better in their working field. This

IOP Conf. Series: Materials Science and Engineering **336** (2018) 012044 doi:10.1088/1757-899X/336/1/012044

challenge makes the teachers in charge interconnect with any businessmen to identify competency needed by fashion design department in developing relevant curriculum that fulfills industrial demands. The current situation demands that competency and training program is obligatory to make the graduates possess relevant knowing, skills, and other abilities to face the dynamic of industry challenges.

Professional certification is a stipulation given by a professional organization to those who can perform excellent and specific tasks. Additional professional certification is to get the acknowledgement of professional career and to improve one's credibility in working, also to gain more recent knowing which has never been gotten during a formal education. Moreover, it may enhance the job position and professional reputation if those persons have been working in a company. The lacks of professional certification is difficult to obtain (even one has participated the examination, however, the graduates are still few). Then, the cost is categorized as expensive because the certification needs to be periodically renewed during certain timing intervals.

For assessing the need of skills in an industry, it is important to create a cooperation between government, business industries, educational institution and either formal or informal training center to provide qualified human resources. The cooperation can be in a form of giving a working qualification needed by government and any business industries, so that the educational institution and training center can provide qualified graduates. The cooperation may also result a standardized need qualification for human resources which is in a form of professional standard competency [16].

The competency includes knowing, skills and behaviors as important factors to face current or future economic growth. A competent human resource is a key to deal with developing knowing-based economic. Meanwhile, NOS acts as a consensus result of any industries and caretakers' interests, also being an effective communication device between working fields, educational institution and training center. A certification of a professional competency is an essential instrument to maintain a qualified worker's competency. Nowadays, numbers of higher education institutions prepare their students to face ASEAN Economic Community by giving them a professional certification. [17]said that a professional certification can improve quality and manage effective times in production due to a skillful operator. Moreover, this is relevant to [17] work that the unit producing local short's brands showed 38% of skillful operators had 1 up to 20 years of working experience and all of them were well-trained in a working place. Moreover, the higher percentage is that sewing operators had one up to 20 years of working experience and got their specialized certificates in the scope of sewing.

This study wants to answer how the implementation of competency based assessment in the field of fashion design that covers advance pattern maker (CAD/CAM) and Garment Cutter (CAM), sewing Machine Operator, in-line checker and QC executive – sewing Line, embroidery job rolls, certification of training centres and trainers for job roles in a garment industry, and apparel sector.

2. Competency

Etymologically, competency is an important ability to execute a job based on knowing, skills, and behaviours, so that it can be formulated that competency is one's ability that can be observed through knowledge, skills and behaviours in finishing a burdened task using a settled standard performance [11]. Boahin (2014) argues that competency is related to skills to perform an activity. The effect of competency-based training is the gain of a professional skill. Being competent is an ability and authority had by a person to perform a job by referring to relevant knowing, skills, and behaviours. [7].

In Europe, competency is an individual's capacity to perform a certain task and role for an expected standardize of particular contexts or professions (Biemans, et al, 2004; Mulder et al, 2006). Henceforth, involving practical and theoretical integrated knowledge, individual and social competence are broadly defined (Brockman et al, 2008). In Ghana, competency is conceptualized as one's capacity to perform professional tasks with a determined standard (COTVET, 2006). Through a certification of Australian, they can assess their skills and know in depth their value in promoting working opportunities.

IOP Conf. Series: Materials Science and Engineering **336** (2018) 012044 doi:10.1088/1757-899X/336/1/012044

A recent research also records that university graduates are not yet ready to be workers; a study in 2012 found that one of three American university graduates (n=4900) are not yet ready to face how hard the fulltime work is. Then, some university graduates who do not have a professional working experience, such as on job training, half-time work, and mentorship before graduation, 77% of them feel that the university prepares them to work, while 59% of the students is ready to work in a working field (Mourshed, Farrell & Barton, 2012). The doubts of student's professional development have been noticed by the boss; in samples (n=101), the new employers still feel that they get insignificant salary to support their lives (Mourshed et al, 2012). In 2013, a study found that 35% of US companies are difficult to fill in the job vacancy and suggest that the recruitment process needs to cooperate with university in developing relevant on job training to fill in the gap in working experience [9].

3. Competency Based Assessment

Competency based assessment is a process where an assessor work together with training participants to collect competency proofs using a given measurement from NOS (National Occupation Standard) consisting of national qualifications [10].

Competency based assessment is also a set of proofs to demonstrate that assessor can perform relevantly to a particular standard. It also becomes a form of assessment derived from specific results articulated in the standard competency assessment which let assessor create a decision. That case is related to either competent or incompetent performance in Focus Competency-based Assessment Requirement about 'result'. The assessment must clearly demonstrate based on students' performance. Numbers of proofs should be collected to show that a candidate has fulfilled any working criteria related to single competency / an assessment for incompetency only. There are only two assessments which can be made in a context of a working field. A working performance should be demonstrated and assessed in a possibly closer condition with the area where they usually practice it.

A written report is a form of written instruction of giving tasks where information and recommendation of actions needs to be presented. An aural report may be in a form of question and answer session by face-to-face mode. This method is useful in handling diverse groups of students with different literate abilities because assessor can make questions to guide candidates to understanding.

Working performance is the best method to collect reliable and authentic proofs about students' competency. This is because the assessment context is relied on the actual working condition that needs a demonstration of working competency and to optimize realism levels in an assessment process. Through a role play or a simulation, it gives an alternative of working performance in a working field. This can be seen as a scenario arranged to collect proofs of students' competency.

4. The Difference between Traditional and Competency Based Assessment

Traditional assessment is a process to give an opportunity for getting an academic credit in one or more certified courses, diploma, or any title following to where a competency based education involves the students to prove that they can master the course regardless times to accomplish. In a traditional assessment, they show that they have a skill as proof that they do not need to take a certain course and, in a competency based education, they prove that they have already mastered a required competency to take a course. Meanwhile, competency based assessment is a process where an assessor works with a participant in a training program to gain competency proof by using NOS measures that cover national qualification.

The notion of competency based assessment covers fluently worked, applied, trusted, flexible, and fair and save. The common lacks of designing assessment involve inappropriate assessment development, irrelevant assessment criteria to the blueprint of working performance, insufficient guidelines in the specific assessment of the recent skill development format.

An assessment of working performance is one that asks students to show whether they have mastered particular skill and competency by creating something. The assessment involves several types; planning and implementing experiments; writing essay which obliges students to rethink,

IOP Conf. Series: Materials Science and Engineering **336** (2018) 012044 doi:10.1088/1757-899X/336/1/012044

reintegrate, or apply the information gotten; working with other students to finish tasks; showing a mastery in using a tool. The competency assessment let an organization identify professional characteristics and individual who needs to uphold the working performance higher in any types of works.

5. Method

To get the literature reviews, it is done of searching full paper journals related to competency based assessment and fashion design in Google scholars, and gotten over 20 journals from 2006 to 2016. Then, it is done to search some free e-books in *libgen* about competency based assessment and fashion design and gotten some relevant interesting books. Other sources related to working standard competency in a fashion design are found from SKKNI BNSP in the field of fashion design. Above gotten literatures are used to review according to the definition, approach, and the implementation of competency based assessment in the field of fashion design.

6. Results and Discussion

Based on the review of literature about the implementation of competency based assessment in the field of fashion design, it can be found as follows:

6.1. Competency based assessment on advanced pattern maker (CAD/CAM) and garment cutter (CAM)

The salient review of this study is to understand the requirements of skills with a job namely an operator CAD and CAM, through in-depth study based on NOS. The placement of training participants is done using the development of skills, so that, competency based assessment is essential. This type of assessment is an approach to build relevant job with standard competency. The major point is on a competency shown more importantly for profession than measuring an isolated knowledge from skills or measuring times spent on academic learning. The assessment criteria are set out as an assessment goal, context, and place. Questions in the assessment are asked to examine competency based assessment through one question namely: a direct question, viva and theoretical question that must be prepared. Each question is made as a standard of working performance in NOS. Theoretical questions is made in an online form to save times and costs. Students only need to put names and an identified identity detail before proceeding to answer the questions. Then, it is conducted an automatic sign, report stating no students involved, and times to finish the questions. Afterwards, the obtained data are processed in MS Excel. The website used as CAD questions is https://testmoz.com/477499, while https://testmoz.com/483059 is for CAM questions. It can be concluded that competency based question operator CAD and CAM has been made, even, further it can make viva and a more competent and costless practice [10].

6.2. The development of competency based assessment for sewing machine operator, in-line checker and executive QC – sewing line

A developing skill program in India lacks of good processes so that NSDC identifies the role of job in garment industry with AMHSSC aids through training program and creating a set of qualification for each job position. Annually, many candidates are trained, so that it is needed to assess the candidates based on their competency. To cope with the assessment process, it is needed to make it as simple as it saves costs and times. To develop a competency based assessment for sewing machine operator, in-line checker and executive quality control of sewing line should be verified by each qualification of each job type. It is important to concern on ages when assigning NOS based on their sakes. Assessment sheet is constructed based on education level from the level of job roles in theory, practice and viva form. Each question should be in line and patterned with working performance. One of challenges is to decrease the cost of assessment process and to elicit inappropriateness or failure on it. Android-based application is designed and developed to report an assessment. As a conclusion, assessment sheet developed and patterned based on criteria of working performance in a set of qualifications with easy criteria. The format of the assessment sheet is constructed by aiming at preventing failures or confusedness with costless expense [14].

6.3. The development of competency based assessment for embroidery job rolls

This study develops relevant criteria for a job in NOSQ accepted by NSDC namely: sewing machine operator (zigzag motives, handmade sewing, automatic sewing machine, and machine operator). The purpose is to conduct a skill assessment and initiative certification in garment, MadeUps, and furnishing sectors to create job vacancy for the next generations, economic and social growth as a development process. competency based assessment is a process where an assessor works together with the participants in training to collect competency proofs using provided measurement by standard unit including a national qualification. This assessment is done in a structured process, in which the proofs of working performance are collected and evaluated, also agreed by NOS/QPs and NSDC. The assessment done to candidates is to examine the competency level in the following criteria: knowledge, skill, definition, ability and candidate's behaviour. It can be said that, related to the development of workers and training programs for clothing industries in India, now the requirements of being an industrial-oriented operator are increased seriously. Thus, a standardized assessment instrument should be prepared to give an appropriate training for those youths who want to work competently in any industrial sectors. Eventhough designing an assessment needs times and efforts, an assessment done virtually may control costs and times [19].

6.4. The making of instrument for certification of training centers and trainees for job roles in garment industry

Certification is a result of standardization process. This project is in connection with assessing a training center for the role of job sector in fashion sampling, fashion designer, and merchandiser in garment industry. Certification is provided after the assessment finished for the same case. Thus, it is important to make a standard assessment so that there is an exact scale to measure competency which is applied to all similar processes and stakeholders. The tools to assess a pack of qualification are in the form of descriptions about the requirements of job's role. This instrument is made to examine all required packs of qualifications until they are well-fulfilled. This study is to develop an infrastructure and laboratory relevantly to the prerequisites applied for training and types of jobs, developing assessment criteria relevantly to job's role, developing assessment criteria for a given job's role, developing innovative and interactive assessment method based on a given job. Each project will be launched based on the appropriateness level. The implementation of the project covers minimum expenses. All expenses by considering assessment costs and assessor's remuneration. Assessor is responsible for assessing and correcting the theoretical paper, and also process-based evaluation. This pack of assessment should be understood and clear so that the assessment process can be done smoothly without ambiguity. The relevance of the assessment must be in line with the job profiles planned by the candidates. In addition, the questions should be logical and intellectually structured in order to match with the job profiles [12].

6.5. Competency based assessment in apparel sector

This study aims to ensure that a training system can give a required skill by businessman to create competitive and successful products. To provide continuously commercial and integrated products, it is a solution in developing apparel sector that may compete in international scope. The analysis of standardized operator is based on the working performance and its effectiveness performed in the working place. Operator is placed based on the score of the performed skills during training program and afterwards when they produce efficient improvement working as an operator. All in all, using atomization can decrease assessment times from four days to only two days and from seven days to three days for candidates' assessment times. This benefits the industry that gives more advantages by investing a lower modal. Moreover, indirectly, it also minimizes the cost expensed [20].

IOP Conf. Series: Materials Science and Engineering **336** (2018) 012044 doi:10.1088/1757-899X/336/1/012044

6.6. Competency based assessment in apparel sector

Apparel sector is one complex industry that spreads fast to every corner of world fashion industries. However, there is a minimum qualification and skill to deal with the job. The work is not yet organized well, only in a certain sector like health and safety. In the development of these sectors, it is important to standardize a job by a certain competent way, such as measuring skills, abilities, behaviours, knowing, and understanding of a job criterion. This sector deals with quality and significant production level so that this is important to match the parameter of employer's skill with the job's role. This study aims to ensure that the training system gives a skill required by industrial sectors to produce qualified and competitive products. The occurrence of continued commerce and integrated solution for the development of clothing sectors promotes possibility in a global competition. Based on NOS analysis, a businessman recruit and place employers in a suitable working place, as well as the training place; giving quality guarantee and supervisor, knowledge about making a report, standardized ability to understand observation results to determine component condition, and ability to communicate a finding after a field inspection. As a conclusion, there is a big gap between a necessary skill and a competency in apparel sector. The gap occurs as a result of competency lacks had by training partner and the assessment. By inviting more candidates to come in the company, the gap can be elicited. It is also necessary to develop a set of questions for the making of basic clothe pattern; sewing machine operator and technician in garment industry. In addition, an assessment instrument is mandatory to be constructed and implemented [15].

7. Conclusion

The implementation of competency based assessment in the field of fashion design through measuring tool, facility, and instrument for garment, apparel, embroidery, CAD, and CAM sectors by patterning performer's criteria relevantly to the qualifications applied. The purpose is to measure one's knowledge, behaviour, and skill in a practical activity in order to perform more competently and effectively regarding times, accesses and costs.

References

[1] Arasinah Kamis, etc, 2014, Validity and Reliability of Clothing Fashion Design (CFaDC) Competency Instrument, Malaysia Middle-East Journal of Scientific Research 19 (Innovation Challenges in Multidiciplinary Research & Practice): 89-97, 2014 Universiti Putra Malaysia, 43400 UPM Serdang, Selangor, Malaysia.

[2] Barbara J. Frazier and Wanda K. Cheek, 2016, An Industry View of Competencies for Entry-Level Merchandising Jobs: Application of the ITAA Meta-Goals, Clothing and Textiles Research Journal 2016, Vol. 34(2) 79-93 sage

[3] Carol A. Falender and Edward P. Shafranske, 2007, Competence in Competency-Based Supervision Practice: Construct and Application Pepperdine University Professional Psychology: Research and Practice Copyright 2007 by the American Psychological Association 2007, Vol. 38, No. 3, 232–240

[4] Develop Competency-Based Assessment Plans Singapore Workforce Development Agency – Quality Assurance Division Develop Competency-Based Assessment Plans Version 1.1 (14 October 2012)

[5] Developing a Competency Based Appraisal Plan The Singapore Manpower Development Agency - Quality Assurance Division Developing a Competency Based Appraisal Plan Version 1.1 (October 14, 2012)

[6] Handbook of Clinical Psychology Competencies, 2010, Volumes 1-3 Springer Science & Business Media. Copyright. edited by Jay C Thomas, Michel Hersen complex interactive clusters of integrated

[7] Boahin, Peter & Hofman, WH Adriaan, 2014, Perceived effects of competency-based training on the acquisition of professional skills, International Journal of Educational Development 36 (2014) 81–89

IOP Conf. Series: Materials Science and Engineering **336** (2018) 012044 doi:10.1088/1757-899X/336/1/012044

[8] Jagranjosh, 2014, Recognising the importance of skill development and entrepreneurship in the country,PM Modi has created a ministery headed by "Sarbananda Sonowa" an MP from 'Lakhimpur constituency of Assam has been has been made the minister of state for "Skill development /&Entrepreneurship (Independent charge)

[9] Joy M. Kozar and Kim Y. Hiller Connell, 2014 The fashion internship experience: identifying learning outcomes in preparing students for the real world, International Journal of Fashion Design, Technology and Education.

[10] Kanchan kumari, 2015, Develop competency based assessment on Advance pattern maker (CAD/CAM) & Garment Cutter (CAM) http://www.nift.ac.in/ Downloads/ Abstarct_ BFT_2015.pdf

[11] KKNI, 2010, Direktorat Jendral Perguruan Tinggi, Kementrian Pendidikan dan Kebudayaan Republik Indonesia.

[12] Prakas, shamal, 2015, Creation Of Tools For Certification Of Training Centres And Trainees For Job Roles In The Garment Industry, Department of Fashion Technology National Institute of Fashion Technology, Mumbai

[13] Prashant Verma, "2015, Competency Based Assessment In Apparel Sector, Department of Fashion Technology National Institute of Fashion Technology, Mumbai

[14] Manish Kumar, 2015, Development of Competency based Assessment for Sewing Machine Operator, In-line Checker and QC Executive – Sewing Line, Department of Fashion Technology National Institute of Fashion Technology, Mumbai

[15] Mouzakitis, 2010; Baqadir, Patrick & Burns, 2011; Minghat & Ruhizan, 2010; RMK ke-10, 2011-2015; Kahirol & Nor Liza, 2007 dalam Arasinah Kamis, Ab. Rahim Bakar, Ramlah Hamzah, Soaib Asimiran Keperluan Kompetensi Pengetahuan Rekaan Fesyen Pakaian Untuk Menceburkan Diri dalam Industri Fesyen2:4 (2014) 25-33 | www.sainshumanika.utm.my | e-ISSN ISSN: 2289-6996

[16] Skkni Sektor Industri Tekstil Dan Barang Tekstil Bidang Garmen Bidang Custom Made Sub Bidang Custom Made Wanita.

[17] Neelam Agrawal Srivastava and Anandita Chatterjii, 2014, Skills Requirement Analysis At The Production Floor In Ready Made Garment Industry Asian Journal Of Management Sciences & Education Issn: 2186-845x ISSN: 2186-8441 Print Vol. 3 No. 1, January 2014 N: 2186-845X ISSN: 2186-8441 Print

[18] Nikhil Sharma dan Ravi Deo Chandra, 2015, Development of National Occupational Standards using Competency Based Assessments, Department of Fashion Technology National Institute of Fashion Technology, Mumbai

[19] Samapriya Banerjee, 2015, Development of competency based assessment for embroidery job rolls Department of Fashion Technology National Institute of Fashion Technology, Mumbai

[20] Sandeep Kumar, 2015, Competency Based Assessment In Apparel Sector, Department of Fashion Technology National Institute of Fashion Technology, Mumbai

[21] Ben10pku, 2016, Pentingnya Sertifikasi Profesi, Kompasiana, dilihat tanggal 9 Agustus 2017, http://www.kompasiana.com/ben10pku/pentingnya-sertifikasi-profesi_56d67eb43493736107a2ff87

[22] Disnaker, 2016, Pentingnya Setifikasi Profesi, di Era Kompetisi Global, dilihat 12 Agustus 2017, http://disnakertrans.sumutprov.go.id/index.php/9-index-berita/57-ini-pentingnya-sertifikasi-profesi-di-era-kompetisi-global

[23] Nurfuada, Rifa N., 2015, Pentingnya Sertifikasi Profesi Jelang MEA, dilihar 14 Agustus 2017, http://news.okezone.com/read/2015/06/16/65/1166303/pentingnya-sertifikasi-profesi-jelang-mea

[24] Bidik, 2017, Lulusan SMK harus Miliki Sertifikasi Profesi, dilihat tanggal 11 Juli 2017, http://bidik.co.id/lulusan-smk-harus-miliki-sertifikasi-profesi/13 052017A reference

[25] Wesselink, Renate & Mulder, Martin, 2006, Competence Based VET in The Netherlands, Wageningen University.

IOP Conf. Series: Materials Science and Engineering **336** (2018) 012045 doi:10.1088/1757-899X/336/1/012045

Developing Student Worksheet Based On Higher Order Thinking Skills on the Topic of Transistor Power Amplifier

Luckey Sardia Ratna Kusuma*, Lusia Rakhmawati , Wiryanto

Department of Electrical Engineering, Universitas Negeri Surabaya, Indonesia

*key.luck.key@gmail.com

Abstract. The purpose of this study is to develop a student worksheet about the transistor power amplifier based on higher order thinking skills include critical, logical, reflective, metacognitive, and creative thinking, which could be useful for teachers in improving student learning outcomes. Research and Development (R & D) methodology was used in this study. The pilot study of the worksheet was carried out with class X AV 2 at SMK Negeri 5 Surabaya. The result showed satisfies aspect of validity with 81.76 %, and effectiveness (students learning outcomes is classically passed out with percentage of 82.4 % and the students gave positive responses to the student worksheet of each statement. It can be concluded that this worksheet categorized good and worthy to be used as a source of learning in the learning activities.

1. Introduction

Globalization is a very real phenomenon that is transforming the world economic system including nearly all aspects of life. With the emergence of a new development model, particularly in the highly industrialized economies, knowledge and information take on increasing importance[4]. Thus, the era of globalization has tremendous concomitant implications for knowledge, education and learning. Educational aspect is intended to advance the quality of education in Indonesia as contained in enacted a new Law No. 20 of 2013 on *National Education System.*

The Law sets forth aims and functions of education, and national standards and the principles of the education system. It applies to all educational institutions that exist or may be created in the Republic of Indonesia. Its underlying philosophy is that education is a national movement in order that different stakeholders participate in the development of education as a life-long process. As such, the Law recognizes active partnership with local government and other local authorities in the process of education decentralization.

In order to realize the goal of national education in Indonesia that then formed a new curriculum in the 21st century, the 2013 curriculum[3]. The application of Curriculum 2013 at the Vocational High School (SMK) refers to the paradigm of 21st century learning curriculum which leads to changes of the Education Unit Level Curriculum (KTSP) to the 2013 Curriculum. These changes include the change of teaching to learning, from a teacher-centered learning into a student-centered learning, and from passive learning to active learning.

All the teaching and learning systems are updated in accordance with the 2013 curriculum. Not only the system of teaching and learning, but also learning device was overhauled to re-adjust to the learning system used. Learning device in question includes syllabi, lesson plans, worksheets, evaluations, modules, and other learning tools that support the learning process. Using the 2013 curriculum, the process of learning is the process of Higher Order Thinking (HOT).

Higher order thinking skills include critical, logical, reflective, metacognitive, and creative thinking[1]. They are activated when individuals encounter unfamiliar problems, uncertainties, questions, or dilemmas. Successful applications of the skills result in explanations, decisions, performances, and products that are valid within the context of available knowledge and experience and that promote continued growth in these and other intellectual skills. Higher order thinking skills are grounded in lower order skills such as discriminations, simple application and analysis, and cognitive strategies and are linked to prior knowledge of subject matter content. Appropriate teaching strategies and learning environments facilitate their growth as do student persistence, self-monitoring, and open-minded, flexible attitudes.

This definition is consistent with current theories related to how higher order thinking skills are learned and developed[2]. Although different theoreticians and researchers use different frameworks to describe higher order skills and how they are acquired, all frameworks are in general agreement concerning the conditions under which they prosper.

On the subject Basic Electronics Engineering at SMK Negeri 5 Surabaya, the 2013 curriculum has been implemented properly by using cooperative learning model as recommended by the education minister[8]. Applying bi-polar transistor as the power amplifier is the basic competencies that have not been obtained by the students of class X AV in the process of teaching and learning in the classroom, and worksheets to the basic competence is also not available from the teachers of the concerned in SMK Negeri 5 Surabaya, to the researchers want to use such material as the material development of worksheets in a study conducted with the hope of helping teachers to make students' better understanding of the material. The purpose of this study is: Determine the validity, practicality, and effectiveness of student worksheet based on HOTS which the 2013 curriculum implementation in the basic competence to apply bi-polar transistors as power amplifier in SMK Negeri 5 Surabaya.

2. Methodology

This research study was carried during the academic year of 2013-2014 in the class X Audio Video department at SMK Negeri 5 Surabaya. Research and Development (R & D) methodology was used in this study [6].

A worksheet based on HOT, which could be useful for student to support their awareness and knowledge of bi-polar transistors as power amplifier, was developed about factors effecting solubility. While the worksheet was developed, these next steps were taken:

1. A topic was determined for preparing the worksheet.
2. Which HOT could be gained from the worksheet were confirmed.
3. A draft of the worksheet was prepared.
4. Experts' input was solicited and received for consideration.
5. The worksheet was revised according to experts' recommendations.
6. The worksheet was applied as a pilot study to bi-polar transistors as power amplifier

In addition, we used FJ King et al approach for teaching strategies. Lessons involving higher order thinking skills require particular clarity of communication to reduce ambiguity and confusion and improve student attitudes about thinking tasks. Modelling of thinking skills, examples of applied thinking, and adaptations for diverse student needs were part of Lesson plans. Scaffolding (giving students support at the beginning of a lesson and gradually requiring students to operate independently) helps students develop higher order learning skills. However, too much or too little support can hinder development. Useful learning strategies include rehearsal, elaboration, organization, and metacognition.

Lessons specifically designed to teach specific learning strategies. Direct instruction (teacher-centered presentations of information) used sparingly. Presentations were short (up to five minutes) and coupled with guided practice to teach sub skills and knowledge. Teacher- and/or student-generated questions about dilemmas, novel problems, and novel approaches elicit answers that have not been learned already.

Sincere feedback providing immediate, specific, and corrective information informs learners of their progress[5]. Small group activities such as student discussions, peer tutoring, and cooperative learning effective in the development of thinking skills. Activities involve challenging tasks, teacher encouragement to stay on task, and ongoing feedback about group progress. Computer-mediated communication and instruction provide access to remote data sources and allow collaboration with students in other locations.

Valid assessment of higher order thinking skills requires that students be unfamiliar with the questions or tasks they are asked to answer or perform and that they have sufficient prior knowledge to enable them to use their higher order thinking skills in answering questions or performing tasks. Psychological research suggests that skills taught in one domain can generalize to others. Over long periods of time, individuals develop higher order skills (intellectual abilities) that apply to the solutions of a broad spectrum of complex problems.

Three item/task formats are useful in measuring higher order skills: (a) selection, which includes multiple-choice, matching, and rank-order items; (b) generation, which includes short-answer, essay, and performance items or tasks; and (c) explanation, which involves giving reasons for the selection or generation responses[7].

In addition classroom teachers recognize the importance of having students develop higher order skills yet often do not assess their students' progress. Several performance-based models are available to assist them in teaching and assessing these skills. Comprehensive statewide assessment of higher order skills is feasible but would be expensive.

3. Results

The results from this study are the student worksheet based on higher order thinking skill, which is used as a source of learning in the learning process. Details of worksheet development are as follows and part of the worksheet can be seen in Figure 1.

a. Defining the problem
b. Determining solution
c. evaluating solutions
d. applying the solution
    (1) Formulating Problems
    (2) Formulate Hypothesis
    (3) Determining Variable
    (4) Designing Step Trial
    (5) Conducting Experiments
    (6) Write the Experimental Results
e. Analyze

This study was conducted to determine the feasibility of student worksheet. It can be seen from the results of the validation, the results of student responses, and student learning outcomes-based the 2013 curriculum for basic electronics engineering subjects on the basis of competence apply bipolar transistor as power amplifier.

Evaluation phases done by evaluate learning process used student worksheet and analyze data of practicality and effectiveness. Validation student worksheet is categorized into three criteria: appearance, language, and content. Based on validation calculations to the overall criteria indicates that the worksheets are valid with a percentage value of 82 %. From these categories can be concluded that the worksheet suitable as a source of learning in the learning activities.

IOP Conf. Series: Materials Science and Engineering **336** (2018) 012045 doi:10.1088/1757-899X/336/1/012045

A. Problems

How to create a simple mini amplifier to work properly to generate great gain (output greater than input)

1. **Definition of Problems**

   Based on the problem review given above, define the problem definition by applying a simple mini amplifier to the electronics device to apply bi-polar transistors as a power amplifier.

   ....

   ....

2. **Determining Solutions**

   From the above problem, find some alternative solutions that might solve the problem.

   ....

   ....

3. **Evaluating Solutions**

   From these solutions, determine the best solution to solve the predetermined problem.

   ....

   ....

4. **Implementing the Solution**

   a. Formulating Problems

   → Question research that contains two variables, namely variable manipulation and response variables

   Make a problem formula related to the problems that have been determined!

   ....

   ....

   b. Formulating Hypotheses

   → temporary answers (guesses) to the results of the experiment to be performed

   Based on the problem formulation, write a hypothesis!

   ....

   ....

   c. Define Variables

   Based on the problem formulation, determine the experimental variables!

   (1) Variable Manipulation → variables that deliberately made a difference

   ....

   ....

   (2) Control Variables → variables are created equal

   ....

   ....

   (3) Variable Response → variable of manipulation variable manipulation result

   ....

   ....

   d. Designing an Experiment Step

   To test the truth of the hypothesis you have written down, you can do an experiment. Arrange the experiment design based on the formulation of the problem you have created. These tools and materials will make it easier for you to set up a test step.

   - Experiment Steps

   Write down the experiment steps that you will do in the fields provided below.

   ....

   ....

   e. Conducting an Experiment

   Experiment in accordance with the experimental steps you have made in accordance with the circuit drawing below!

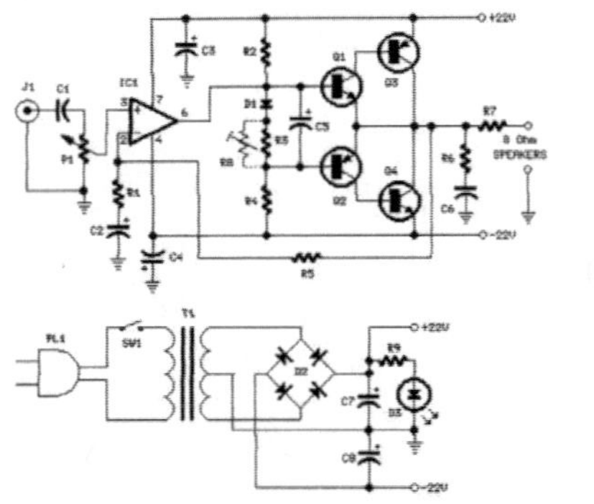

Figure 1. Mini Amplifier Circuit Scheme

   f. Experiment Results

   Based on the experiments you have done, write down the results of the experiments you obtained.

   ....

   ....

5. **Analyze**

   After experimenting, analyze the results of the experiments (shortcomings and strengths) by writing it down below.

   ....

   ....

B. Conclusions

Based on the experiments that have been done, write down a conclusion that you get!

....

Figure 1. Part of worksheet development of bi-polar transistors as power amplifier

In addition the results of student responses of 17 students who work on worksheets, there are three students who received a score of 70; 6 students who received grades of 80; and 8 students received grades of 90, which means that amounted to 17.65% from 17 students obtained a score of 70; amounting to 35.29% of the 17 students received grades of 80; and amounted to 47.06% from 17 students received grades 90. The total overall value of these was in 1390, with an average value of 81.76. Finally, practicality quality using student response questionnaire based on actual score is 81.5 %, it means student worksheet is practice.

The measurement of higher order thinking skills requires that students are unfamiliar with the questions or tasks they are asked to answer or perform and that they have sufficient prior knowledge to enable them to apply their higher order thinking skills to answer the question or solve the problem. Meeting these requirements is problematic in dealing with either real-life or subject-matter domain problems. Student profiles of correct answers can help to determine the extent to which the overall assessment is novel for individual students.

References

[1] Arikunto, Suharsimi. 2006. *Dasar-Dasar Evaluasi Pendidikan Edisi Revisi*. Jakarta: Rineka Cipta.

[2] Erlina dkk. 2011. "*Pengembangan Lembar Kerja Siswa dengan Pendekatan Makroskopis-Mikroskopis-Simbolik pada Materi Ikatan Kimia*". Jurnal Penelitian Universitas Tanjungpura Pontianak.

[3] Lismawati. 2010. Pengoptimalan Penggunaan Lembar Kerja Siswa. Jakarta: Rineka Cipta.

[4] Masjhudi dkk. 2012. *Pengembangan Lembar Kegiatan Siswa (LKS) dengan Model Siklus Belajar 5E Berbasis Konstruktivistik pada Materi Sistem Sirkulasi Manusia untuk Kelas XI SMA*. Jurnal Penelitian Universitas Negeri Malang.)

[5] Samanthis, Alunanda. 2013. *Pengembangan Perangkat Pembelajaran Menggunakan Model Project Based Learning pada Standar Kompetensi Memperbaiki Radio Penerima di SMK Negeri 3 Surabaya*. Skripsi Pendidikan Teknik Elektro Universitas Negeri Surabaya.

[6] P. D. S. Ahmet Mentes and A. P. D. Aykut H. Turan 2012 Assessing The Usability of University Websites: an Empirical Study on Namik Kemal University," *TOJET: The Turkish Online Journal of Educational Technology* **11** *(3)*, pp. 61-69

[7] Sugiyono. 2010. *Metode Penelitian Pendidikan (Pendekatan Kuantitatif, Kualitaif, dan R&D).* Bandung: Alfabeta.

[8] Zahara, Rita. 2013. *Pengembangan Lembar Kerja Siswa (LKS) Praktikum Berbasis Inkuiri Terbimbing pada Subpokok Materi Hubungan Hasil Kali Kelarutan dan Pengendapan.* Jurnal Penelitian Universitas Pendidikan Indonesia.

IOP Conf. Series: Materials Science and Engineering **336** (2018) 012046 doi:10.1088/1757-899X/336/1/012046

The Development of Learning Management System Using Edmodo

Joko[*], Gayuh Septia Wulandari

Department of Electrical Engineering, Universitas Negeri Surabaya

[*]unesa_joko@yahoo.com

Abstract. The development of Learning Management System (LMS) can be used as an online learning media by managing the teacher in delivering the material and giving a task. This study aims to: 1) to know the validity of learning devices using LMS with Edmodo, 2) know the student's response to LMS implementation using Edmodo, and 3) to know the difference of the learning outcome that is students who learned by using LMS with Edmodo and Direct Learning Model (DLM). This research method is quasi experimental by using control group pretest-posttest design. The population of the study was the student at SMKN 1 Sidoarjo. Research sample X TITL 1 class as control goup, and X TITL 2 class as experimental group. The researcher used scale rating to analyze the data validity and students' respon, and t-test was used to examine the difference of learning outcomes with significant 0.05. The result of the research shows: 1) the average learning device validity use Edmodo 88.14%, lesson plan validity is 92.45%, pretest-posttest validity is 89.15%, learning material validity is 84.64%, and affective and psychomotor-portfolio observation sheets validity is 86.33 included very good criteria or very suitable to be used for research; 2) the result of students' response questionnaire after taught by using LMS with Edmodo 86.03% in very good category and students agreed that Edmodo can be used in learning; and 3) the learning outcome of LMS by using Edmodo with DLM are: a) there are significant difference of the student cognitive learning outcome which is taught by using Edmodo with the student who use DLM. The average of student learning outcome that is taught LMS using Edmodo is 81.69 compared to student with DLM outcome 76.39, b) there is difference of affective learning outcome that is taught LMS using Edmodo compared to student using DLM. The average of student learning outcomeof affective that is taught LMS by using Edmodo is 83.50 compared students who use DLM 80.34, and c) there is difference of student psychomotor learning outcome that is taught with LMS using Edmodo compared student who use DLM. The average of student learning outcome that is taught with LMS using Edmodo is 85.60 compared to student who uses DLM 82.31.

1. Introduction

The learning process using internet media is different from the process of learning face to face with the teacher, so learning using internet media need to consider the risk, system, structure, schedule, and cost. One solution to overcome the risks in terms of cost is to use the Learning Management System (LMS). LMS has a scope of course material delivery, assessment, monitoring, and communication. LMS integrate material content, pedagogical and professional competence, science and technology with the expectation of quality and optimal learning outcomes.

To support LMS can use Edmodo. Edmodo is a communication medium and discussion between teachers and students efficiently, and the design is similar to facebook. Teachers can upload teaching materials, assignments, quizzes, and grades. Students can download teaching materials, assignments, quizzes, and send assignments. The use of Edmodo also facilitates interaction and discussion among fellow students about the topic that has been determined, and can be monitored by teachers. Teachers and students can access anytime and anywhere without dotage by distance, place, and time to obtain a source of information quickly and accurately.

Learning using LMS with Edmodo is expected to help the learning process of students, students are not easily bored in learning so that the learning outcomes are more optimal. Learning LMS using Edmodo also facilitate teachers in the delivery of subject matter because the subject matter can be packed more practically and become alternative media replacement paper.

The purpose of this research are: 1) to know the validity level of LMS learning design using Edmodo developed, 2) to know the student's response to learning by LMS using Edmodo, and 3) to know the difference of learning result of students with LMS using Edmodo compared to learning using Direct Learning Model (DLM).

Learning is a complex process that happens to everyone and lasts a lifetime, from the time they are babies to the grave later. One sign that a person has learned is a change of behavior within himself (Sadiman, 2005: 2). These behavioral changes concern both cognitive and psychomotor changes as well as values and attitudes (Musfiqon, 2012: 3). In learning there is certainly a process of learning, except to learn by self-taught. According to Rusman (2012: 1) learning is a system consisting of several components that are interconnected with each other. These components include objectives, materials, methods and evaluation. Internet is a giant library of the world, because in the internet there are billions of sources of information so that it can use the information in accordance with the needs.

LMS is a system that helps learning activities and manages e-Learning platform content content on e-Learning (Effendi and Zhuang, 2005: 85). LMS is a system that supports the implementation of electronic learning (e-Learning) by providing learning materials, learning process instructions undertaken by students, evaluation materials, and display the learning process (Prawiradilaga, 2004: 179).

LMS can serve as a software or software for learning, documentation, reports of a learning activity, online learning activities (connected to the internet), and online training materials (Ellis, 2009: 3). While Edmodo by Gatot (2013: 3) is a social media platform that is often described as facebook for school and can function more in accordance with needs. Edmodo is an app that appeals to teachers and students with social elements similar to facebook, but there is a greater value in online education applications. Edmodo makes it easier for teachers to monitor student interactions within the LMS because of the open course. Edmodo is almost like an easy facebook operation. This is different from learning with DLM.

Implementation of DLM in need of face-to-face directly between teacher and student, teacher explanation, question and answer, teacher-led training, follow-up training, evaluation, and continued with assignment. In such learning, the learning process tends to be more teacher-centered. In integrating LMS learning tools with Edmodo for optimum results, the result of learning device integration needs to be validated by the instructional design expert, the learning media expert, content expert, and evaluation expert. Likewise in learning with DLM, then the learning tool must also be validated. Response is any individual behavior that is a response to the stimulus. The human individual participates as the controller between the stimulus and the response so that which determines the form of individual response to the stimulus is the individual's own stimuli and factors

(Hamalik, 2011: 15). The responses are students' opinions of interest, feelings of happiness and present, and ease of understanding the components: 1) the material or the content of the lesson, 2) the teaching material format, 3) the pictures, 4) the activities in the student worksheet, 5) the learning atmosphere, and 6) the way teachers teach and the learning approaches used (Trianto, 2011: 242). Learning outcomes are the abilities possessed by students as a result of learning acts that can be observed from the student's performance. Learning outcomes are patterns of action, values, understandings, attitudes, appreciation and skills (Suprijono, 2009: 5). Learning outcomes consist of three domains, namely learning outcomes of cognitive, affective, and psychomotor.

2. Method

This research use research and development research type (R & D) referring Sugiyono (2015: 407) in developing LMS with Edmodo subjects Basic Elementary and Electronics class X. While research method using quasi experiment with exsperimental design non-equivalent control gruop pretest-posttest (Sugiono, 2015: 111). Fig. 1 shows the design of non-equivalent control group pretest-posttest.

$$O_1 \quad - \quad O_2$$
$$O_3 \quad X \quad O_4$$

Fig. 1 Experimental Design

X : Treatment in experimental group with LMS using Edmodo
O1: Pretest at control group
O2: Postest control class after learning using DLM
O3: Pretest at exsperimental group
O4: Posttest at exsperimental group

The research population is students SMKN 1 Sidoarjo. Determination of research sample using stratified random sampling technique. The sample of the study was selected students of class X TITL 2 experimental group and class X TITL 1 control group. Determination of experimental group and control group is using drawing technique. Number of students in each class is 36 students.

The research was conducted in Audio Room of SMK Negeri 1 Sidoarjo East Java Indonesia in the even semester of academic year 2016/2017. Data collection techniques consist of: 1) validation sheet, to collect validation data of learning device, and student's response to learning by LMS using Edmodo, 2) pretest-posttest to collect initial capability data and cognitive domain learning outcomes, 3) observation to collect the results of learning affective sphere, and 4) observation & portfolio to collect data of psychomotor domain learning outcomes.

The research instruments consist of: 1) validation sheet, to know the result of learning device validation developed, 2) questionnaire of student response, to know the student response to learning by LMS using Edmodo, 3) item pretest-posttest to know the initial ability and learning result cognitive domain, 4) observation sheet to know the affective domain learning result, and 5) observation sheet and portfolio to know the psychomotor domain learning result.

Steps to develop learning tools and integrate on the LMS with Edmodo adapting R & D Sugiono's research (2015: 407), covering: 1) analyzing the potential and problems of students, educators and technicians, media, tool resources, and student learning outcomes and standards (4) validate product design, 5) revise the product design, 6) conduct limited product testing, 7) revise the product, 8) test the experimental class, and 9) collect and analyze data on evaluation results and student responses.

Experts design validation results and student responses are further analyzed. The learning outcomes of the experimental class cognitive sphere were compared with the cognitive classroom learning outcomes of control class students. For the learning result of the affective and psychomotor spheres of the experimental class and control class each compared to the Minimal Mastery Criteria (MCC) specified.

The researcher used scale rating to analyze the data validity refers to Table 1. Determination of percentage of learning device eligibility obtained from the validator results using the following formula.

IOP Conf. Series: Materials Science and Engineering **336** (2018) 012046 doi:10.1088/1757-899X/336/1/012046

$$Percentage = \frac{\sum answer\ validator}{\sum highest\ score\ validator}\ x\ 100\%$$

Table 1 Percentage of Rating Scale and Assesment Criteria

No	Quality Assessment	Percentage (%)
1	Very Valid / Very Worthy	> 81.5 - 100
2	Valid / Worthy	62.5 – 81.5
3	Less Valid / Less Worthy	43.5 – 62.5
4	Invalid / Ineligible	25.0 – 43.5

(Adapted from Widoyoko, 2012: 112)

The researcher used scale rating to analyze the data students' response. The determination of the level of response categories of students to the LMS using Edmodo basic subjects of electricity and electronics refer to Table 2.

Table 2 Percentage of rating Scale

No	Quality Assessment	Percentage (%)
1	Very good	> 81.5 – 100
2	Good	62.5 – 81.5
3	Less good	43.5 – 62.5
4	Not good	25.0 – 43.5

To test the difference of learning result of cognitive sphere of experiment and control group using independent t-test sample with significant level of 0.05. Before performing the test t-test requirements analysis, namely homogeneity and normality test. In addition, the students' initial abilities of the experimental and control groups were not significantly different.

The first hypothesis of this study is the learning tool LMS with Edmodo developed in the category valid or feasible to use. Second hypothesis, student response to learning LMS with Edmodo in good category. While the third hypothesis is that there are significant differences in cognitive learning outcomes of students using LMS with Edmodo compared with student learning outcomes using DLM.

The indicators of success of this study are as follows: 1) validation of learning design in valid category or suitable to use if average percentage of rating result ≥ 62.5 - 81.5%, 2) student response to LMS learning with Edmodo in good category if average of percentage of rating result ≥ 62.5 - 81.5 %, and 3) the significance level of t-test result 0.05 for the cognitive domain learning outcomes, mastery of affective and psychomotor learning achievement of students with MMC ≥ 75.

3. Results and Discussion

Results

In the learning occurs interaction between students and students, and the interaction between students with teachers. The interaction that occurs is influenced by the instructional model applied by the teacher in the learning process. The interaction that occurs in DLM tends to be more limited, both the interaction between the teacher and the students because in DLM the learning process is more controlled by the teacher when compared to the interaction that occurs in the LMS with Edmodo.

Learning interactions between teachers and students use the LMS with Edmodo, as shown in Fig. 2.

Fig. 2 Student Interaction with Teacher

The learning device of LMS with Edmodo conducted validation is: 1) learning implementation plan, 2) subject matter, and 3) item pretest-postest. Validated aspects of learning planning validated aspects include indicators, objectives, material selection, selection of learning resources, selection of learning media, learning model, learning scenario, assessment, and literature. While the aspects of the subject matter that is validated include the layout and layout, illustrations and material content, language, and literature.

In the subject matter of cognitive domain, the validated aspect is: a) the material, ie the conformity of the item with the indicator, the item according to the competence, the letters and the picture is well and neatly arranged, the difficulty level of the problem, b) the construction, clear and unambiguous, the main formulation of the item and the choice of answers is a necessary statement only, the item does not give a clue the answer key, the choice of homogeneous and logical answers, the picture is clear and functioning, the length of the answer is relatively the same, and c) the language, using the language according to Indonesian and communicative rules.

The validated aspect on the affective observation sheet, the psychomotor-portfolio observation sheet is the indicator linkage with the objectives, the conformity of the statement with the purpose, using the language according to Indonesian and communicative rules, and the appropriateness of the assessment rubric. A summary of the validation results of learning device is shown in Table 3.

TABLE 3 SUMMARY OF LEARNING DEVICE VALIDATION RESULTS

No	Learning Device	Average Rating Result (%)
1	Lesson plan	92.45
2	Learning materials	84.64
3	Pretest-Posttest	89.15
4	Affective and psychomotor-portfolio observation sheets	86.33
	Average	88.14

The average of learning device rate validation is 88.14%, then based on Table 1 is in very valid or very feasible category, so the learning device of LMS with Edmodo developed is very feasible to be tested in experimental group students. To find out student response in learning using LMS with Edmodo students were given questionnaire of student response after learning. The summary of student response to LMS with Edmodo from media, material, and access aspects is shown in Table 4.

TABLE 4 SUMMARY OF STUDENT RESPONSE RESULTS

No	Response to Aspects	Average Rating Result (%)
1	Media	84.30
2	Materials	86,86
3	Access Edmodo	86,94
	Rerata	86.03

The average result of the student's response rating is 86.03%, then based on Table 2 is in very good or very like category, it means that students have a very good response to LMS with Edmodo. The results of the affective learning outcome of the results of observations at the time students do practice tests. Aspects that are assessed obedience do practical work, responsibility creativity, honesty, and discipline, While the learning outcome psychomotor that is assessed is the skill of using the tools and materials, the attitude in work, the sequence of doing the job, the accuracy of data-analysis-conclusion, the speed of work, the compatibility of shape & size, and the compliance of reports. Initial ability (preteset) student of experiment group average 39.94. Histogram of students' pretest of experimental group showed Fig 3.

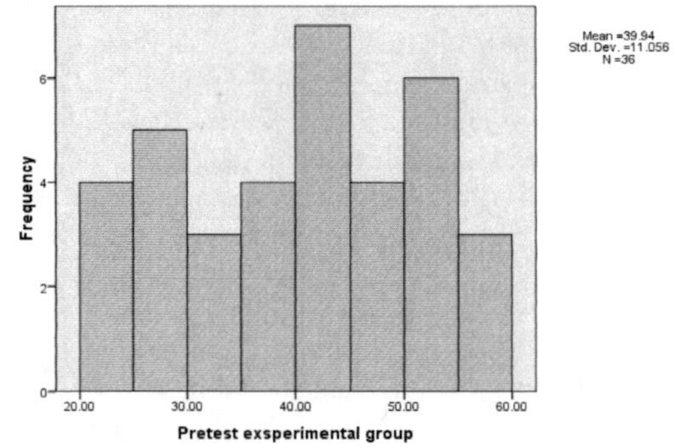

Fig. 3 Histogram Pretest Experiment Group

Pretest results of control group average 37.00. Histogram of students' pretest of control group shown Fig. 4.

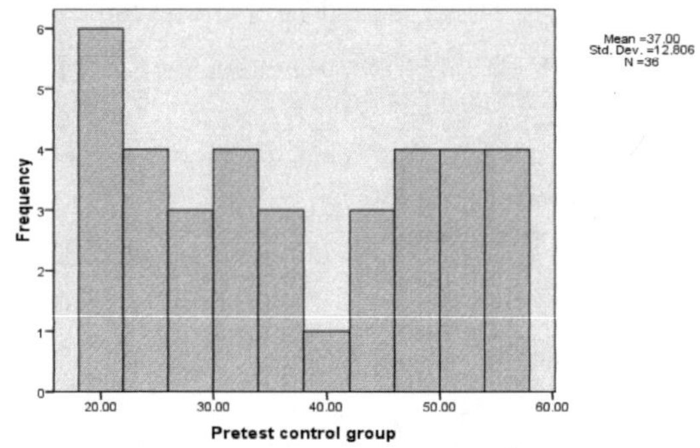

Fig. 4 Histogram Pretest Results Control Group

Learning outcome of experimental group average 81.69. Histogram result of student learning outcome (posttest) experiment group shown Fig. 5.

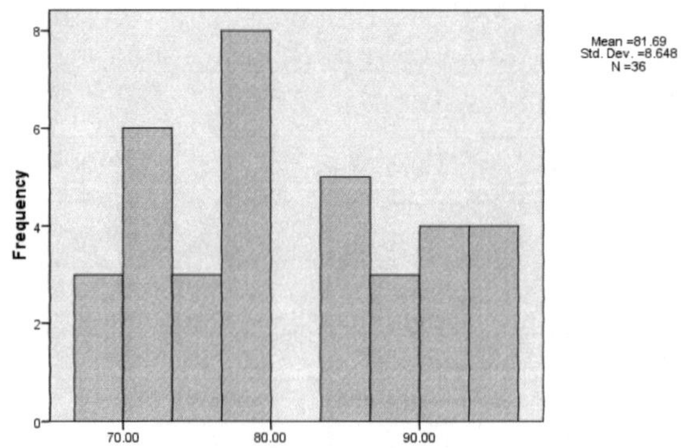

Fig. 5 Histogram posttest of Exsperimental Group

Learning outcome control group average 76.39. Histogram learning outcome control group shown Fig. 6.

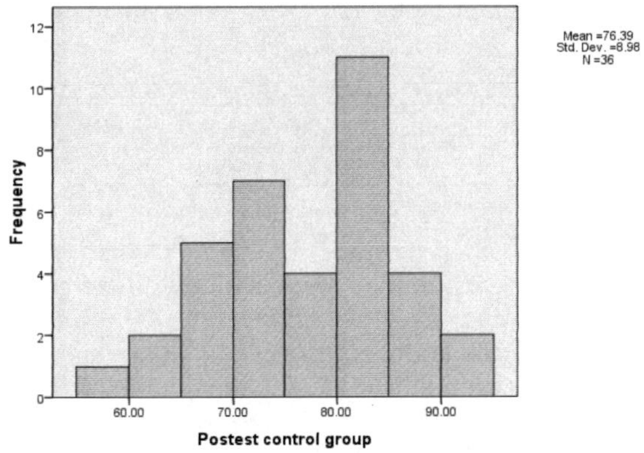

Fig. 6 Histogram Learning Outcome of Control Group

To know the normality of pretest data and learning outcaome data of experiment group and control group was done normality test. A summary of the results of normality test of pretest data and learning outcome data experimental group and control group is shown Table 5.

TABLE 5 TEST OF NORMALITY

	Kolmogorov-Smirnov[a]		
	Statistic	df	Sig.
Pretest exsperimental group	.143	36	.060
Pretest control group	.138	36	.080
Posttest exsperimental group	.133	36	.106
Posttest control group	.128	36	.141

The significance of pretest and learning outcome data experimental group and control group sig. > 0.05 conducted normal data.

To find out homogeneity of pretest data and learning data experiment group and control group conducted homogeneity test. The summary of the homogeneity test of pretset data and learning outcome data experimental and control group shown Table 6

TABEL 6 TEST OF HOMOGENEITY OF VARIANCES

	Levene Statistic	df1	df2	Sig.
Pretest data	1.798	1	70	.184
Posttest adata	.142	1	70	.708

Value sig. > 0.05 conducted variant of pretest data and posttest data experimental group and control group is homogeneous.

Because of pretest data experimental group and control groups normal and homogeneous, the t-test can be performed to test for differences in the pretest of the experimental group and the control group to proceed. The summary of the pretest results of the experimental group and control group shown Table 7.

TABLE 7 INDEPENDENT SAMPLES T-TEST PRETEST DATA

		t-test for Equality of Means		
		t	df	Sig. (2-tailed)
Pretest data	Equal variances assumed	1.044	70	.300

Value of sig. 0.30 > 0.05, meaning there is no significant difference between pretest data of experimental group and control group.

Since there were no significant differences in pretest data between experimental and control groups, normal and homogeneous data, the t-test to determine learning outcome differences between the experimental and control groups could be continued. A summary of the t-test results of the experimental and control group learning outcome shown Table 8.

TABEL 8 INDEPENDENT SAMPLES T-TEST POSTTEST DATA

		t-test for Equality of Means		
		t	df	Sig. (2-tailed)
Posttest data	Equal variances assumed	2.716	70	.008

The avalue sig. 0.008 <0.05, meaning there is a significant difference in the learning outcome of the experimental group and the control group.

Student learning outcome of affective domain of experimental was taught using LMS with Edmodo average 83.50 ≥75 or over MMC, as well as learning outcome of affective domain of control group which taught using DLM average 80.34 ≥75 or beyond MMC. The results show that the experiment group's higher learning outcomes are higher than the control group's learning outcomes.

Student learning outcomes psychomotor domain experimental groups who learn to use the LMS with Edmodo average 85.60 ≥75 or beyond MMC. While the results of student learning outcomes psychomotor domain control group that learn with DLM averages 82.31 ≥75 or beyond MMC. The learning outcomes psychomotor doamain of the experimental group were higher than the control group.

4. Discussion

The result of the student's response rating is in the category of very good or very fond of the LMS with Edmodo. This is because the media and materials that exist in the LMS with Edmodo more interesting, the present, the ease in understanding the components, and easy accessed. This will make students more interested, more happy against LMS with Edmodo.

There are significant differences in learning outcomes of the cognitive domain of the student group learning to use the LMS with Edmodo and higher than the student group learning using DLM. This can happen because the LMS with Edmodo can help students use technology and learn more independently than students who learn to use DLM. In addition, in the LMS with Edmodo the students' learning motivation also increases, not quickly bored, and the learning process is not bound by room, distance, time and place.

The learning outcomes The affective domain learning outcomes of the group of students learning to use the LMS with Edmodo were also higher than the group of students studying on DLM. This is because the learning using LMS with Edmodo students more independent and more utilizing technology, and more obedient in completing or performing tasks, more creative, more disciplined, and more responsible. In contrast to DLM, students are more dependent on teachers so that students' creativity, discipline, and responsibilities tend to be less developed.

The learning outcomes of the psychomotor domain of the group of students learning to use the LMS with Edmodo are better than the students' psychomotor learning outcomes learning on DLM. This is more because students who learn to use the LMS with Edmodo are more accustomed to practicum or practice. The material being studied is not new for the students as it has been supported by other subject matter that Edmodo has introduced and can support the practice activities, they are familiar and easily operate it in searching the supporting material for the practice.

5. Conclusion

Based on the results of the analysis, results of research and discussion, it can be concluded: 1) learning design developed (learning implementation plan, lesson material, evaluation of learning outcomes), and integrated in LMS with Edmodo in the category very valid or very suitable for use in basic learning electricity electronics, 2) student responses to learning using LMS with Edmodo in very good category, and 3) learning outcomes of the cognitive domain of the group of students who were taught using LMS with Edmodo higher and significantly different than the group of students who were taught using DLM. The learning result of the affective and psychomotor domains of the students group was taught using LMS with Edmodo higher than the learning using DLM.

References

[1] Arsyad, Azhar, *Media Pembelajaran,* Jakarta: PT. Raja Grafindo Persada, 2015.

[2] Effendy, Empy & Zhuang, Hartono, *E-Learning Konsep dan Aplikasi*, Yogyakarta: CV. Sandi Offset, 2005.

IOP Conf. Series: Materials Science and Engineering **336** (2018) 012046 doi:10.1088/1757-899X/336/1/012046

[3] Supramono, Gatot, *Perencanaan Media Pembelajaran*, Jakarta: Kencana Prenadamedia Group, 2013.
[4] Hamalik, Oema, Kurikulum dan Pengembangan, Jakarta: Bumi Aksara, 2011.
[5] Kelly, R. L, *Profitable & Sustainable Primary Industries: Soil Biology Basics*, New South Wales: Department of Primary Industries, 2005.
[6] Musfiqon, *Pengembangan Media dan Sumber Pembelajaran*, Jakarta: PT. Presti Pustakaraya, 2005.
[7] Prawiradilaga, Dewi Salma & Siregar, Eveline, *Mozaik Teknologi Pendidikan*, Jakarta: Prenada Media, 2004.
[8] Rusman, *Model-model Pembelajaran*, Jakarta: PT. Raja Grafindo Persada, 2012.
[9] Sugiyono, *Metode Penelitian Pendidikan Pendekatan Kuantitatif, Kualitatif, dan R&D*, Bandung: Alfabeta, 2015.
[10] Trianto, *Mendesain Model Pembelajaran Inovatif- Progresif. Konsep Landasan, dan Implementasinya pada Kurikulum Tingkat Satuan Pendidikan (KTSP),* Jakarta: Kencana Prenada Media Group, 2012.
[11] Widoyoko, E. P., *Teknik Penyusunan Instrumen Penelitian*,Yogyakarta: Pustaka Belajar, 2012.

IOP Conf. Series: Materials Science and Engineering **336** (2018) 012047 doi:10.1088/1757-899X/336/1/012047

Design of Smart Educational Robot as a Tool For Teaching Media Based on Contextual Teaching and Learning to Improve the Skill of Electrical Engineering Student

M S Zuhrie[1*], I Basuki[2], I G P Asto B[3], L Anifah[4]

[1,2] Department of Electrical Engineering, Faculty of Engineering, Universitas Negeri Surabaya, Indonesia.
[3,4] Department of Computer Engineering, Faculty of Engineering, Universitas Negeri Surabaya, Indonesia.

*zuhrie@unesa.ac.id

Abstract. The development of robotics in Indonesia has been very encouraging. The barometer is the success of the Indonesian Robot Contest. The focus of research is a teaching module manufacturing, planning mechanical design, control system through microprocessor technology and maneuverability of the robot. Contextual Teaching and Learning (CTL) strategy is the concept of learning where the teacher brings the real world into the classroom and encourage students to make connections between knowledge possessed by its application in everyday life. This research the development model used is the 4-D model. This Model consists of four stages: Define Stage, Design Stage, Develop Stage, and Disseminate Stage. This research was conducted by applying the research design development with the aim to produce a tool of learning in the form of smart educational robot modules and kit based on Contextual Teaching and Learning at the Department of Electrical Engineering to improve the skills of the Electrical Engineering student. Socialization questionnaires showed that levels of the student majoring in electrical engineering competencies image currently only limited to conventional machines. The average assessment is 3.34 validator included in either category. Modules developed can give hope to the future are able to produce Intelligent Robot Tool for Teaching.

1. Introduction

The development of robotics in Indonesia has been very encouraging. The barometer is the success of the Indonesian Robot Contest. In the contest no less than 40 major universities in Indonesia took part. But the development of such robots is limited to the contest and has not been developed to address the issues more real, especially in the industrial world.

The focus of this research is the design of the device in the course of the smart educational robot through the firefighter home robot that is booming and will be implemented in annual of the Indonesia Robot Contest and the industrial in order to advance the world of robotics Indonesia. Robots of this type developed by developed countries such as America, Japan, the United Kingdom the cost is very high. Currently, the robot actually uses high tech, because the control is fully controlled by a microprocessor as a substitute for human and we strive to develop the Low-Cost Technology. [1]

The focus of this research is a teaching module manufacturing, planning mechanical design, control system through microprocessor technology and maneuverability of the robot. The first phase of the study is expected to produce a smart educational robot that is reliable both in terms of technology and in terms of economy.

The principal issue is how the curriculum and the learning modules are implemented in accordance with the approach of industrial needs [2]. In order to support competency-based curriculum needs of the industry, has also been compiled modules both manual and interactive, oriented to the achievement of competence as well as accommodate the working life skills. With regard to the problems that will arise is what kind of learning modules appropriate to the characteristics of the attainment of the goal of learning the world of work as well as the achievement of competence in the field of life skills by using the main teaching materials in the form of modules and learning media. [3]

To answer these problems in this study will be compiled in the form of learning devices in the course of the smart educational robot as a tool for teaching based on Contextual Teaching and Learning to the needs of industry. Expected results of this study will be able to bridge the needs of energy, particularly of graduates majoring in Electrical Engineering Universitas Negeri Surabaya which will work as a vocational teacher or plunge into the world of industry.

2. Theoretical Framework

2.1. Educational Robot

The educational robot is a broad term that refers to a collection of activities, instructional programs, physical platforms, educational resources and pedagogical philosophy. The primary of objective of the educational robot is to provide a set of experience to facilitate the student's development of knowledge, skills, and attitudes for the design, analysis, application and operation of robots. The term robot here is used quite broadly and may include articulated robots, mobile robots or autonomous vehicles of any scale.

2.2. Contextual Teaching and Learning Strategy

Contextual Teaching and Learning is a learning concept that helps teachers connect the contents of the subject matter with real-world situations and motivate students to make connections between their knowledge application in about his life as a member of the family and community.[6]

Meanwhile, contextual teaching and learning is the concept of learning where the teacher brings the real world into the classroom and encourage students to make connections between knowledge possessed by its application in everyday life.[7]

US Department of Education and the National School-to-Work Office is a concept that helps teachers to connect teaching materials with situations in the real world and motivates students to make connections between knowledge and practice into their lives as family members, citizens, and labor.[8]

2.3. Development Learning Tool 4-D Model

The steps of 4D model are (1) define stage, the purpose of this stage is to determine and define the conditions of learning. This phase has five basic steps include: front-end analysis, learner analysis, concept analysis, task analysis, and formulation of learning objectives, (2) design stage, the design of prototype learning device is produced. The result of this phase is usually a preliminary design of the learning device depending on requirements. Device components used are very diverse, such as student book, modules, teacher guide, student activity sheets, lesson plans, tests of student learning outcomes, and learning media, (3) develop stage, this stage aims to produce learning tools that have been revised based on the input of experts. The next step is a trial with an appropriate number of students in a class of real (not unlimited). This activity was conducted to determine how effective learning tools developed when applied to the learning process. The effectiveness of the learning device can be seen through the observation, for example, the activities of teachers and students, the ability to manage learning and student achievement test, (4) disseminate stage, this stage is the stage of deployment and uses learning device. Learning tools that have been tested and revised earlier, duplicated and distributed for use on learning on a larger scale.[1]

3. Research Methods

This research was conducted by applying the research design development with the aim to produce a tool of learning in the form of the smart educational robot as a tool for teaching based on Contextual Teaching and Learning at the Department of Electrical Engineering to improve the skills of Electrical Engineering student.[4]

Procedure development in research carried out through several stages as follows:

- Identifying the various issues surrounding the smart educational robot used in the robot contests through the study of literature and data mining to be used as reference for the formulation of theory, simulation and applications approach;
- Analyze and formulate the identification of various issues surrounding the smart educational robot relevant to developed in the Department of Electrical Engineering Analyze and formulate the identification of various issues surrounding the smart educational robot relevant to developed in the Department of Electrical Engineering;
- Make analysis and draw up scenarios content and design of learning and infrastructure needed for the development of equipment, teaching staff, laboratory personnel, and the establishment of appropriate evaluation system material needs robotics;
- Formulate indicators of success for teaching and learning process oriented to the achievement of the performance demands of the professional needs;
- Implementation of the manufacture of smart educational robot as a tool for teaching based on Contextual Teaching and Learning as a means to integrate the theory, simulations and applications suitable with various issues surrounding the learning needs and smart educational robot for contest;
- Conduct field trials on smart educational robot course tools of learning based on Contextual Teaching and Learning in the Department of Electrical Engineering oriented to the mastery of the competency needs;
- Analyzing the results of field trials and make improvements to validate test procedures and test results;
- Doing the final revision of the smart educational robot as a tool for teaching based on Contextual Teaching and Learning, the results of field test and validation.[4]

4. Data Collection and Analysis Techniques Data

Information on the results of field surveys and discussions in order to implement the standards needs of the Indonesia Robot Contest and the learning device is analyzed with descriptive techniques. Likewise, data from focus group discussions in order to identify and formulate the essential topics as teaching materials for integrated robotic courses in the form of theory, simulation, and applications by using qualitative descriptive analysis techniques. Qualitative descriptive analysis techniques emphasis on making the evaluation and synthesis of the conclusions resulting from the activities. Synthesis and conclusions on the outcome of this study were formulated through forums workshops and focus group discussions.[5]

5. Results And Discussion

5.1. Data from the Indonesia Robot Contest 2009-2015

From the data of the Indonesia Robot Contest during the period 2009-2015 can be seen that the modules that have been developed have reached the fourth stage of the research methods of development that disseminate methods. So that, the module is perfect, ready to be duplicated and distributed as learning devices subject to intelligent robot courses. In Indonesia Robot Contest, material-based teaching module compiled specifically for the contest is not only electrical or electronic engineering majors alone, but multi-disciplines, and the object of research being developed not to be the latest technology and appropriate technology alone, but include the development of learning tools.

5.2. Data Socialization Result in Student Learning Tool

Based on the data about the socialization of the learning is done on students majoring in electrical engineering at the above resulted in the following data. For questions 1, 2, and 3 all respondents consisting of 10 students from representatives of various courses in the department of electrical engineering (100%) answered no understanding of the application of smart educational robot in the industry, or about smart educational robot equipment. Respondents also argued robotics competence that is not taught in the department of electrical engineering. This shows that the image of the competence of the department of electrical engineering is currently only limited to the classic electronic circuit, which ranges on flip-flop, NAND gate, AND gate only, not offensive competence in the industrial smart educational robot. Yet today the development of machinery industry has many uses industrial intelligent robot.

To question no. 4 of smart educational robot teaching module of r of respondents who answered draw was as much as 9 students (90%) of the 10 student-vocational and vocational representatives argue unattractive as much as one student (10%). As for the question no. 5 and no. 6 on the use of computer tools and modules of respondents who think "interesting" as many as 10 students (100%) and argue "unattractive" as none (0%). This shows the positive response of the device in the form of learning modules and devices are developed lesson

To question no. 7 all respondents as many as 10 people (100%) found the module that was developed to facilitate the understanding of the matter. All respondents (100%) believed that they feel happy and motivated by learning to use teaching modules and tool of learning. This indicates that the module can be motivating Electrical Engineering student and assist Electrical Engineering student in understanding the material.

5.3. Data from the validation of learning by teaching staff (lecturers)

A module that has been created and then validated on 5 validators consisting of faculty learning experts, education experts, engineering experts, and grammarians. And the average assessment is 3.34 validator included in either category. So that the module can be used in trial 2 that experiments performed in the course of the smart educational robot.

6. Conclusions

Based on the results of data analysis and discussion, the researchers can draw conclusions as follows:

This is a module developed research results and monitoring research officer with the team for 6 years continues to develop research-based robotics contest, with the hope of the future be able to a tool of learning in the form of intelligent robot modules and kit based on Prj Contextual Teaching and Learning.

The results of student responses during learning tools developed socialization shows that the image of competence for the department of electrical engineering is currently still limited to classic electronic circuit, which ranges on flip flop, NAND gate, AND gate only, only, not offend industrial robots. Yet today the development of industrial automation has many uses industrial robots. Therefore we need a learning device that can support learning robotics competence in the department of electrical engineering. From the results of student responses also showed a positive response to the module of intelligent robot based on Contextual Teaching and Learning developed.

Module that has been created and then validated on 5 validator consisting of faculty learning experts, education experts, engineering experts and grammarians. And the average assessment is 3.34 validator included in either category. So that the module can be used in trial 2 that experiments performed in the course of.

IOP Conf. Series: Materials Science and Engineering **336** (2018) 012047 doi:10.1088/1757-899X/336/1/012047

References

[1] Finch, Curtis R. & John R Crunkilton. 1983. curriculum development in vocational and technical education, london: allyn and bacon, inc.

[2] Blank,We. (1982). handbook for developing competency based training program. englewood cliffs; prentice hall.

[3] Adviso f, bernardo 1990 Development of the national training council as the coordinating body for technical and vocational trainng, jakarta: depdikbud

[4] Norton, R.E 1985 Dacum handbook. columbus: the national centre for research in vocational the ohio state Electrical Engineering

[5] Kulik Ja. Meta-analytic studies of findings on computer-based instruction. in: baker el, o'neil hf, eds. technology assessment in education and training. hillsdale, nj: erlbaum, 1994:52-66.

[6] Nurhadi, dan Agus Gerrad Senduk. 2003. Kontekstual dan Penerapannya Dalam KBK. Malang: Universitas Negeri Malang

[7] Berns, R. and Erickson, P. 2001 An Interactive Web based Model for the Professional Development of Teachers in Contextual Teaching and Learning. Bowling Green State University. Available at: http://www.bgsu.edu/ctl (accessed on August 22nd , 2017)

[8] A. Blanchard, Contextual Teaching and Learning, Educational Services, 2001.